Lecture Notes in Mathematics

Edited by A. Dold and B. Eckmann

767

Makoto Namba

Families of Meromorphic Functions on Compact Riemann Surfaces

Springer-Verlag
Berlin Heidelberg New York 1979

Author

Makoto Namba
Mathematical Institute
Tohoku University
Aoba, Sendai, 980
Japan

AMS Subject Classifications (1970): 14HXX, 32A46, 32CXX, 32GXX

ISBN 3-540-09722-8 Springer-Verlag Berlin Heidelberg New York
ISBN 0-387-09722-8 Springer-Verlag New York Heidelberg Berlin

Library of Congress Cataloging in Publication Data
Namba, Makoto, 1943-
Families of meromorphic functions on compact Riemann surfaces.
(Lecture notes in mathematics; 767)
Bibliography: p.
Includes index.
1. Riemann surfaces. 2. Functions, Meromorphic. I. Title. II. Series: Lecture notes in mathematics (Berlin); 767.
QA3.L28 no. 767 [QA333] 510'.8s [515'.982] 79-24604
ISBN 0-387-09722-8

Printing and binding: Beltz Offsetdruck, Hemsbach/Bergstr.
2141/3140-543210

To my teachers

Professor S. Sasaki and Professor M. Kuranishi

PREFACE

These notes were taken originally from the course held at Tohoku University in 1976. The aim of the course was to present a study of meromorphic functions on compact Riemann surfaces from the deformation theoretic point of view. In 1977, I had an opportunity to visit the University of Göttingen as a Humboldt fellow. Through many discussions with the staffs in Grauert's seminar, a lot of improvements became possible. I would like to express my deep gratitude to Professor Grauert and the staff in Göttingen, Professor Schneider, Dr. Commichau and so on. I also express my thanks to Alexander von-Humboldt Foundation for supporting me during my stay in Göttingen.

Finally, but not the least, I would like to express my sincere thanks to my colleagues, Professors Kuroda and Oda and Messrs. Ishida and Imayoshi for many valuable suggestions and constant encouragement; and to Miss Tokuko Sasaki for her beautiful job of typing these notes.

Sendai, Japan July 1979. Makoto Namba

Table of Contents

INTRODUCTION

" Examples are most important
in deformation theory. "
K. Kodaira

The purpose of the present lecture notes is to present a study of meromorphic functions on compact Riemann surfaces from the deformation theoretic point of view. For a compact Riemann surface V, the field $\mathbb{C}(V)$ of all meromorphic functions on V is nothing but the set $Hol(V, \mathbb{P}^1)$ of all holomorphic maps of V into $\mathbb{P}^1$, the complex projective line. By a general theorem of Douady [13], it has a complex space structure, whose underlying topology is the compact-open topology. $\mathbb{C}(V)$, as a field, has been studied by many mathematicians. One of our purposes is to study $\mathbb{C}(V) = Hol(V, \mathbb{P}^1)$ as a Douady space. For this purpose, we use many known facts about Jacobi varieties and symmetric products. As general references, two beautiful books, Gunning [28] and Mumford [59], are mainly used.

In Chapter 0, we review the deformation theory of compact complex manifolds by Kodaira-Spencer and Kuranishi. The existence theorem of Kuranishi [48], [49] of versal families of compact complex manifolds is of fundamental importance in deformation theory. The idea of his proof [49] is the leading principle of the present lecture notes.

In Chapter 1, we give a general theory on $Hol(V, \mathbb{P}^1)$. $Hol(V, \mathbb{P}^1)$ is written as the disjoint union of open subspaces $R_n(V)$, the set of all meromorphic functions of mapping order n. The structure of $R_n(V)$ for $n \geqq 2g$ (g = the genus of V) is simple. Making use of it, we determine the structure of the complex space $R_n(V)/(Aut(\mathbb{P}^1)\times Aut(V))$ when V is a complex 1-torus. This last space can be considered as the <u>moduli space of elliptic functions of order</u> n <u>on</u> V. This result is generalized in Chapters 3 and 4.

In Chapter 2, we concentrate our study to $R_n(V)$ for $n \leqq g$. It is a difficult problem to determine the integers n with non-empty $R_n(V)$ and the structure of $R_n(V)$ for such n. We give two simple theorems on this problem. One of them asserts that if m and n are positive integers such that (1) m and n are relatively prime and (2) $(m-1)(n-1) \leqq g-1$, then at least one of $R_m(V)$ and $R_n(V)$ is empty. As a corollary, we assert that if p is a prime number with non-empty $R_p(V)$ and if a positive integer n satisfies $(p-1)(n-1) \leqq g-1$, then (1) $R_n(V)$ is empty for $n \not\equiv 0 \pmod p$ and (2) $R_n(V) \cong R_{n/p}(\mathbb{P}^1)$ for $n \equiv 0 \pmod p$. This corollary is well known for a hyperelliptic V and $p = 2$.

Next, we advance to study the complex space $R_n = \cup_{t\in T_g} R_n(V_t)$, (disjoint union), where T_g is the Teichmüller space of compact Riemann surfaces of genus g ($\geqq 2$) and V_t is the compact Riemann surface corresponding to $t \in T_g$. For this purpose, we give two different methods. One of them is to make use of a general theory of (relative) Douady spaces of holomorphic maps of compact complex manifolds.

In Chapter 3, we give a construction of the (relative) Douady space $\cup_{s\in S} \mathrm{Hol}(V_s, W_s)$. Since Douady's original construction is very difficult, we give another easy method due to Namba [61], whose idea comes from Kuranishi [49]. Our method gives us some local informations about the space. Using it, we prove that $R_n = \cup_{t\in T_g} R_n(V_t)$ is non-singular and of dimension $2n+2g-2$. Hence, the qoutient space $R_n/\mathrm{Aut}(\mathbb{P}^1)$ is also non-singular and of dimension $2n+2g-5$.

In the last section of this chapter, we discuss the deformation theory of holomorphic maps developped by Horikawa [32], Miyajima [54] and Kouchiyama [47].

Some of the results of this and the following chapters are used in the previous chapters. Hence, logically they should have been presented earlier. But we arranged this way, because we hope that our arrangement makes the theory easier to understand.

In Chapter 4, we study families of effective divisors and linear systems on projective manifolds. Again, we construct our spaces using Kuranishi's idea. We introduce the notion of semi-regularity for linear systems and give a theorem which generalizes the usual semi-regularity theorem for divisors by Kodaira-Spencer [45].

In Chapter 5, we apply the results in Chapter 4 to the case of compact Riemann surfaces. We have the complex space

$$\mathbb{G}^r_n(V) = \{g^r_n \mid g^r_n \text{ is a linear system on } V \text{ of degree } n \text{ and of dimension } r\}$$

A linear system $g^r_n \in \mathbb{G}^r_n(V)$ is semi-regular if and only if, for independent divisors $D_\nu \in g^r_n$, $0 \leqq \nu \leqq r$, the linear subspaces

$$H^0(V, \mathcal{O}(K_V - D_\nu)) = \{\omega \in H^0(V, \mathcal{O}(K_V)) \mid (\omega) \geqq D_\nu\}, \quad 0 \leqq \nu \leqq r,$$

are independent in $H^0(V, \mathcal{O}(K_V))$. (K_V = the canonical bundle of V.) Semi-regularity theorem asserts in this case that if g^r_n is semi-regular, then it is a non-singular point of $\mathbb{G}^r_n(V)$ and

$$\dim_{g^r_n} \mathbb{G}^r_n(V) = (r+1)(n-r) - rg .$$

We also have the complex space

$$\mathbb{G}^r_n = \cup_{t \in T_g} \mathbb{G}^r_n(V_t), \quad \text{(disjoint union)}.$$

We introduce the notion of weak semi-regularity. If $g^r_n \in \mathbb{G}^r_n$ is weakly semi-regular, then it is a non-singular point of $\mathbb{G}^r_n$ and

$$\dim_{g^r_n} \mathbb{G}^r_n = (r+1)(n-r) - rg + 3g - 3 .$$

We prove that every pencil $g^1_n \in \mathbb{G}^1_n$ is weakly semi-regular. Thus $\mathbb{G}^1_n$ is non-singular and of dimension $2n+2g-5$. This proves again the non-singularity of the spaces $R_n/\mathrm{Aut}(\mathbb{P}^1)$ and R_n.

At the end of §5.2, we construct the global moduli space of non-degenerate holomorphic maps of compact Riemann surfaces of genus g

($g \geqq 2$) into $\mathbb{P}^r$. In particular ($r = 1$), we construct the <u>global moduli space of algebraic functions of order</u> n <u>and genus</u> g ($g \geqq 2$). It is a normal complex space of dimension $2n+2g-5$.

We put

$$T_g(n) = \{t \in T_g \mid V_t \text{ has a meromorphic function of order } n\} .$$

Our final result is

<u>Theorem</u>. Let p be a prime number such that $(p-1)^2 \leqq g-1$. Then

(1) $T_g(p)$ is an open subspace of a closed complex subspace of T_g and has dimension $2p+2g-5$.

(2) $T_g(p)$ is singular at $t \in T_g(p)$ if and only if $\dim H^0(V_t, \mathcal{O}([2D_\infty(f)])) > 3$, where $f \in R_p(V_t)$ and $D_\infty(f)$ is the polar divisor of f.

<u>Corollary</u>.

(1) (Rauch [71]) If $g \geqq 2$, then $T_g(2)$, the hyperelliptic locus, is a non-singular closed complex subspace of T_g of dimension $2g-1$.

(2) If $g \geqq 4$, then $T_g(3)$, the locus of trigonal compact Riemann surfaces, is non-singular and of dimension $2g+1$.

(3) If p is a prime number such that $p \geqq 5$ and $(p-1)(2p-3) \leqq g-1$, then $T_g(p)$ is non-singular.

The theory presented here is far from being complete. Many problems remain unsolved. I would be very happy if some of the readers get interested in solving them.

Chapter 0. Review of the deformation theory of compact complex manifolds.

0.1. Families of compact complex manifolds.

The contents of this chapter are mainly taken from the survey article Namba [67]. By a complex space, we mean a reduced, Hausdorff, complex analytic space, unless otherwise stated. A complex manifold is, by definition, a connected non-singular complex space. A 1-dimensional complex manifold is traditionally called a Riemann surface. In the present lecture notes, we need only some basic knowledges about complex spaces. The reader may consult Gunning-Rossi [29], Narasimhan [68] or Fischer [18].

Let X and S be complex spaces and let $\pi : X \longrightarrow S$ be a surjective proper holomorphic map. The triple (X,π,S) is called a family of compact complex manifolds if (1) every fiber $V_s = \pi^{-1}(s)$, $s \in S$, is connected and (2) f is smooth, i.e., there are an open covering $\{X_i\}$ of X, open subsets U_i of $\mathbb{C}^d$ (d is independent of the indices i), open subsets S_i of S and holomorphic isomorphisms (i.e., biholomorphic maps) $\eta_i : X_i \longrightarrow U_i \times S_i$, such that the diagram

$$\begin{array}{ccc} X_i & \xrightarrow{\eta_i} & U_i \times S_i \\ & {\scriptstyle\pi}\searrow \quad \swarrow{\scriptstyle proj} & \\ & S_i & \end{array}$$

is commutative for each i. In this case, S is called the parameter space of the family (X,π,S). By the definition, each fiber V_s is a compact complex manifold of dimension d. We often write $\{V_s\}_{s\in S}$ instead of (X,π,S). If S is connected, then we say that V_t is a deformation of V_s for any $s,t \in S$.

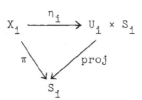

Example 0.1.1. Let $\mathbb{H} = \{\omega \in \mathbb{C} \mid Im(\omega) > 0\}$ be the complex upper

half plane. For $\omega \in \mathbb{H}$, we consider the complex 1-torus $T_\omega = \mathbb{C}/(\mathbb{Z}+\mathbb{Z}\omega)$. Then $\{T_\omega\}_{\omega\in\mathbb{H}}$ is a family of complex 1-tori. In fact, for any $m,n\in\mathbb{Z}$, let

$$g_{m,n} : (z,\omega) \in \mathbb{C}\times\mathbb{H} \longmapsto (z+m+n\omega,\omega)\in \mathbb{C}\times\mathbb{H}$$

be an automorphism of $\mathbb{C}\times\mathbb{H}$. Then the group $G = \{g_{m,n}\}_{m,n\in\mathbb{Z}}$ acts properly on $\mathbb{C}\times\mathbb{H}$ without fixed point. The quotient manifold $(\mathbb{C}\times\mathbb{H})/G$ with the induced projection $\pi : (\mathbb{C}\times\mathbb{H})/G \longrightarrow \mathbb{H}$ defines the family $\{T_\omega\}_{\omega\in\mathbb{H}}$. $T_{\omega'}$ is biholomorphic to T_ω if and only if $\omega' = (a\omega+b)/(c\omega+d)$ for some integers a,b,c and d with $ad-bc = 1$.

Example 0.1.2. Let $g \geq 2$ be an integer. Let T_g be the Teichmüller space of compact Riemann surfaces of genus g. It is a complex manifold of dimension $3g-3$. For $t\in T_g$, let V_t be the compact Riemann surface corresponding to t. Then $\{V_t\}_{t\in T_g}$ is a family of compact Riemann surfaces. It is called the Teichmüller family of compact Riemann surfaces of genus g. It is known that this family has the following properties:

(1) For any compact Riemann surface V of genus g, there is a point $t\in T_g$ such that V is biholomorphic to V_t.

(2) For any point $o\in T_g$, there is an open neighborhood U of o in T_g such that V_t is not biholomorphic to V_o for any $t\in U-\{o\}$.

(3) T_g is biholomorphic to a bounded, holomorphically convex domain in $\mathbb{C}^{3g-3}$.

(See Teichmüller [81], Rauch [72], Ahlfors [1], Bers [4] and Grothendieck [26, Exp.17].)

Now, let $(X,\pi,S) = \{V_s\}_{s\in S}$ be a family of compact complex manifolds. For a point $o\in S$, put $V_o = V$. Let X_i, U_i, S_i, η_i, etc., be as above. We consider only small deformations of V. Hence, we may assume that the set $I = \{i\}$ of indices is a finite set and $S_i = S$ for all $i \in I$. Let $(z_i) = (z_i^1, \cdots, z_i^d)$ be a coordinate system in $U_i \subset \mathbb{C}^d$. We identify U_i with $X_i \cap V$. Then $\mathcal{U} = \{U_i\}_{i \in I}$ is a finite open covering of V. We may assume that each U_i is Stein. If $U_i \cap U_k$ is non-empty, then the holomorphic isomorphism

$$\eta_{ik} : \eta_k(X_i \cap X_k) \longrightarrow \eta_i(X_i \cap X_k)$$

defined by $\eta_{ik} = \eta_i \eta_k^{-1}$ is written as

$$\eta_{ik}(z_k, s) = (g_{ik}(z_k, s), s),$$

where $g_{ik} : \eta_k(X_i \cap X_k) \longrightarrow U_i$ is a holomorphic map.

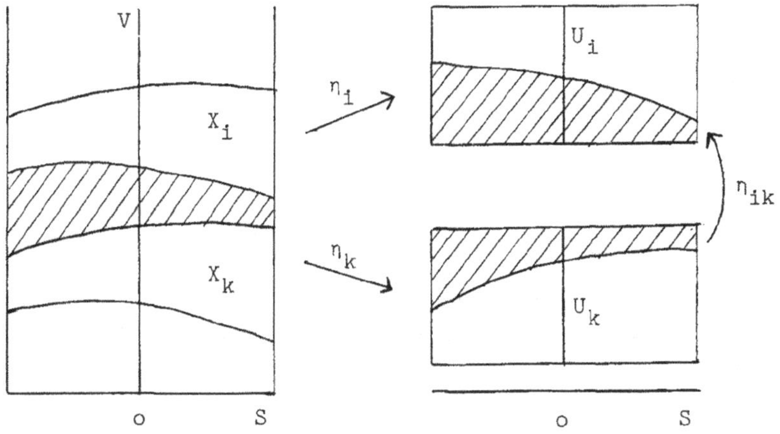

Let Ω be an ambient space of S such that $\dim \Omega = \dim T_o S$, where $T_o S$ is the Zariski tangent space to S at o. (Such Ω exists. See Gunning-Rossi [29, p.153].) Let $(s^1, \cdots, s^q)$ be a coordinate system in Ω. Let S be defined in Ω as the set of zeros of holomorphic functions $v_1, \cdots, v_p$:

$$S = \{ s \in \Omega \mid v_1(s) = \cdots = v_p(s) = 0 \}.$$

Then it is easy to see that

(1) $v_\alpha(o) = 0$ and $(\partial v_\alpha / \partial s^\beta)_o = 0$ for $1 \leqq \alpha \leqq p$ and $1 \leqq \beta \leqq q$.

Now, if $X_i \cap X_j \cap X_k$ is non-empty, then $\eta_{ik} = \eta_{ij}\eta_{jk}$ on $\eta_k(X_i \cap X_j \cap X_k)$. It is written as

$$g_{ik}(z_k,s) = g_{ij}(g_{jk}(z_k,s),s) \quad \text{for} \quad (z_k,s) \in \eta_k(X_i \cap X_j \cap X_k).$$

Fix a point $z_k \in U_i \cap U_j \cap U_k$. Then there are an open neighborhood U_o of z_k in $U_i \cap U_j \cap U_k$ and an open neighborhood Ω_o of o in Ω such that g_{ik}, g_{ij} and g_{jk} are extended holomorphically to $U_o \times \Omega_o$. Moreover, there are vector valued holomorphic functions a^α, $1 \leqq \alpha \leqq p$, on $U_o \times \Omega_o$ such that

$$g_{ik}(z'_k,s) = g_{ij}(g_{jk}(z'_k,s),s) + \sum_\alpha a^\alpha(z'_k,s)v_\alpha(s)$$

for $(z'_k,s) \in U_o \times \Omega_o$. Taking the partial derivative at (z_k,o) with respect to s and using (1), we get

$$(\partial g_{ik}/\partial s)_{(z_k,o)} = (\partial g_{ij}/\partial z_j)_{(z_j,o)}(\partial g_{jk}/\partial s)_{(z_k,o)} + (\partial g_{ij}/\partial s)_{(z_j,o)},$$

where $z_j = g_{jk}(z_k,o)$. This means that $\theta(\mathcal{U}) = \{(\partial g_{ik}/\partial s)_{(z_k,o)}\}$ is a 1-cocycle for the sheaf $\Theta = \mathcal{O}(TV)$ of germs of holomorphic vector fields on V, with respect to the covering $\mathcal{U} = \{U_i\}_{i\in I}$, i.e.,

$$\theta(\mathcal{U}) \in \check{H}^1(\mathcal{U},\Theta) \cong H^1(V,\Theta), \quad \text{(Leray's theorem)}.$$

($\check{H}$ means the Čech cohomology group.) We show that, as an element of $H^1(V,\Theta)$, $\theta(\mathcal{U})$ is independent of the choice of the covering $\mathcal{U} = \{U_i\}$. In fact, let $\{X'_\alpha\}_{\alpha\in A}$ be another finite covering of V by open subsets X'_α of X with the same property as $\{X_i\}_{i\in I}$. Put $U'_\alpha = X'_\alpha \cap V$ and $\mathcal{U}' = \{U'_\alpha\}_{\alpha\in A}$. Then, $\mathcal{U}$ and $\mathcal{U}'$ are refinements of $\mathcal{U}'' = \mathcal{U} \cup \mathcal{U}' = \{U_i\} \cup \{U'_\alpha\}$ (disjoint union). It is clear that the restrictions of $\theta(\mathcal{U}'')$ to $\mathcal{U}$ and $\mathcal{U}'$ are $\theta(\mathcal{U})$ and $\theta(\mathcal{U}')$, respectively. Hence $\theta(\mathcal{U}) = \theta(\mathcal{U}'') = \theta(\mathcal{U}') \in H^1(V,\Theta)$.

This element is denoted by $\rho_o(\partial/\partial s)$ and is called the <u>infinitesimal deformation at</u> $o \in S$ <u>to the direction</u> $\partial/\partial s \in T_oS$. It was found by Kodaira-Spencer [43]. The linear map

$$\rho_o : T_oS \longrightarrow H^1(V, \Theta)$$

is called the <u>Kodaira-Spencer map at</u> $o \in S$ <u>of the family</u> $\{V_s\}_{s\in S}$. $\{V_s\}_{s\in S}$ is said to be <u>effectively parametrized at</u> $o \in S$ if ρ_o is injective.

Let $(X',\pi',S') = \{V'_{s'}\}_{s'\in S'}$ and $(X,\pi,S) = \{V_s\}_{s\in S}$ be two families of compact complex manifolds. A <u>morphism of</u> $\{V'_{s'}\}_{s'\in S'}$ <u>to</u> $\{V_s\}_{s\in S}$ is, by definition, a pair $(h,\tilde{h})$ of holomorphic maps $h : S' \longrightarrow S$ and $\tilde{h} : X' \longrightarrow X$ such that (1) the diagram

$$\begin{array}{ccc} \tilde{h} : X' & \longrightarrow & X \\ \pi' \downarrow & & \downarrow \pi \\ h : S' & \longrightarrow & S \end{array}$$

commutes and (2) the induced map $\tilde{h}_{s'} : V'_{s'} \longrightarrow V_{h(s')}$ is a holomorphic isomorphism for any $s' \in S'$. Thus, we can define an <u>isomorphism of families</u>. Let $(X,\pi,S) = \{V_s\}_{s\in S}$ be a family and $h : S' \longrightarrow S$ be a holomorphic map. Let $h^*X = S' \times_S X$ be the fiber product and $\hat{h} : h^*X \longrightarrow X$ and $\pi^* : h^*X \longrightarrow S'$ be the projections:

$$\begin{array}{ccc} \hat{h} : h^*X & \longrightarrow & X \\ \pi^* \downarrow & & \downarrow \pi \\ h : S' & \longrightarrow & S . \end{array}$$

Then, it is easy to see that (h^*X,π^*,S') is a family of compact complex manifolds. Moreover, the pair $(\hat{h},h)$ is a morphism of (h^*X,π^*,S') to (X,π,S). (h^*X,π^*,S') is called the <u>induced family over</u> h. Making use of local coordinates, we can easily show that, if $(h,\tilde{h})$ is a morphism of (X',π',S') to (X,π,S), then there is an isomorphism $(id,\tilde{g})$ of (X',π',S') to (h^*X,π^*,S') such that $(h,\hat{h})(id,\tilde{g}) = (h,\tilde{h})$. (id = the identity map : $S' \longrightarrow S'$.)

Let $(h,\tilde{h})$ be a morphism as above. For a point $o' \in S'$, put $o = h(o')$ and $i = \tilde{h}_{o'} : V'_{o'} \longrightarrow V_o$. Let $i^*\Theta_o$ be the inverse image sheaf by the map i. ($i^*\Theta_o = \mathcal{O}(i^*TV_o)$, the sheaf of germs of holomorphic sections of the pull back i^*TV_o of the holomorphic tangent bundle TV over i.) Then, the linear maps

$$i_* : H^1(V'_{o'}, \Theta_{o'}) \longrightarrow H^1(V'_{o'}, i^*\Theta_o)$$

$$i^* : H^1(V_o, \Theta_o) \longrightarrow H^1(V'_{o'}, i^*\Theta_o),$$

induced by i, are linear isomorphisms. Making use of local coordinates, we can easily show the following

Lemma 0.1.3. The diagram

$$\begin{array}{cccc} (dh)_{o'} : & T_{o'}S' & \longrightarrow & T_oS \\ & \rho_{o'}\downarrow & & \downarrow\rho_o \\ i^{*-1}i_* : & H^1(V_{o'}, \Theta_{o'}) & \longrightarrow & H^1(V_o, \Theta_o) \end{array}$$

commutes. ($(dh)_{o'}$ = the differential of the map h at o'.)

Now, let $(X,\pi,S) = \{V_s\}_{s\in S}$ be a family of compact complex manifolds. Let $o \in S$. The family is said to be complete at o if, for any family $(X',\pi',S') = \{V'_{s'}\}_{s'\in S'}$ of compact complex manifolds with a point $o' \in S'$ and a holomorphic isomorphism $i : V'_{o'} \xrightarrow{\approx} V_o$, there are an open neighborhood U' of o' in S' and a morphism $(h,\tilde{h})$ of $\{V'_{s'}\}_{s'\in U'}$ to $\{V_s\}_{s\in S}$ such that $h(o') = o$ and $\tilde{h}_{o'} = i$. $\{V_s\}_{s\in S}$ is said to be complete if it is complete at every point of S. $\{V_s\}_{s\in S}$ is said to be versal at $o \in S$ if (1) it is complete at $o \in S$ and (2) the differential $(dh)_{o'}$ of the above holomorphic map h at o' is uniquely determined, i.e., if $(h_1,\tilde{h}_1)$ is another such morphism, then $(dh)_{o'} = (dh_1)_{o'}$.

By Lemma 0.1.3, it is easy to see that, if a family is complete at $o \in S$ and effectively parametrized at o, then it is versal at o.

$\{V_s\}_{s\in S}$ is said to be <u>universal at</u> $o\in S$ if (1) it is complete at $o\in S$ and (2) the above holomorphic map h itself is uniquely determined as a mapping germ at o'.

In short, if $\{V_s\}_{s\in S}$ is complete at $o\in S$, then it contains all small deformations of V_o. If it is versal at $o\in S$, then it is the smallest among complete families. The universality is of course stronger than the versality. The families in Examples 0.1.1 and 0.1.2 are universal at every point of the parameter spaces.

The following theorem is of fundamental importance in the deformation theory of compact complex manifolds.

<u>Theorem 0.1.4. (Kuranishi).</u> For any comact complex manifold V, there is a complete family $\{V_s\}_{s\in S}$ of compact complex manifolds with a point $o\in S$ such that (1) $V_o = V$, (2) it is effectively parametrized at o (hence it is versal at o) and (3) further, if $H^0(V,\Theta) = 0$, then it is universal at every point of S. Moreover, the parameter space S is given as follows: There are an open neighborhood U of 0 in $H^1(V,\Theta)$ and a holomorphic map $\alpha : U \longrightarrow H^2(V,\Theta)$ such that (4) $S = \{\xi\in U \mid \alpha(\xi) = 0\}$, (5) $o = 0$, (6) $(d\alpha)_0 = 0$ and (7) the Kodaira-Spencer map ρ_o is equal to $(di)_o$, where $i : S \hookrightarrow U$ is the inclusion map.

<u>Corollary 0.1.5. (Kodaira-Nirenberg-Spencer [46]).</u> Let V be a compact complex manifold with $H^2(V,\Theta) = 0$. Then there is a complete family $\{V_s\}_{s\in S}$ with a point $o\in S$ such that (1) $V_o = V$, (2) it is effectively parametrized at o and (3) S is non-singular at o and $\dim_o S = \dim H^1(V,\Theta)$.

<u>Corollary 0.1.6. (Kodaira-Spencer [44]).</u> Let $\{V_s\}_{s\in S}$ be a family of compact complex manifolds. Assume that, for a non-singular point $o\in S$, the Kodaira-Spencer map ρ_o is surjective (resp. isomor-

phic). Then the family is complete (resp. versal) at o.

Remark 0.1.7. Theorem 0.1.4 was first proved in a complicated way by Kuranishi [48]. Later on, he gave another simpler proof [49], whose detailed and rigorous reformulation was given in his Montreal lectures [50]. Other proofs of Kuranishi's theorem were given by Douady [12] and Commichau [10]. The family and the parameter space in Theorem 0.1.4 are sometimes called the Kuranishi family and the Kuranishi space, respectively. They are uniquely determined up to (non-canonical) holomorphic isomorphism as germs of complex spaces. Unfortunately, they may not in general be universal at o as examples show (see Example 0.1.10 below).

Remark 0.1.8. The corresponding theorem for compact (not necessarily reduced) complex spaces was proved by Grauert [24] and Douady [15]. The corresponding theorem for germs of (not necessarily reduced) complex spaces with isolated singularities was proved by Donin [11].

Now, we give a very short sketch of Kuranishi's proof of Theorem 0.1.4. Let $(X,\pi,S) = \{V_s\}_{s\in S}$ be a family of compact complex manifolds. If S is connected, then (X,π,S) is a differentiable fiber bundle over S (see Kuranishi [50]). Hence $\{V_s\}_{s\in S}$ is regarded as a family of (isomorphism classes of) complex structures on a fixed differentiable manifold V. Taking S sufficiently small, let $g : V \times S \longrightarrow X$ be a diffeomorphism. (A differentiable function on a complex space is a function which is locally extended to differentiable functions on ambient spaces. Thus we can define a differentiable map of complex spaces.) We use the same notations as above. Let $U_i' \Subset U_i$ be Stein open and $\{U_i'\}_{i\in I}$ cover V again. ($A \Subset B$ means that the closure $\bar{A}$ is compact and is contained in B.) Put, for $(z_i,s) \in U_i' \times S$,

$$\eta_i \circ g(z_i,s) = (\zeta_i(z_i,s),s).$$

Then $\zeta_i(z_i,s)$ is a differentiable function of (z_i,s) such that $\zeta_i(z_i,o) = z_i$. For $s \in S$, we define a vector valued (0,1)-form $\phi(s)$ on $\mathbf{V}$ as follows:

$$\phi(s) = \sum_{\alpha,\beta} \phi^{\alpha}_{i\beta}(s) \frac{\partial}{\partial z^{\alpha}_i} \otimes d\bar{z}^{\beta}_i,$$

$$\partial\zeta^{\gamma}_i(z_i,s)/\partial\bar{z}^{\beta}_i - \sum_{\alpha}\phi^{\alpha}_{i\beta}(s)\partial\zeta^{\gamma}_i(z_i,s)/\partial z^{\alpha}_i = 0, \quad 1 \leqq \beta, \gamma \leqq d.$$

Then, $\phi(s)$ is globally defined and represents the almost complex structure on $\mathbf{V}$ corresponding to V_s. ($\phi(o) = 0$.) Since V_s is in fact a complex manifold, $\phi(s)$ must satisfy the integrability condition:

$$\bar{\partial}\phi(s) - \tfrac{1}{2}[\phi(s),\phi(s)] = 0.$$

(The bracket [,] is the Lie bracket, defined in a way similar to the Lie bracket of vector fields.)

Now, Kuranishi considered the subset of all complex structures in the space of all almost complex structures on $\mathbf{V}$. Locally (around a fixed complex structure V on $\mathbf{V}$), it is defined by the equation:

$$\bar{\partial}\phi - \tfrac{1}{2}[\phi,\phi] = 0,$$

where ϕ is a vector valued (0,1)-form with a small 'norm'. Taking a subset of complex structures which are transversal to the orbit of Diffeo($\mathbf{V}$), the diffeomorphism group of $\mathbf{V}$, he got the Kuranishi space S. Using Hodge theory, he showed that S is of finite dimension and is obtained as stated in Theorem 0.1.4.

<u>Theorem 0.1.9. (Wavrik [84])</u>. Let $\{V_s\}_{s\in S}$ be the Kuranishi family of $V = V_o$. Assume that ρ_o is a linear isomorphism. Then the family is universal at o if and only if $\dim H^0(V_s, \Theta_s)$ is constant around o.

<u>Example 0.1.10. (Morrow-Kodaira [55], Suwa [80])</u>. Let $(Z_0:Z_1)$ be a homogeneous coordinate system of $\mathbb{P}^1$, the complex projective line.

Put $\zeta = Z_0/Z_1$. Put $U_1 = U_2 = \mathbb{C}$. For a non-negative integer m, let

$$M^{(m)} = (U_1 \times \mathbb{P}^1) \cup (U_2 \times \mathbb{P}^1)/\sim ,$$

where the equivalence relation $\sim$ is defined by:

$$(z_1,\zeta_1) \in U_1 \times \mathbb{P}^1 \sim (z_2,\zeta_2) \in U_2 \times \mathbb{P}^1,$$

if and only if $\zeta_1 = z_2^m\zeta_2$ and $z_1z_2 = 1$. Then $M^{(m)}$ is a compact complex manifold of dimension 2. It is called a <u>Hirzebruch</u> <u>surface</u>. Note that $M^{(m)}$ is a $\mathbb{P}^1$-bundle over $\mathbb{P}^1$. ($M^{(0)} = \mathbb{P}^1 \times \mathbb{P}^1$.) we can show that

(0) $\dim H^0(M^{(m)}, \Theta) = \begin{cases} m+5 & (m > 0) \\ 6 & (m = 0), \end{cases}$

(1) $\dim H^1(M^{(m)}, \Theta) = \begin{cases} m-1 & (m > 0) \\ 0 & (m = 0), \end{cases}$

(2) $\dim H^2(M^{(m)}, \Theta) = 0.$

For $t = (t_1,\cdots,t_{m-1}) \in \mathbb{C}^{m-1}$, let

$$M_t = (U_1 \times \mathbb{P}^1) \cup (U_2 \times \mathbb{P}^1)/\sim ,$$

where the equivalence relation $\sim$ is defined by:

$$(z_1,\zeta_1) \in U_1 \times \mathbb{P}^1 \sim (z_2,\zeta_2) \in U_2 \times \mathbb{P}^1,$$

if and only if $\zeta_1 = z_2^m\zeta_2 + t_1z_2 + t_2z_2^2 + \cdots + t_{m-1}z_2^{m-1}$ and $z_1z_2 = 1$. Then it is easy to see that $M_0 = M^{(m)}$ and $\{M_t\}_{t\in\mathbb{C}^{m-1}}$ is a family of compact complex manifolds whose Kodaira-Spencer map at 0 is a linear isomorphism. Hence, by Corollary 0.1.6, the family is versal at 0. But it is not universal at 0 for $m \geq 2$. (This can be shown directly or by using Theorem 0.1.9.)

If $m = 3$, then every M_t, $t \in \mathbb{C}^2 - \{0\}$, is biholomorphic to $M^{(1)}$ which is <u>not</u> biholomorphic to $M_0 = M^{(3)}$. Such a phenomenon is called <u>jumping</u> <u>structures</u>. This phenomenon does not occur for families of compact Riemann surfaces. Kodaira-Spencer [43] first found this phenomenon for a family of Hopf surfaces.

If $m = 4$, then (1) every M_t, $t \in \mathbb{C}^3$ - {$t_1 t_3$-plane}, is biholomorphic to $M^{(0)}$, (2) every M_t, $t \in$ {$t_1 t_3$-plane} - {0}, is biholomorphic to $M^{(2)}$ and (3) $M_0 = M^{(4)}$. Such a stratification holds for any m. See Suwa [80] for detailed arguments.

In almost all examples, the Kuranishi space S of V is non-singular and the Kodaira-Spencer map ρ_o at the reference point $o \in S$ is a linear isomorphism of $T_o S$ onto $H^1(V, \Theta)$. We say that V is <u>unobstructed</u> if this is the case. Otherwise, V is said to be <u>obstructed</u>. If $H^2(V, \Theta) = 0$, then V is unobstructed by Corollary 0.1.5. But it is a usual phenomenon that V is unobstructed, even if $H^2(V, \Theta) \neq 0$. It seems pretty difficult to give examples of obstructed compact complex manifolds. Such examples were given by Kodaira-Spencer [43], Kas [34], Mumford [56], Burns-Wahl [9], Horikawa [32,III], [33], Nakamura [60], and so on. Here, we explain only the example given by Kodaira-Spencer.

<u>Example 0.1.11. (c.f., Kodaira-Spencer [43])</u>. Let $q \geqq 2$. Let

$$R = \{(a_1, \cdots, a_q, b_1, \cdots, b_q, c_1, \cdots, c_q) \in \mathbb{C}^{3q} \mid$$

$$a_j b_k = a_k b_j,\ a_j c_k = a_k c_j,\ b_j c_k = b_k c_j \quad \text{for} \quad 1 \leqq j, k \leqq q\}.$$

Then R is a cone which defines a non-singular projective variety $\tilde{R}$ in $\mathbb{P}^{3q-1}$. In fact, $\tilde{R}$ is biholomorphic to $\mathbb{P}^2 \times \mathbb{P}^{q-1}$. Hence R is an (irreducible) subvariety of $\mathbb{C}^{3q}$ of dimension $q+2$ with a unique singular point 0, the origin. We can easily show that $\dim T_0 R = 3q$. For $x = (a_1, \cdots, a_q, b_1, \cdots, b_q, c_1, \cdots, c_q) \in R$, we define automorphisms $g_\beta(x)$, $1 \leqq \beta \leqq 2q$, of $\mathbb{P}^1$ by

$$g_\beta(x) = \text{the identity map for } 1 \leqq \beta \leqq q,$$

$$g_{q+\beta}(x) = \exp((a_\beta\zeta^2 + b_\beta\zeta + c_\beta)\partial/\partial\zeta) \quad \text{for} \quad 1 \leqq \beta \leqq q,$$

where exp is the exponential map of the complex Lie group $\mathrm{Aut}(\mathbb{P}^1)$ and ζ is an inhomogeneous coordinate in $\mathbb{P}^1$.

Let

$$S = \{ s \mid s \text{ is a } (q \times q)\text{-matrix such that } \det(\mathrm{Im}(s)) > 0 \},$$

where $\mathrm{Im}(s)$ is the imaginary part of s. Let I be the $(q \times q)$-identity matrix and put $\omega(s) = (I,s)$, a $(q \times 2q)$-matrix. We denote by $\omega_\beta(s)$ the β-th column vector of $\omega(s)$. Let $\tilde{g}_\beta$, $1 \leqq \beta \leqq 2q$, be the automorphisms of $\mathbb{C}^q \times \mathbb{P}^1 \times S \times R$ defined by

$$\tilde{g}_\beta : (z,\zeta,s,x) \longmapsto (z+\omega_\beta(s), g_\beta(x)\zeta, s, x).$$

Let Γ be the group of automorphisms of $\mathbb{C}^q \times \mathbb{P}^1 \times S \times R$ generated by $\tilde{g}_\beta$, $1 \leqq \beta \leqq 2q$. It is abelian. Then Γ acts on $\mathbb{C}^q \times \mathbb{P}^1 \times S \times R$ properly without fixed point. The quotient complex space $X = (\mathbb{C}^q \times \mathbb{P}^1 \times S \times R)/\Gamma$ with the induced projection $\pi : X \longrightarrow S \times R$ defines a family of compact complex manifolds. For any point $s \in S$, $\pi^{-1}(s,0)$ is (biholomorphic to) $T_s \times \mathbb{P}^1$, where T_s is a complex q-torus. In general, the fiber $\pi^{-1}(s,x)$ is a $\mathbb{P}^1$-bundle over T_s. We can prove that $(X,\pi,S \times R)$ is the Kuranishi family of $\pi^{-1}(s,0) = T_s \times \mathbb{P}^1$ for any $s \in S$.

A compact complex manifold V is said to be <u>rigid</u> if the Kuranishi space of V is just one point. By Kuranishi's theorem (Theorem 0.1.4),

<u>Theorem 0.1.12. (Frölicher-Nijenhuis [19], Kodaira-Spencer [43])</u>. If $H^1(V,\Theta) = 0$, then V is rigid.

<u>Example 0.1.13</u>. The d-dimensional complex projective space $\mathbb{P}^d$

is rigid (see Bott [8]). $\mathbb{P}^1 \times \mathbb{P}^1$ is rigid (see Example 0.1.10).

0.2. Families of submanifolds and holomorphic maps.

Let W be a complex manifold. Douady [13] proved that the set $D(W)$ of all (not necessarily reduced) compact complex subspaces of W has a (not necessarily reduced) complex space structure. (This is true even if W is a not necessarily reduced complex space.) $D(W)$ is called the Douady space of W. His proof uses theory of Banach analytic spaces and is very difficult.

If we consider only the set $S(W)$ of all compact complex submanifolds of W, then we can give a complex space structure on $S(W)$ by a simpler method which is an analogy of Kuranishi's proof on his Theorem 0.1.4. ($S(W)$ is an open subspace of $D(W)$.)

A family (X,π,S) of compact complex manifolds is called a family of compact complex submanifolds of W if (1) X is a closed complex subspace of $W \times S$ and (2) π is the restriction to X of the projection $W \times S \longrightarrow S$. In this case, each fiber $\pi^{-1}(s)$ is identified with a compact complex submanifold V_s of W. We sometimes write $\{V_s\}_{s\in S}$ instead of (X,π,S), as before.

Example 0.2.1. (Kodaira-Spencer [43]). Let $W = \mathbb{P}^{d+1}$. Let $(Z_0:Z_1:\cdots:Z_{d+1})$ be a homogeneous coordinate system in $\mathbb{P}^{d+1}$. For an integer $h \geq 1$, the set of all homogeneous polynomials of degree h of $Z_0, Z_1, \cdots, Z_{d+1}$ forms a vector space of dimension $N+1 = \binom{d+1+h}{h}$. The monomials

$$Z_0^h,\ Z_0^{h-1}Z_1,\ \cdots,\ Z_{d+1}^h$$

form a basis of it. For a point $s = (s_0:s_1:\cdots:s_N) \in \mathbb{P}^N$, we associate

$$f(Z,s) = s_0 Z_0^h + s_1 Z_0^{h-1} Z_1 + \cdots + s_N Z_{d+1}^h .$$

Put

$$V_s = \{Z \in \mathbb{P}^{d+1} \mid f(Z,s) = 0\} ,$$

$$S = \{s \in \mathbb{P}^N \mid V_s \text{ is non-singular and connected}\} .$$

Then, S is Zariski-open in $\mathbb{P}^N$ and $\{V_s\}_{s\in S}$ is a family of non-singular hypersurfaces of degree h in $\mathbb{P}^{d+1}$.

Let $\{V_s\}_{s\in S}$ be a family of compact complex submanifolds of W. For a point $o \in S$, let $\{W_i\}_{i\in I}$ be a finite collection of open subsets of W with coordinate systems

$$(w_i,z_i) = (w_i^1,\cdots,w_i^m,z_i^1,\cdots,z_i^d)$$

in W_i such that (1) V_o is covered by $\{W_i\}$ and (2) $U_i = V_o \cap W_i$ is defined in W_i by the equation : $w_i = 0$. Then there are an open neighborhood S' of o in S and a vector valued holomorphic function $\phi_i(z_i,s)$ on $U_i \times S'$ such that $V_s \cap W_i$ is defined in W_i by the equation : $w_i = \phi_i(z_i,s)$, for any $s \in S'$. Note that $\phi_i(z_i,o) = 0$ for all $z_i \in U_i$. Now, let

$$w_i = f_{ik}(w_k,z_k)$$

$$z_i = g_{ik}(w_k,z_k)$$

be the coordinate transformation in $W_i \cap W_k$. Then

$$F_{ik}(z_k) = (\partial f_{ik}/\partial w_k)(o,z_k)$$

is the transition matrix of the <u>normal bundle</u> F of $V = V_o$ in W. The functions $\phi_i(z_i,s)$ must satisfy the following compatibility condition:

$$f_{ik}(\phi_k(z_k,s),z_k) = \phi_i(g_{ik}(\phi_k(z_k,s),z_k),s).$$

Taking the partial derivative at (z_k,o) with respect to s, we get

$$F_{ik}(z_k)(\partial\phi_k/\partial s)_{(z_k,o)} = (\partial\phi_i/\partial s)_{(z_i,o)},$$

where $z_i = g_{ik}(z_k, o)$. This means that $\{(\partial\phi_i/\partial s)_{(z_i,o)}\}$ is an element of $H^0(V, \mathcal{O}(F))$, where $\mathcal{O}(F)$ is the sheaf of germs of holomorphic sections of F. (Henceforth, for a holomorphic vector bundle B, we denote by $\mathcal{O}(B)$ the sheaf of germs of holomorphic sections of B.) It is easy to see that this element does not depend on the choice of the covering $\{W_i\}_{i\in I}$ of V and is denoted by $\sigma_o(\partial/\partial s)$. It is called the infinitesimal displacement at $o \in S$ to the direction $\partial/\partial s$. It was found by Kodaira [41]. The map

$$\sigma_o : T_oS \longrightarrow H^0(V, \mathcal{O}(F))$$

is a linear map, called the characteristic map at $o \in S$. $\{V_s\}_{s\in S}$ is said to be injectively parametrized at $o \in S$ if σ_o is injective.

A family $\{V_s\}_{s\in S}$ of compact complex submanifolds of W is said to be maximal at $o \in S$, if, for any family $\{V'_{s'}\}_{s'\in S'}$ of compact complex submanifolds of W with a point $o' \in S'$ such that $V'_{o'} = V_o$, there are an open neighborhood U of o' in S' and a holomorphic map f of U into S such that (1) $f(o') = o$ and (2) $V_{f(s')} = V'_{s'}$ for all $s' \in U$.

$\{V_s\}_{s\in S}$ is said to be maximal if it is maximal at every point of S.

The following theorem corresponds to Kuranishi's Theorem 0.1.4. Our proof is also an analogy of the proof Kuranishi [49]. In Chapter 3, we give its proof in the case of holomorphic maps.

Theorem 0.2.2. (Namba [61]). Let W be a complex manifold. For any compact complex submanifold V of W, there is a maximal family $\{V_s\}_{s\in S}$ of compact complex submanifolds of W with a point $o \in S$ such that (1) $V_o = V$ and (2) it is injectively parametrized at every point of S. Moreover, the parameter space S is given as follows: there are an open neighborhood U of 0 in $H^0(V, \mathcal{O}(F))$ (F = the normal bundle of V in W.) and a holomorphic map $\alpha : U \longrightarrow H^1(V, \mathcal{O}(F))$

such that (3) $S = \{\xi \in U \mid \alpha(\xi) = 0\}$, (4) $o = 0$, (5) $(d\alpha)_o = 0$ and (6) $\sigma_o = (di)_o$, where $i : S \hookrightarrow U$ is the inclusion map.

Corollary 0.2.3. (Kodaira [41]). Let V be a compact complex submanifold of W with $H^1(V, \mathcal{O}(F)) = 0$. ($F$ = the normal bundle of V in W.) Then there is a maximal family $\{V_s\}_{s\in S}$ of compact complex submanifolds of W with a point $o \in S$ such that (1) $V_o = V$, (2) it is injectively parametrized at every point of S and (3) S is non-singular at o and $\dim_o S = \dim H^0(V, \mathcal{O}(F))$.

Corollary 0.2.4. (Kodaira [41]). Let $\{V_s\}_{s\in S}$ be a family of compact complex submanifolds of W. Assume that, for a non-singular point o of S, σ_o is surjective. Then the family is maximal at o.

For Example, the family in Example 0.2.1 is a maximal family. In almost all examples, the complex space S in Theorem 0.2.2 is non-singular and the characteristic map σ_o is a linear isomorphism of T_oS onto $H^0(V, \mathcal{O}(F))$. We say that V is unobstructed relative to W if this is the case. Otherwise, V is said to be obstructed relative to W. If $H^1(V, \mathcal{O}(F)) = 0$, then V is unobstructed relative to W by Corollary 0.2.3. But it is a usual phenomenon that V is unobstructed relative to W even if $H^1(V, \mathcal{O}(F)) \neq 0$. Examples of obstructed submanifolds were given by Zappa [86] and Mumford [56]. In Chapter 2, we give such examples.

For a complex manifold W, let $S(W)$ be the set of all compact complex submanifolds of W. For V_1 and V_2 in $S(W)$, put

$$d(V_1,V_2) = \sup\{\, d_W(x,V_2) \mid x \in V_1\} + \sup\{\, d_W(V_1,y) \mid y \in V_2\}\ ,$$

where d_W is a metric on W. Then d is a metric on $S(W)$.

Let $\{V_s\}_{s\in S}$ be the family in Theorem 0.2.2. Then, the correspondence

$$j : s \in S \longmapsto V_s \in S(W)$$

is injective (by the construction of the family). We take j as a "local coordinate system" in $S(W)$. Then, by the maximality, these local data are patched up to give a global complex space structure on $S(W)$. We can prove that the space thus defined is Hausdorff, using the fact that j is a continuous map from S to $(S(W),d)$. (However, I do not know weather the topology of the space $S(W)$ thus defined is equal to that of $(S(W),d)$.)

For a point $s \in S(W)$, let V_s be the compact complex submanifold of W corresponding to s. Then, we get the following theorem, which is a special case of Douady [13].

Theorem 0.2.5. For a complex manifold W, the set $S(W)$ of all compact complex submanifolds of W admits a complex space structure. Moreover, $\{V_s\}_{s\in S(W)}$ is a family of compact complex submanifolds of W and has the following universal property : for any family $\{V'_{s'}\}_{s'\in S'}$ of compact complex submanifolds of W, there is a unique holomorphic map f of S' into $S(W)$ such that $V'_{s'} = V_{f(s')}$ for all $s' \in S'$.

Let $\{V_s\}_{s\in S}$ be a family of compact complex submanifolds of W. Take a point $o \in S$. We give a relation between the Kodaira-Spencer map ρ_o and the characteristic map σ_o. Put $V = V_o$. Then, there is an exact sequence of vector bundle homomorphisms:

$$0 \longrightarrow TV \longrightarrow (TW)|V \longrightarrow F \longrightarrow 0 ,$$

where $(TW)|V$ is the restriction to V of the holomorphic tangent bundle TW of W. Hence, we have the following exact sequence of cohomology groups:

$$\cdots \longrightarrow H^0(V, \mathcal{O}(TW|V)) \longrightarrow H^0(V, \mathcal{O}(F)) \xrightarrow{\delta} H^1(V, \Theta) \longrightarrow \cdots$$

($\Theta = \mathcal{O}(TV)$). Then it is easy to see that the following lemma holds:

Lemma 0.2.6. The following diagram commutes:

$$\begin{array}{ccc} T_oS & \xrightarrow{\ \rho_o\ } & H^1(V, \Theta) \\ & {\scriptstyle \sigma_o}\searrow \quad \nearrow{\scriptstyle \delta} & \\ & H^0(V, \mathcal{O}(F)). & \end{array}$$

Hence

Proposition 0.2.7. Assume that, for a non-singular point $o \in S$, σ_o and δ are surjective. Then $\{V_s\}_{s \in S}$ is complete at o.

For example, the family in Example 0.2.1 is a complete family unless $(n,h) = (2,4)$. If $(n,h) = (2,4)$, then every V_s has small deformations of non-algebraic K3-surfaces.

Next, we consider families of holomorphic maps. Let V and W be compact complex manifolds. Then the set

$$\mathrm{Hol}(V,W) = \{ f \mid f : V \longrightarrow W \text{ is a holomorphic map} \}$$

is a complex space. In fact, by identifying $f \in \mathrm{Hol}(V,W)$ with its graph $\Gamma_f \subset V \times W$, $\mathrm{Hol}(V,W)$ is regarded as an open subspace of $S(V \times W)$. The underlying topology is eventurely the compact-open topology. The normal bundle of Γ_f in $V \times W$ is canonically isomorphic to f^*TW, the pull back of TW over f.

We say that $f \in \mathrm{Hol}(V,W)$ is unobstructed if the graph Γ_f is unobstructed relative to $V \times W$. Otherwise, f is said to be obstructed. If $H^1(V, \mathcal{O}(f^*TW)) = 0$, then f is unobstructed by Corollary 0.2.3. But it is a usual phenomenon for a holomorphic map to be unobstructed, even if $H^1(V, \mathcal{O}(f^*TW)) \neq 0$.

We state the following theorem without proof.

Theorem 0.2.8. Let V and W be compact complex manifolds.

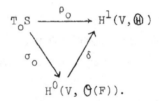

Assume that W has a local affine structure. (i.e., W has a coordinate covering whose coordinate transformations are affine maps.) Then every holomorphic map of V into W is unobstructed.

For example, every holomorphic map into a complex torus is unobstructed.

Akahori-Namba [2] gave examples of obstructed holomorphic maps. We talk about them in Chapter 2.

Next, the automorphism group Aut(V) of a compact complex manifold V is a complex Lie group. This fact was first proved by Bochner-Montgomery [7]. It is an open and closed subspace of Hol(V,V).

The complex Lie groups Aut(W) and Aut(W) × Aut(V) act (holomorphically) on Hol(V,W) by the composition of maps as follows:

$$(b,f) \in \mathrm{Aut}(W) \times \mathrm{Hol}(V,W) \longmapsto bf \in \mathrm{Hol}(V,W),$$

$$(b,a,f) \in \mathrm{Aut}(W) \times \mathrm{Aut}(V) \times \mathrm{Hol}(V,W) \longmapsto bfa^{-1} \in \mathrm{Hol}(V,W).$$

We may say that the orbit space Hol(V,W)/(Aut(W) × Aut(V)) is the <u>moduli space of holomorphic maps of</u> V <u>into</u> W. However, it is not in general even a Hausdorff space. (e.g., $V = W = \mathbb{P}^1$.)
Put

$$\mathrm{Open}(V,W) = \{ f \mid f \text{ is an open holomorphic map of } V \text{ onto } W \}.$$

Assume that it is non-empty. Then Open(V,W) is an open subspace of Hol(V,W). The groups Aut(W) and Aut(W) × Aut(V) act on Open(V,W). We can prove that the action on Open(V,W) of Aut(W) is free and proper. Hence, by Holmann's theorems (Holmann[30], [31]),

<u>Theorem 0.2.9. (Namba [64]).</u>
(1) The orbit space Open(V,W)/Aut(W) admits a complex space structure such that the projection

$$\mathrm{Open}(V,W) \longrightarrow \mathrm{Open}(V,W)/\mathrm{Aut}(W)$$

is a principal Aut(W)-bundle.

(2) If Aut(V) is compact, then Open(V,W)/(Aut(W) × Aut(V)) has a complex space structure.

This last space Open(V,W)/(Aut(W) × Aut(V)) may be called the moduli space of open holomorphic maps of V onto W. In Chapter 3, we give a proof of Theorem 0.2.9.

One of the simplest (non-trivial) example of Hol(V,W) is the case:

V = a compact Riemann surface,

W = $\mathbb{P}^1$, the complex projective line,

and this is the main object of our present study.

Chapter 1. Structure of $\mathrm{Hol}(V, \mathbb{P}^1)$.

1.1. Preliminary remarks on $\mathrm{Hol}(V, \mathbb{P}^1)$.

Let V be a compact Riemann surface of genus g. A <u>family of meromorphic functions on</u> V <u>with the parameter space</u> S means a holomorphic map

$$\mathfrak{F} : V \times S \longrightarrow \mathbb{P}^1,$$

where S is a complex space. We sometimes write $\{f_s\}_{s\in S}$ instead of $\mathfrak{F}$, where

$$f_s(P) = \mathfrak{F}(P,s) \quad \text{for} \quad (P,s) \in V \times S.$$

Let $\mathrm{Hol}(V, \mathbb{P}^1)$ be the Douady space of all holomorphic maps of V into $\mathbb{P}^1$ whose underlying topology is the compact-open topology. Then the map

$$\tilde{\mathfrak{F}} : (P,f) \in V \times \mathrm{Hol}(V, \mathbb{P}^1) \longmapsto f(P) \in \mathbb{P}^1$$

is holomorphic. Hence it is a family of meromorphic functions on V. This family is characterized by the following universal property (see Theorem 0.2.5): For any family $\{f_s\}_{s\in S}$, there is a unique holomorphic map $h : S \longrightarrow \mathrm{Hol}(V, \mathbb{P}^1)$ such that $f_s = h(s)$ for all $s \in S$.

For a positive integer n, we denote by $R_n(V)$ the subset of $\mathrm{Hol}(V, \mathbb{P}^1)$ of all meromorphic functions of order n. Here, by the <u>order of a meromorphic function</u> f, we mean the mapping order of $f : V \longrightarrow \mathbb{P}^1$. Note that $R_n(V)$ may be empty for some $n \leqq g$. On the other hand, every $R_n(V)$, for $n \geqq g+1$, is non-empty. (Consider non-Weierstrass points.)

$\mathrm{Hol}(V, \mathbb{P}^1)$ is then written as the disjoint union

$$\mathrm{Hol}(V, \mathbb{P}^1) = \mathrm{Const} \cup R_1(V) \cup R_2(V) \cup \cdots,$$

where Const is the set of all constant functions on V and is biholomorphic to $\mathbb{P}^1$.

<u>Lemma 1.1.1</u>. Every $R_n(V)$ is open and closed in $\mathrm{Hol}(V, \mathbb{P}^1)$.

<u>Proof</u>. It is enough to show that Const and all $R_n(V)$ are open. Let ω be a non-vanishing continuous real 2-form on $\mathbb{P}^1$. Then

$$\int_{V\times f} \tilde{\mathcal{F}}^*\omega = \int_V f^*\omega = \mathrm{ord}(f) \int_{\mathbb{P}^1} \omega ,$$

where $\mathrm{ord}(f)$ is the order of f. (If f is constant, then we put $\mathrm{ord}(f) = 0$.) The left hand side is a continuous function of $f \in \mathrm{Hol}(V, \mathbb{P}^1)$. Hence $\mathrm{ord}(f)$ is locally constant. <u>Q.E.D.</u>

Let $\mathcal{F} = \{f_s\}_{s\in S}$ be a family of meromorphic functions on V. Let $o \in S$. The characteristic map at o (see §0.2):

$$\sigma_o : T_oS \longrightarrow H^0(V, \mathcal{O}(f_o^*T\mathbb{P}^1))$$

is defined in this case by: $\sigma_o(\partial/\partial s) = (\partial\mathcal{F}/\partial s)_o$.

By Theorem 0.2.2, $\mathrm{Hol}(V, \mathbb{P}^1)$ is locally (around a point $o = f \in \mathrm{Hol}(V, \mathbb{P}^1)$) identified with

$$\tilde{S} = \{\, \xi \in U \mid \alpha(\xi) = 0 \,\} ,$$

where U is an open neighborhood of 0 in $H^0(V, \mathcal{O}(f^*T\mathbb{P}^1))$ and α is a holomorphic map of U into $H^1(V, \mathcal{O}(f^*T\mathbb{P}^1))$, such that $o = 0$ and $(d\alpha)_o = 0$. Moreover, the characteristic map

$$\sigma_f : T_f\mathrm{Hol}(V, \mathbb{P}^1) \longrightarrow H^0(V, \mathcal{O}(f^*T\mathbb{P}^1))$$

defined above with respect to the universal family $\tilde{\mathcal{F}}$ is identified with the natural inclusion map

$$(di)_o : T_o\tilde{S} \hookrightarrow H^0(V, \mathcal{O}(f^*T\mathbb{P}^1)).$$

A meromorphic function f on V is said to be <u>unobstructed</u> if

(1) f is a non-singular point of $\mathrm{Hol}(V, \mathbb{P}^1)$ and (2) σ_f is a linear isomorphism of $T_f\mathrm{Hol}(V, \mathbb{P}^1)$ onto $H^0(V, \mathcal{O}(f^*T\mathbb{P}^1))$. Otherwise, f is said to be <u>obstructed</u>. (See §0.2.)

Let $(Z_0:Z_1)$ be the standard homogeneous coordinate system in $\mathbb{P}^1$. Put $\infty = (0:1)$, the point of infinity. Then we have

$$T\mathbb{P}^1 = [2\infty],$$

where $[2\infty]$ is the line bundle determined by the divisor $2\infty = \infty + \infty$. Hence

$$f^*T\mathbb{P}^1 = f^*[2\infty] = [2D_\infty(f)],$$

where $D_\infty(f)$ is the <u>polar divisor</u> of f. Thus we get

<u>Proposition 1.1.2</u>. For a meromorphic function f on V, assume that $H^1(V, \mathcal{O}([2D_\infty(f)])) = 0$. Then f is unobstructed.

Henceforth, for a vector bundle F and a divisor D on V, we denote by $h^\nu(F)$ and $h^\nu(D)$ the dimensions of $H^\nu(V, \mathcal{O}(F))$ and $H^\nu(V, \mathcal{O}([D]))$, respectively. ([D] is the line bundle determined by D.)

Now, for $f \in R_n(V)$,

$$h^0(2D_\infty(f)) - h^1(2D_\infty(f)) = 2n+1-g,$$

by Riemann-Roch theorem. Thus, by the above description of the local structure of $\mathrm{Hol}(V, \mathbb{P}^1)$ and by Proposition 1.1.2, we get the following two propositions. (We give other direct proofs of them in §1.2.)

<u>Proposition 1.1.3</u>. Let $f \in R_n(V)$. Then

$$2n+1-g \leqq \dim_f R_n(V) \leqq \dim T_f R_n(V) \leqq h^0(2D_\infty(f)).$$

<u>Proposition 1.1.4</u>. If $n \geqq g$, then every $f \in R_n(V)$ is unobstructed. Hence, if $n \geqq g$, then $R_n(V)$ is non-singular and of dimension

2n+1-g.

Remark 1.1.5. $R_g(V)$ is empty if and only if V is hyperelliptic (elliptic) and g is odd (see Chapter 2).

One of the simplest examples is

Example 1.1.6. Let $V = \mathbb{P}^1$. Then $R_n(\mathbb{P}^1)$ is non-singular and of dimension $2n+1$. In fact, $f \in R_n(V)$ is expressed as

$$f(Z_0:Z_1) = (a_0Z_0^n + a_1Z_0^{n-1}Z_1 + \cdots + a_nZ_1^n : \\ b_0Z_0^n + b_1Z_0^{n-1}Z_1 + \cdots + b_nZ_1^n),$$

where the resultant $R(a;b) = R(a_0,\cdots a_n;b_0,\cdots,b_n)$ of the homogeneous polynomials $a_0Z_0^n+\cdots+a_nZ_1^n$ and $b_oZ_0^n+\cdots+ b_nZ_1^n$ is not zero. Hence there is a bijection

$$f \in R_n(\mathbb{P}^1) \longmapsto (a:b) = (a_0:\cdots:a_n:b_0:\cdots:b_n) \in \mathbb{P}^{2n+1} - \Delta,$$

where

$$\Delta = \{(a:b) \in \mathbb{P}^{2n+1} \mid R(a;b) = 0\} .$$

We can easily show that it is biholomorphic.

Now, we prove

Proposition 1.1.7. Let V and W be compact Riemann surfaces and let $u : V \longrightarrow W$ be a surjective holomorphic map of the mapping order e. Then the map

$$j_u : f \in R_n(W) \longmapsto fu \in R_{ne}(V)$$

is a holomorphic imbedding. (fu is the composition of the maps.)

Proof. It is clear that j_u is injective. For any $f \in R_n(W)$, we get the following commutative diagram:

$$\begin{array}{ccc} T_f R_n(W) & \xrightarrow{(dj_u)_f} & T_{fu} R_{ne}(V) \\ {\scriptstyle \sigma_f}\downarrow & & \downarrow{\scriptstyle \sigma_{fu}} \\ H^0(W, \mathcal{O}(f^* T\mathbb{P}^1)) & \xrightarrow{u^*} & H^0(V, \mathcal{O}((fu)^* T\mathbb{P}^1)), \end{array}$$

where $(dj_u)_f$ is the differential of j_u at f and u^* is the induced linear map. We can easily show that u^* is injective. Hence $(dj_u)_f$ is injective. Q.E.D.

The automorphism group $\mathrm{Aut}(\mathbb{P}^1)$ of $\mathbb{P}^1$ acts on $R_n(V)$ by the composition of maps.

By Theorem 0.2.9, the orbit space $R_n(V)/\mathrm{Aut}(\mathbb{P}^1)$ is a complex space and the projection $\tilde{\omega} : R_n(V) \longrightarrow R_n(V)/\mathrm{Aut}(\mathbb{P}^1)$ is a principal $\mathrm{Aut}(\mathbb{P}^1)$-bundle.

Lemma 1.1.8. Let $f \in R_n(V)$. Then the characteristic map σ_f induces an injective linear map

$$\hat{\sigma}_f : T_{\tilde{\omega}(f)}(R_n(V)/\mathrm{Aut}(\mathbb{P}^1)) \longrightarrow H^0(V, \mathcal{O}(f^* T\mathbb{P}^1))/f^* H^0(\mathbb{P}^1, \mathcal{O}(T\mathbb{P}^1)).$$

Proof. We have the following exact-commutative diagram:

$$\begin{array}{ccccccccc} & & 0 & & 0 & & & & \\ & & \downarrow & & \downarrow & & & & \\ 0 & \to & T_e \mathrm{Aut}(\mathbb{P}^1) & \longrightarrow & T_f R_n(V) & \longrightarrow & T_{\tilde{\omega}(f)}(R_n(V)/\mathrm{Aut}(\mathbb{P}^1)) & \to & 0 \\ & & {\scriptstyle \sigma_e}\downarrow & & {\scriptstyle \sigma_f}\downarrow & & {\scriptstyle \hat{\sigma}_f}\downarrow & & \\ 0 & \to & H^0(\mathbb{P}^1, \mathcal{O}(T\mathbb{P}^1)) & \to & H^0(V, \mathcal{O}(f^* T\mathbb{P}^1)) & \to & \dfrac{H^0(V, \mathcal{O}(f^* T\mathbb{P}^1))}{f^* H^0(\mathbb{P}^1, \mathcal{O}(T\mathbb{P}^1))} & \longrightarrow & 0 \\ & & \downarrow & & & & & & \\ & & 0 & & & & & & \end{array}$$

(e is the identity of $\mathrm{Aut}(\mathbb{P}^1)$). Chasing the diagram, we see that $\hat{\sigma}_f$ is injective. Q.E.D.

Proposition 1.1.9. Let V and W be compact Riemann surfaces and let $u : V \longrightarrow W$ be a surjective holomorphic map of order e. Then the map

$$\hat{j}_u : \tilde{\omega}(f) \in R_n(W)/\mathrm{Aut}(\mathbb{P}^1) \longmapsto \tilde{\omega}(fu) \in R_{ne}(V)/\mathrm{Aut}(\mathbb{P}^1)$$

is a holomorphic imbedding.

Proof. It is easy to see that $\hat{j}_u$ is well defined, holomorphic and injective. Note that the diagram

$$\begin{array}{ccc} T_{\tilde{\omega}(f)}(R_n(W)/\mathrm{Aut}(\mathbb{P}^1)) & \xrightarrow{d(\hat{j}_u)} & T_{\tilde{\omega}(fu)}(R_{ne}(V)/\mathrm{Aut}(\mathbb{P}^1)) \\ \downarrow \hat{\sigma}_f & & \downarrow \hat{\sigma}_{fu} \\ \dfrac{H^0(W, \mathcal{O}(f^*T\mathbb{P}^1))}{f^*H^0(\mathbb{P}^1, \mathcal{O}(T\mathbb{P}^1))} & \xrightarrow{\hat{u}^*} & \dfrac{H^0(V, \mathcal{O}((fu)^*T\mathbb{P}^1))}{(fu)^*H^0(\mathbb{P}^1, \mathcal{O}(T\mathbb{P}^1))} \end{array}$$

commutes, where $\hat{u}^*$ is the linear map induced by u^*. Hence, by Lemma 1.1.8, it suffices to show that $\hat{u}^*$ is injective. But, we have the following exact-commutative diagram:

$$\begin{array}{ccccccccc} & & & & 0 & & & & \\ & & & & \downarrow & & & & \\ 0 & \rightarrow & H^0(\mathbb{P}^1, \mathcal{O}(T\mathbb{P}^1)) & \rightarrow & H^0(W, \mathcal{O}(f^*T\mathbb{P}^1)) & \longrightarrow & \dfrac{H^0(W, \mathcal{O}(f^*T\mathbb{P}^1))}{f^*H^0(\mathbb{P}^1, \mathcal{O}(T\mathbb{P}^1))} & \rightarrow & 0 \\ & & \| & & \downarrow u^* & & \downarrow \hat{u}^* & & \\ 0 & \rightarrow & H^0(\mathbb{P}^1, \mathcal{O}(T\mathbb{P}^1)) & \rightarrow & H^0(V, \mathcal{O}((fu)^*T\mathbb{P}^1)) & \rightarrow & \dfrac{H^0(V, \mathcal{O}((fu)^*T\mathbb{P}^1))}{(fu)^*H^0(\mathbb{P}^1, \mathcal{O}(T\mathbb{P}^1))} & \rightarrow & 0 \end{array}$$

Chasing the diagram, we see that $\hat{u}^*$ is injective. Q.E.D.

In the following sections, we shall use known facts about symmetric product S^nV and the Jacobi variety $J(V)$ of V. We recall here their definitions and basic properties. (See, e.g., Gunning [28] for detail.)

Let V be a compact Riemann surface of genus g. The n-th symmetric product S^nV of V is, by definition, the quotient space $V^n/\mathfrak{S}_n$, where V^n is the Cartesian product of V and $\mathfrak{S}_n$ is the symmetric group of permutations on $\{1,2,\cdots,n\}$ acting on V^n as follows:

$$((P_1,\cdots,P_n),\sigma) \in V^n \times \mathfrak{S}_n \longmapsto (P_{\sigma(1)},\cdots,P_{\sigma(n)}) \in V^n.$$

S^nV is a n-dimensional compact complex manifold and is the set of all effective divisors on V of degree n.

Let K_V be the canonical bundle of V. Then $H^0(V,\mathcal{O}(K_V))$ is the set of all holomorphic 1-forms on V. Take a basis $\{\omega_1,\cdots,\omega_g\}$ of it. For a 1-cycle γ on V, the vector

$$u(\gamma) = (\textstyle\int_\gamma \omega_1,\cdots,\int_\gamma \omega_g) \in \mathbb{C}^g$$

is called a period. Then the set Γ of all periods forms a free subgroup of $\mathbb{C}^g$. It can be shown that, if $\gamma_1,\cdots,\gamma_{2g}$ is a basis of 1-cycles on V, then the periods $u(\gamma_1),\cdots,u(\gamma_{2g})$ form a basis of Γ and are linearly independent vectors over $\mathbb{R}$ (the field of real numbers). Hence $\mathbb{C}^g/\Gamma$ is a complex g-torus. We call it the Jacobi variety of V and denote by $J(V)$. It is known that $J(V)$ is an Abelian variety (i.e., is embedded in a projective space) and is canonically identified with the set of all (isomorphism classes of) holomorphic line bundles on V of degreee 0.

Take a point $P_o \in V$ and fix once and for all. We call it the base point. The Jacobi map $\phi : S^nV \longrightarrow J(V)$ is, by definition, the holomorphic map:

$$\phi : P_1+\cdots+P_n \longmapsto \left(\sum_{k=1}^{n}\int_{P_o}^{P_k}\omega_1,\cdots,\sum_{k=1}^{n}\int_{P_o}^{P_k}\omega_g \right) \text{ (mod periods).}$$

It can be shown that, for $n = 1$,

$$\phi : V \longrightarrow J(V)$$

is a holomorphic embedding. For $n = g$,

$$\phi : S^g V \longrightarrow J(V)$$

is a bimeromorphic map, (Jacobi inversion). Abel's theorem asserts that every fiber of $\phi : S^n V \longrightarrow J(V)$ is a set of all mutually linearly equivalent effective divisors, i.e., a complete linear system, and is therefore biholomorphic to a complex projective space. (Its dimension may depend on the choice of the fiber.)

If we identify $J(V)$ with the set of all holomorphic line bundles on V of degree 0, then the Jacobi map ϕ is given by

$$\phi : P_1+\cdots+P_n \longmapsto [P_1+\cdots+P_n-nP_o].$$

1.2. Construction of $R_n(V)/\mathbb{C}^*$.

We regard $\mathbb{C}^* = \mathbb{C} - \{0\}$ as a subgroup of $\mathrm{Aut}(\mathbb{P}^1)$. We construct the complex space $R_n(V)/\mathbb{C}^*$ concretely. This was suggested by Oda.

A meromorphic function $f \in R_n(V)$ is determined up to constant multiple by its zero divisor $D_0(f)$ and its polar divisor $D_\infty(f)$. The divisors $D_0(f)$ and $D_\infty(f)$ are linearly equivalent and have no point in common. Conversely, for any two linearly equivalent effective divisors D_1 and D_2 having no point in common, there is a meromorphic function f on V such that $D_1 = D_0(f)$ and $D_2 = D_\infty(f)$. This consideration leads to the construction of $R_n(V)/\mathbb{C}^*$ as follows:

Let $S^n V$ and $J(V)$ be the n-th symmetric product and the Jacobi variety of V, respectively. We fix a point $P_o \in V$ once and for all. Let K_V be the canonical bundle of V and let $\{\omega_1,\cdots,\omega_g\}$ be a basis of $H^0(V, \mathcal{O}(K_V))$. Let

$$\phi : S^n V \longrightarrow J(V)$$

be the Jacobi map with respect to the base point P_o and $\{\omega_1,\cdots,\omega_g\}$ That is to say:

$$\phi : P_1+\cdots+P_n \in S^nV \longmapsto \left(\sum_k\int_{p_o}^{p_k}\omega_1,\cdots,\sum_k\int_{p_o}^{p_k}\omega_g\right) \quad \text{(mod periods).}$$

Let

$$S^nV \times_{J(V)} S^nV = \{(D_1,D_2) \in S^nV \times S^nV \mid \phi(D_1) = \phi(D_2)\}$$

be the fiber product over $J(V)$. Put

$$B = \{(D_1,D_2) \in S^nV \times S^nV \mid D_1 \text{ and } D_2 \text{ have at least one point in common}\}.$$

Then B is an irreducible subvariety of $S^nV \times S^nV$ of codimension 1 (see Gunning [28, p.130]). Now, put

$$Q_n(V) = (S^nV \times_{J(V)} S^nV) - B.$$

Theorem 1.2.1. $R_n(V)/\mathbb{C}^*$ is biholomorphic to $Q_n(V)$.

Proof. Define a map

$$\alpha : R_n(V) \longrightarrow Q_n(V)$$

by $\alpha(f) = (D_0(f),D_\infty(f))$. It is factored through the projection

$$\tau : R_n(V) \longrightarrow R_n(V)/\mathbb{C}^*$$

so that $\alpha = \hat{\alpha}\tau$, for some map

$$\hat{\alpha} : R_n(V)/\mathbb{C}^* \longrightarrow Q_n(V).$$

By the above consideration, $\hat{\alpha}$ is a bijection. We show that $\hat{\alpha}$ is a biholomorphic map.

We first show that $\hat{\alpha}$ is holomorphic. Equivalently, we prove that α is holomorphic. For this purpose, it is enough to show that the map

$$s \in S \longmapsto (D_0(f_s), D_\infty(f_s)) \in S^nV \times S^nV$$

is holomorphic for any family $\mathcal{F} = \{f_s\}_{s \in S}$ of meromorphic functions of order n on V. The idea of the proof comes from the proof of Weierstrass Preparation theorem (see Gunning-Rossi[29]).

Let $o \in S$ be any fixed point. We may write

$$D_0(f_o) = \nu_1 P_1^o + \cdots + \nu_l P_l^o,$$

$$D_\infty(f_o) = \mu_1 Q_1^o + \cdots + \mu_m Q_m^o,$$

where P_i^o and Q_j^o are mutually distinct points of V and ν_i and μ_j are positive integers such that

$$\Sigma\nu_i = \Sigma\mu_j = n.$$

We denote by $B(P,\varepsilon)$ (resp. $\overline{B}(P,\varepsilon)$) the open (resp. closed) ε-ball with the center $P \in V$ with respect to a metric on V. Taking ε sufficiently small, we may assume that 2ε-balls $B(P_i^o, 2\varepsilon)$ and $B(P_j^o, 2\varepsilon)$ are mutually disjoint. Let z_i be a local coordinate in $B(P_i^o, 2\varepsilon)$ such that $z_i(P_i^o) = 0$.

Shrinking S if necessary and taking ε sufficiently small, we may assume that $\mathcal{F}$ is a holomorphic function on $(\cup B(P_i^o, 2\varepsilon)) \times S$. In fact, otherwise we may choose a sequence of points in $\mathcal{F}^{-1}(\infty)$ converging to, say, (P_1^o, o). Hence $f_o(P_1^o) = \infty$, a contradiction.

Let C_i be the boundary of $\overline{B}(P_i^o, \varepsilon)$. We put

$$\delta = \min\{|\mathcal{F}(P,o)| = |f_o(P)| \;\big|\; P \in C_i,\ i = 1, \cdots, l\}.$$

Then δ is a positive number.

Again, shrinking S if necessary and taking ε sufficiently small, we may assume that there is an ambient space Ω of S and that the holomorphic functions

$$\mathcal{F} : B(P_i^o, 2\varepsilon) \times S \longrightarrow \mathbb{C},\ i = 1, \cdots, l,$$

are extended to holomorphic functions

$$\tilde{\mathfrak{F}}_i : B(P_i^o, 2\varepsilon) \times \Omega \longrightarrow \mathbb{C}, \; i = 1, \cdots, l \; .$$

Since each C_i is compact, we may assume that

$$|\tilde{\mathfrak{F}}_i(P,s) - f_o(P)| < \delta \quad \text{for} \quad (P,s) \in C_i \times \Omega.$$

By Rouche's theorem, for each fixed $s \in \Omega$, the equation $\tilde{\mathfrak{F}}_i(P,s) = 0$ has just ν_i roots

$$P_{i1}(s), \cdots, P_{i\nu_i}(s)$$

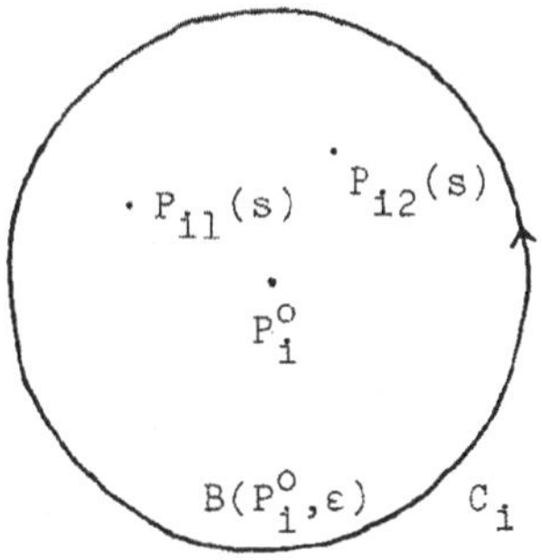

in $B(P_i^o, \varepsilon)$. Note that

$$P_{i1}(o) = \cdots = P_{i\nu_i}(o) = P_i^o \; .$$

We denote by $z_{ik}(s)$ the coordinate of the point $P_{ik}(s)$ in $B(P_i^o, 2\varepsilon)$. Then, for $p = 1, \cdots, \nu_i$, we have

$$\sum_k z_{ik}(s)^p = \frac{1}{2\pi\sqrt{-1}} \int_{C_i} \frac{(\partial \tilde{\mathfrak{F}}_i(\zeta,s)/\partial\zeta)\zeta^p d\zeta}{\tilde{\mathfrak{F}}_i(\zeta,s)} .$$

Hence, $\sum_k z_{ik}(s)^p$ is a holomorphic function of $s \in \Omega$. By Gunning [28, p.79], if $s \in S$, then this is one of the coordinates of the point $D_0(f_s)$ in $S^n V$. Hence

$$s \in S \longmapsto D_0(f_s) \in S^n V$$

is a holomorphic map. In a similar way, (considering $1/\mathfrak{F}(P,s)$), we can show that

$$s \in S \longmapsto D_\infty(f_s) \in S^n V$$

is a holomorphic map. Hence $\hat{\alpha}$ is a holomorphic map.

In order to show that $\hat{\alpha}^{-1}$ is also holomorphic, it is enough to show that α has a holomorphic local cross section through any given point $f_o \in R_n(V)$. As before, we write

$$D_0(f_0) = \nu_1 P_1^0 + \cdots + \nu_l P_l^0,$$

$$D_\infty(f_0) = \mu_1 Q_1^0 + \cdots + \mu_m P_m^0.$$

For simplicity, we may choose a basis $\{\omega_1, \cdots, \omega_g\}$ of $H^0(V, \mathcal{O}(K_V))$ and a basis $\{\gamma_1, \cdots, \gamma_{2g}\}$ of 1-cycles on V such that

$$\int_{\gamma_k} \omega_i = 2\pi\sqrt{-1}\delta_{ik} \quad (\delta_{ik} \text{ is the Kronecker's } \delta),$$

$$\int_{\gamma_{k+g}} \omega_i = 2a_{ik},$$

for $i,k = 1,\cdots,g$. Then, there are integers m_k and n_k $(k = 1,\cdots,g)$ such that

$$\sum_i \nu_i \int_{P_0}^{P_i^0} \omega_k - \sum_i \mu_i \int_{P_0}^{Q_i^0} \omega_k = 2m_k\pi\sqrt{-1} - 2\sum_j n_j a_{jk}$$

for $k = 1,\cdots,g$. We put

$$U = Q_n(V) \cap [(\prod_i S^{\nu_i} B(P_i^0,\varepsilon)) \times (\prod_j S^{\mu_j} B(Q_j^0,\varepsilon))],$$

where $S^{\nu_i}B(P_i^0,\varepsilon)$, etc., is the ν_i-th symmetric product of $B(P_i^0,\varepsilon)$. Then, for each $(D_1,D_2) \in U$, we associate a meromorphic function $f_{(D_1,D_2)} \in R_n(V)$ defined by

$$f_{(D_1,D_2)}(P) = \exp\left(\int_{P_0}^{P} (\Sigma n_k \omega_k)\right) \prod_i E_e(P,P_i) / \prod_j E_e(P,Q_j),$$

where

$$D_1 = P_1 + \cdots + P_n \quad \text{and} \quad D_2 = Q_1 + \cdots + Q_n.$$

Here, $E_e(P,Q)$ is the basic function defined by

$$E_e(P,Q) = \vartheta\left(\int_Q^P \hat{\omega} - e\right),$$

where ϑ is the Riemann's ϑ-function, $\hat{\omega} = (\omega_1, \cdots, \omega_g)$ and $e \in \mathbb{C}^g$ is a vector such that $\vartheta(e) = 0$ (see Mumford [59]). Moreover, we may assume that the $g-1$ other zeros $R_1,\cdots,R_{g-1}$ than Q of the equation $E_e(P,Q) = 0$ with respect to P are fixed, i.e., are independent of Q and are outside of any $\bar{B}(P_i^o,\varepsilon)$ and $\bar{B}(Q_i^o,\varepsilon)$ (see Mumford [59, p.66]). We show that

$$(P,D_1,D_2) \in V \times U \longmapsto f_{(D_1,D_2)}(P) \in \mathbb{P}^1$$

is a holomorphic map. In fact, since $B(P_i^o,\varepsilon)$ and $B(Q_j^o,\varepsilon)$ are mutually disjoint, it is clear that $f_{(D_1,D_2)}(P)$ is a holomorphic map of $(P,D_1,D_2) \in V \times U$, unless $P = R_1,\cdots,R_{g-1}$. On the other hand, let R_0 be a point of V which is distinct from $R_1,\cdots,R_{g-1}$ and is outside of any $\bar{B}(P_i^o,\varepsilon)$ and $\bar{B}(Q_j^o,\varepsilon)$. Then the meromorphic 1-form

$$\xi_P = \left(\frac{\partial}{\partial z} \log \frac{E_e(z,P)}{E_e(z,R_0)}\right) dz$$

has simple poles only at P, R_0 with residues ± 1 respectively (see Mumford [59, p.67]). Using ξ_P, the function $f_{(D_1,D_2)}$ is written as

$$f_{(D_1,D_2)}(P) = c_o \exp\left(\int_{P_o}^{P} \Sigma n_k \omega_k + \sum_i \int_{P_o}^{P} \xi_{P_i} - \sum_j \int_{P_o}^{P} \xi_{Q_j}\right),$$

where

$$c_o = \prod_i E_e(P_o,P_i) / \prod_j E_e(P_o,Q_j),$$

(c.f., Bliss [5, p.155]). From this expression, it is clear that $f_{(D_1,D_2)}(P)$ is a holomorphic map of $(P,D_1,D_2) \in V \times U$, unless $P = R_0$. Hence we conclude that $f_{(D_1,D_2)}(P)$ is a holomorphic map of $(P,D_1,D_2) \in V \times U$. From the construction,

$$D_0(f_{(D_1,D_2)}) = D_1 \quad \text{and} \quad D_\infty(f_{(D_1,D_2)}) = D_2.$$

In particular, $f_{(D_0(f_o),D_\infty(f_0))}$ has the same zeros and poles as f_o. Hence there is a non-zero constant c such that

$$f_o = c f_{(D_0(f_o),D_\infty(f_o))}.$$

Now the map

$$(D_1,D_2) \in U \longmapsto cf_{(D_1,D_2)} \in R_n(V)$$

is a holomorphic cross section of the map α through f_o. Q.E.D.

For the rest of this section, we give direct proofs of Proposition 1.1.3 and Proposition 1.1.4.

For a point $(D_1,D_2) \in Q_n(V)$, we take $f \in R_n(V)$ such that $D_0(f) = D_1$ and $D_\infty(f) = D_2$. f has only finite number of points in V as branch locus. Hence there is $b \in \mathrm{Aut}(\mathbb{P}^1)$ such that the divisors $D_0(bf)$ and $D_\infty(bf)$ consist of n distinct points. It is clear that f and bf have biholomorphic neighborhoods in $R_n(V)$. By Theorem 1.2.1, $\tau : R_n(V) \longrightarrow Q_n(V)$ is a principal $\mathbb{C}^*$-bundle. Hence (D_1,D_2) and $(D_0(bf),D_\infty(bf))$ have biholomorphic neighborhoods in $Q_n(V)$. Thus, in order to give estimates of $\dim_{(D_1,D_2)}Q_n(V)$, we may assume that the divisors D_1 and D_2 consist of n distinct points. Let

$$\tilde{\phi} : S^nV \times S^nV \longrightarrow J(V)$$

be the holomorphic map defined by

$$\tilde{\phi}(D,D') = \phi(D) - \phi(D'),$$

where the notation $x-y$ means the group theoretic difference in the complex Lie group $J(V)$. Then

$$Q_n(V) = \{(D,D') \in (S^nV \times S^nV) - B \mid \tilde{\phi}(D,D') = 0\} .$$

Put

$$D_1 = P_1^o+\cdots+P_n^o \quad \text{and} \quad D_2 = Q_1^o+\cdots+Q_n^o.$$

As before, we choose ε so small that $\overline{B}(P_i^o,\varepsilon)$ and $\overline{B}(Q_j^o,\varepsilon)$, $i,j = 1,\cdots,n$, are mutually disjoint. We choose suitable pathes so that

$$\sum_i \int_{P_o}^{P_i^o} \omega_k - \sum_j \int_{P_o}^{Q_j^o} \omega_k = 0, \quad k = 1,\cdots,g.$$

We put

$$W = B(P_1^o,\varepsilon)\times\cdots\times B(P_n^o,\varepsilon)\times B(Q_1^o,\varepsilon)\times\cdots\times B(Q_n^o,\varepsilon).$$

Then it is clear that the map $\widetilde{\phi}$ is locally written as follows:

$$\widetilde{\phi}(P_1,\cdots,P_n,Q_1,\cdots,Q_n)$$

$$= \Big(\sum_i \int_{P_i^o}^{P_i} \omega_1 - \sum_j \int_{Q_j^o}^{Q_j} \omega_1, \cdots, \sum_i \int_{P_i^o}^{P_i} \omega_g - \sum_j \int_{Q_j^o}^{Q_j} \omega_g \Big)$$

for $(P_1,\cdots,P_n,Q_1,\cdots,Q_n) \in W$. In particular, $Q_n(V)$ is locally biholomorphic to

$$S = \{(P_1,\cdots,P_n,Q_1,\cdots,Q_n) \in W \mid \sum_i \int_{P_i^o}^{P_i} \omega_k - \sum_j \int_{Q_j^o}^{Q_j} \omega_k = 0, \quad k = 1,\cdots,g\}.$$

Hence we get

$$2n-g \leqq \dim_{(D_1,D_2)} Q_n(V).$$

Now, the differential of $\widetilde{\phi}$ at (D_1,D_2) is given by

$$(d\widetilde{\phi})_{(D_1,D_2)} = \begin{pmatrix} \omega_1(P_1^o),\cdots,\omega_1(P_n^o),-\omega_1(Q_1^o),\cdots,-\omega_1(Q_n^o) \\ \cdots\cdots\cdots\cdots\cdots\cdots \\ \cdots\cdots\cdots\cdots\cdots\cdots \\ \omega_g(P_1^o),\cdots,\omega_g(P_n^o),-\omega_g(Q_1^o),\cdots,-\omega_g(Q_n^o) \end{pmatrix}$$

Here $\omega_k(P_i^o)$, etc., means the value $h_k(P_i^o)$, where $\omega_k = h_k(z)dz$ is the expression of ω_k in a local coordinate z in $B(P_i^o,\varepsilon)$. Let ${}^t(d\widetilde{\phi})_{(D_1,D_2)}$ be the transpose of the matrix $(d\widetilde{\phi})_{(D_1,D_2)}$. Then, for $(\lambda_1,\cdots,\lambda_g) \in \mathbb{C}^g$,

$${}^t(d\widetilde{\phi})_{(D_1,D_2)}\,{}^t(\lambda_1,\cdots,\lambda_g) = 0$$

if and only if

$$(\Sigma_k \lambda_k \omega_k)(P_i^o) = (\Sigma_k \lambda_k \omega_k)(Q_i^o) = 0, \quad i = 1,\cdots,n.$$

Hence

$$\begin{aligned} \text{rank } (d\tilde{\phi})_{(D_1,D_2)} &= \text{rank}\,{}^t(d\tilde{\phi})_{(D_1,D_2)} \\ &= g - h^1(D_1+D_2) = g - h^1(2D_2). \end{aligned}$$

Thus, by Riemann-Roch theorem,

$$\begin{aligned} \dim T_{(D_1,D_2)}Q_n(M) &\leqq 2n - \text{rank } (d\tilde{\phi})_{(D_1,D_2)} \\ &= 2n - g + h^1(2D_2) = h^0(2D_2) - 1. \end{aligned}$$

Hence we get

$$2n - g \leqq \dim_{(D_1,D_2)} Q_n(V) \leqq \dim T_{(D_1,D_2)} Q_n(V) \leqq h^0(2D_2) - 1.$$

This, together with Theorem 1.2.1, implies Proposition 1.1.3.

Assume that $n \geqq g$. Then

$$\text{rank } (d\tilde{\phi})_{(D_1,D_2)} = g - h^1(2D_2) = g.$$

Hence $\tilde{\phi}$ is of maximal rank at (D_1,D_2). This implies that $Q_n(V)$ is non-singular and of dimension $2n-g$. This, together with Theorem 1.2.1, implies Proposition 1.1.4.

Remark 1.2.2. In order to assert that a given complex space is non-singular at a point, this kind of argument will appear repeatedly in the present lecture notes.

1.3. Structure of $R_n(V)/\mathrm{Aut}(\mathbb{P}^1)$.

The complex space $R_n(V)/\mathrm{Aut}(\mathbb{P}^1)$ will be constructed in Chapter 3 and Chapter 5 by two different methods. In this section, we study the structure of $R_n(V)/\mathrm{Aut}(\mathbb{P}^1)$ by a direct method.

We define holomorphic maps

$$\beta : R_n(V) \longrightarrow J(V),$$

$$\gamma : Q_n(V) \longrightarrow J(V),$$

by

$$\beta(f) = \phi(D_\infty(f)),$$

$$\gamma(D_1, D_2) = \phi(D_2),$$

respectively, where $\phi : S^nV \longrightarrow J(V)$ is the Jacobi map. Then it is clear that the following diagram commutes:

$$\begin{array}{ccc} R_n(V) & \xrightarrow{\ \alpha\ } & Q_n(V) \\ & {\scriptstyle\beta}\searrow \quad \swarrow{\scriptstyle\gamma} & \\ & J(V) & \end{array}$$

where α is the map defined in §1.2. On the other hand, let

$$\tilde{\omega} : R_n(V) \longrightarrow R_n(V)/\mathrm{Aut}(\mathbb{P}^1)$$

be the projection. Then, there is a holomorphic map

$$\psi : R_n(V)/\mathrm{Aut}(\mathbb{P}^1) \longrightarrow J(V)$$

such that the following diagram commutes:

$$\begin{array}{ccc} R_n(V) & \xrightarrow{\ \tilde{\omega}\ } & R_n(V)/\mathrm{Aut}(\mathbb{P}^1) \\ & {\scriptstyle\beta}\searrow \quad \swarrow{\scriptstyle\psi} & \\ & J(V) & \end{array}$$

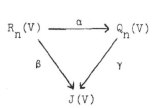

In fact, if $f \in R_n(V)$ and $b \in \mathrm{Aut}(\mathbb{P}^1)$, then $D_\infty(f)$ and $D_\infty(bf)$ are linearly equivalent.

We study the complex space $R_n(V)/\mathrm{Aut}(\mathbb{P}^1)$ using the map ψ. Since α and $\tilde{\omega}$ are surjective, we have

$$\psi(R_n(V)/\mathrm{Aut}(\mathbb{P}^1)) = \beta(R_n(V)) = \gamma(Q_n(V)).$$

<u>Lemma 1.3.1.</u> The image $\psi(R_n(V)/\mathrm{Aut}(\mathbb{P}^1))$ is the set of all points $x \in J(V)$ such that the complete linear system $\phi^{-1}(x)$ has no fixed point.

<u>Proof.</u> Let $x \in \psi(R_n(V)/\mathrm{Aut}(\mathbb{P}^1)) = \gamma(Q_n(V))$. Take a point $(D_1, D_2) \in Q_n(V)$ such that $x = \phi(D_2)$. Then D_1 and D_2 belong to $\phi^{-1}(x)$ and have no point in common.

Conversely, if $\phi^{-1}(x)$ has no fixed point, then any divisor $D \in \phi^{-1}(x)$ is the polar divisor $D_\infty(f)$ of some meromorphic function $f \in R_n(V)$ (see Lemma 1.3.2 below). Then $\beta(f) = x$. <u>Q.E.D.</u>

<u>Lemma 1.3.2. (Walker [83, Chapter 6, Theorem 8.1]).</u> For an effective divisor D, the complete linear system $|D|$ has no fixed point if and only if $D = D_\infty(f)$ for a meromorphic function f.

<u>Proof.</u> If $D = D_\infty(f)$, then D and $D_0(f) \in |D|$ have no common point. Conversely, assume that $|D|$ has no fixed point. If $\dim|D| = 1$, then it is clear that $D = D_\infty(f)$, for some meromorphic function f. Let $\dim|D| > 1$. Then the <u>associated meromorphic map</u>

$$\Phi_{|D|} : V \longrightarrow \mathbb{P}^N \quad (N = \dim|D|)$$

is holomorphic and every element of $|D|$ is obtained as the pullback of a hyperplane section of the image $\Phi_{|D|}(V)$. Let D be obtained by a hyperplane H. Choose another hyperplane H' such that H' does not intersect with $H \cap \Phi_{|D|}(V)$. Then the pencil $\{\lambda H + \lambda' H'\}_{(\lambda:\lambda') \in \mathbb{P}^1}$

gives a pencil of divisors on V, which defines a meromorphic function f such that $D = D_\infty(f)$.

Q.E.D.

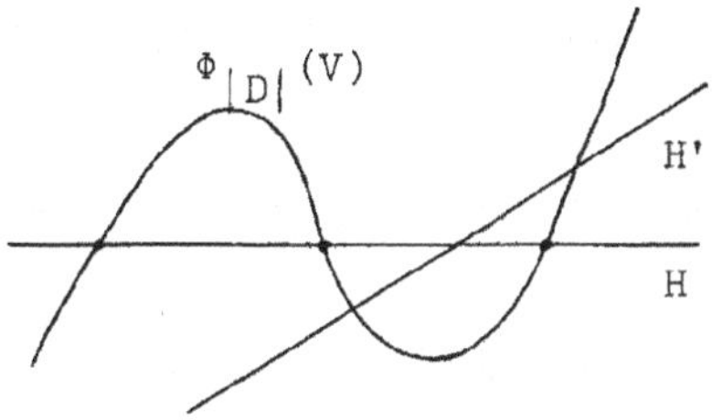

Now, we employ the following notations:

$$G_n^r = \{D \in S^nV \mid h^0(D) \geqq r + 1\},$$

$$W_n = \phi(S^nV),$$

$$W_n^r = \phi(G_n^r),$$

for $n \geqq 1$ and $r \geqq 0$. Note that Gunning [28] writes G_n^{r+1}, W_n^{r+1} for our G_n^r, W_n^r, respectively. We employ these notations in order to fit the notations in Chapter 5. Note that

$$S^nV = G_n^0 \supset G_n^1 \supset \cdots$$

are closed complex subspaces of S^nV and

$$W_n = W_n^0 \supset W_n^1 \supset \cdots$$

are closed complex subspaces of $J(V)$. It is clear that

$$\phi^{-1}(W_n^r) = G_n^r.$$

Note also that

$$\dim W_n = n \quad \text{for} \quad 1 \leqq n \leqq g,$$

$$W_n = J(V) \quad \text{for} \quad n \geqq g.$$

The set $\psi(R_n(V)/\mathrm{Aut}(\mathbb{P}^1))$ is, by Lemma 1.3.1, contained in W_n^1. We put

$$F_n^r = W_{n-1}^r + W_1,$$

where the notation A+B is the group theoretic sum in the complex

Lie group $J(V)$. F_n^r is called the <u>gap subvariety of</u> W_n^r. Note that $W_n^{r+1} \subset F_n^r$. In fact, for any point $x \in W_n^{r+1}$ and any point $P \in V$, the linear system $L = \{E \mid E + P \in \phi^{-1}(x)\}$ is complete and $\phi(L) + \phi(P) = x$.

In the sequel, for two subsets A and B of $J(V)$, we denote by $A \setminus B$ the set theoretic difference, while we denote by $A-B$ the group theoretic difference.

Now, the following lemma is an easy consequence of Gunning [28, p.66].

<u>Lemma 1.3.3</u>. For $r = 1,2,\cdots$,

$$\psi(R_n(V)/\mathrm{Aut}(\mathbb{P}^1)) \cap (W_n^r \setminus W_n^{r+1}) = W_n^r \setminus F_n^r.$$

Now, we describe the fibers of the map ψ. The following lemma is easy to see.

<u>Lemma 1.3.4</u>. Let $f \in R_n(V)$ and $b \in \mathrm{Aut}(\mathbb{P}^1)$. Put $x = \beta(f) = \beta(bf)$. Then the line L connecting $D_0(f)$ and $D_\infty(f)$ in $\phi^{-1}(x)$, as a complex projective space, coincides with the line connecting $D_0(bf)$ and $D_\infty(bf)$ in $\phi^{-1}(x)$. Conversely, for any distinct points D_1 and D_2 on L, there is $b \in \mathrm{Aut}(\mathbb{P}^1)$ such that $D_1 = D_0(bf)$ and $D_2 = D_\infty(bf)$.

Let B be the subvariety of $S^nV \times S^nV$ defined in §1.2.

<u>Lemma 1.3.5</u>. Let L be a line in $\phi^{-1}(x)$ for a point $x \in W_n^1$. Let Δ_L be the diagonal set in $L \times L$. Then, either $L \times L \subset B$ or $(L \times L) \cap B = \Delta_L$.

<u>Proof</u>. Take a point $(D_1, D_2) \in (L \times L) - B$. Then $(D_1, D_2) \in Q_n(V)$. Hence, by Lemma 1.3.4, $(D_1', D_2') \notin B$ for any distinct points D_1' and D_2' on L. <u>Q.E.D.</u>

Let $x \in W_n^1$. We denote by G_x^1 the Grassmann variety of all lines in $\phi^{-1}(x)$. Put

$$H_x = \{ L \in G_x^1 \mid (L \times L) \cap B = \Delta_L \} .$$

Then, by Lemma 1.3.5, H_x is a Zariski open subset of G_x^1.

<u>Lemma 1.3.6</u>. For any point $x \in W_n^1$, $\psi^{-1}(x)$ is biholomorphic to H_x.

<u>Proof</u>. Let $f \in R_n(V)$ be such that $D_\infty(f) \in \phi^{-1}(x)$. We associate with f the line L in H_x connecting $D_0(f)$ and $D_\infty(f)$. Then, this correspondence

$$\delta : f \in \beta^{-1}(x) \longmapsto L \in H_x$$

is clearly holomorphic. Moreover, by Lemma 1.3.4, it induces a bijective holomorphic map

$$\hat{\delta} : \tilde{\omega}(f) = f \pmod{\mathrm{Aut}(\mathbb{P}^1)} \in \psi^{-1}(x) \longmapsto L \in H_x .$$

(We sometimes write $f \pmod{\mathrm{Aut}(\mathbb{P}^1)}$ instead of $\tilde{\omega}(f)$.) In order to show that $\hat{\delta}^{-1}$ is also holomorphic, it is enough to show that δ has a holomorphic local cross section through any given $f \in \beta^{-1}(x)$. Let U be a small open neighborhood of $L = \delta(f)$ in H_x. We may assume that, for any $L' \in U$, two distinct points D_1 and D_2 on L' can be chosen so that they depend holomorphically on L'. As was shown in the proof of Theorem 1.2.1, there exists a meromorphic function $g \in R_n(V)$ depending holomorphically on (D_1, D_2) such that $D_1 = D_0(g)$ and $D_2 = D_\infty(g)$. Thus there exists $g_{L'} \in R_n(V)$ depending holomorphically on $L' \in U$ such that $\delta(g_{L'}) = L'$. In particular, $\delta(g_L) = L$. Hence, by Lemma 1.3.4, there is $b_o \in \mathrm{Aut}(\mathbb{P}^1)$ such that $f = b_o g_L$. Then

$$L' \in U \longmapsto b_o g_{L'} \in \beta^{-1}(x)$$

is a desired holomorphic section. Q.E.D.

Let $x \in W_n^1$. We put $r = \dim \phi^{-1}(x)$. Then $r \geqq 1$. Take any $D \in \phi^{-1}(x)$. Let $\Phi_{|D|} : V \longrightarrow \mathbb{P}^r$ be the meromorphic map associated with $|D|$. (Let $\{\eta_0, \eta_1, \cdots, \eta_r\}$ be a basis of $H^0(V, \mathcal{O}([D]))$. The meromorphic map

$$\Phi_x = \Phi_{|D|} : P \in V \longmapsto (\eta_0(P):\eta_1(P):\cdots:\eta_r(P)) \in \mathbb{P}^r$$

is called the <u>meromorphic map associated with the complete linear system</u> $\phi^{-1}(x) = |D|$.)

Assume that $\phi^{-1}(x) = |D|$ has no fixed point. Then Φ_x is a holomorphic map.

If $r = 1$, i.e., $x \in W_n^1 \setminus W_n^2$, then the holomorphic map Φ_x is a meromorphic function on V such that $\beta(\Phi_x) = x$. In this case, by Lemma 1.3.6, $\psi^{-1}(x)$ is just one point and is equal to Φ_x (mod $\mathrm{Aut}(\mathbb{P}^1)$)

Assume now $r \geqq 2$, i.e., $x \in W_n^r$. Then the image $C_x = \Phi_x(V)$ of Φ_x is a curve in $\mathbb{P}^r$ which is <u>not</u> contained in any hyperplane in $\mathbb{P}^r$. Every divisor in $\phi^{-1}(x) = |D|$ is obtained as the pull back over Φ_x of a hyperplane section of C_x. Hence, we may regard $\phi^{-1}(x)$ as the dual space of $\mathbb{P}^r$. A line L in $\phi^{-1}(x)$ corresponds to a $(r-2)$-dimensional linear subspace S of $\mathbb{P}^r$. The condition:

$$(L \times L) \cap B = \Delta_L$$

corresponds to the condition:

$$C_x \cap S = \phi \quad \text{(empty)}.$$

We denote by G_x^* the Grassmann variety of all $(r-2)$-dimensional linear subspaces in $\mathbb{P}^r$. We put

$$H_x^* = \{S \in G_x^* \mid C_x \cap S = \phi \quad \text{(empty)}\}.$$

Then, H_x^* is a Zariski open subset of G_x^*.

Now, Lemma 1.3.5 implies

Theorem 1.3.7. (1) For $x \in W_n^2$, assume that $\phi^{-1}(x)$ has no fixed point. Then $\psi^{-1}(x)$ is biholomorphic to H_x^*. Moreover, an element $\widetilde{\omega}(f) = f(\mathrm{mod}\ \mathrm{Aut}(\mathbb{P}^1))$ in $\psi^{-1}(x)$ is obtained as the composition of Φ_x and the projection $C_x = \Phi_x(V) \longrightarrow \mathbb{P}^1$ with the center S, the $(r-2)$-dimensional linear subspace of $\mathbb{P}^r$ $(r = \dim \phi^{-1}(x))$, corresponding to f (mod Aut($\mathbb{P}^1$)).

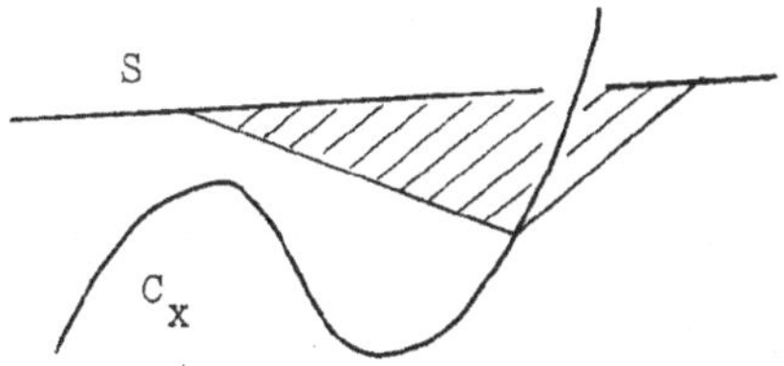

(2) If $x \in W_n^1 \setminus F_n^1$, then $\psi^{-1}(x)$ consists of one point: Φ_x (mod Aut($\mathbb{P}^1$)).

Example 1.3.8. Let $V = \mathbb{P}^1$. Then $J(\mathbb{P}^1)$ is one point. Let $(Z_0:Z_1)$ be as before the standard homogeneous coordinate system in $\mathbb{P}^1$. We put $0 = (1:0)$ and $n0 = 0+\cdots+0$ (n-times), a divisors on $\mathbb{P}^1$. Put

$$L(n0) = \{\text{rational functions whose poles are at most } n0\}.$$

Then $L(n0)$ is a vector space isomorphic to $H^0(\mathbb{P}^1, \mathcal{O}([n0]))$ and has the basis $\{1, x, \cdots, x^n\}$, where $x = Z_1/Z_0$. Hence, the holomorphic map $\Phi_{|n0|}$ is given by

$$\Phi_{|n0|} : x \in \mathbb{P}^1 \longmapsto (1:x:\cdots:x^n) \in \mathbb{P}^n.$$

The image $C_n = \Phi_{|n0|}(\mathbb{P}^1)$ is a non-singular rational curve in $\mathbb{P}^n$. If $n \geqq 2$, then, by Theorem 1.3.7, $R_n(\mathbb{P}^1)/\mathrm{Aut}(\mathbb{P}^1)$ is holomorphic to H^* the Zariski open subspace of the Grassmann variety consisting of all (n-2)-dimensional linear subspace of $\mathbb{P}^n$ which do not intersect with C_n. In particular, $R_2(\mathbb{P}^1)/\mathrm{Aut}(\mathbb{P}^1)$ is biholomorphic to $\mathbb{P}^2 - C_2$, where C_2 is a conic. Thus, $\mathbb{P}^5 - \Delta$ (see Example 1.1.6) is a principal Aut($\mathbb{P}^1$)-bundle over $\mathbb{P}^2 - C_2$.

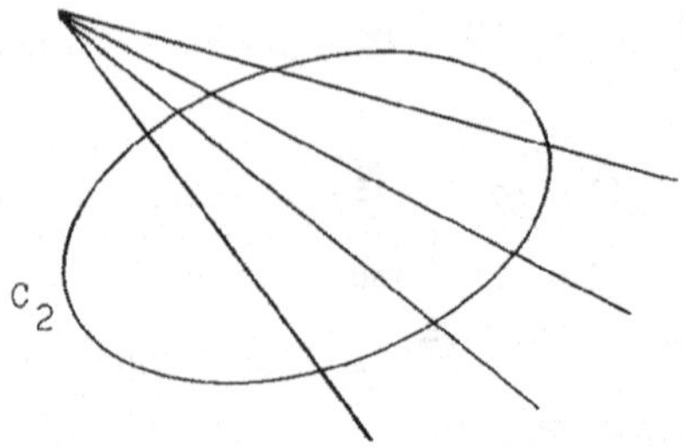

Now, we put

$$R_n^r(V) = \{f \in R_n(V) \mid h^0(D_\infty(f)) \geqq r + 1\}$$

for $r = 1, 2, \cdots$. Then

$$R_n(V) = R_n^1(V) \supset R_n^2(V) \supset \cdots$$

are closed complex subspaces of $R_n(V)$. In fact,

$$R_n^r(V) = \beta^{-1}(W_n^r).$$

It is clear that $\mathrm{Aut}(\mathbb{P}^1)$ acts on each $R_n^r(V) - R_n^{r+1}(V)$.

In order to describe the orbit space $(R_n^r(V) - R_n^{r+1}(V))/\mathrm{Aut}(\mathbb{P}^1)$, we need the following consideration:

The Jacobi variety $J(V)$ is identified with the set of all holomorphic line bundles on V of degree 0. We denote by B_x the line bundle corresponding to $x \in J(V)$. We put

$$F_x = B_x \otimes [nP_o],$$

where P_o is the base point and $nP_o = P_o + \cdots + P_o$ (n-times) is a divisor. The following theorem is a special case of Schuster [77, p.289]. We prove it in Chapter 4.

<u>Theorem 1.3.9</u>. The (disjoint) union $\mathbb{H}^0 = \cup_{x \in J(V)} H^0(V, \mathcal{O}(F_x))$ admits a complex space structure so that canonical projection $\lambda : \mathbb{H}^0 \longrightarrow J(V)$ is a <u>linear</u> <u>fiber</u> <u>space</u> in the sense of Grauert [23].

In particular,

$$\lambda : \lambda^{-1}(W_n^r \setminus W_n^{r+1}) \longrightarrow W_n^r \setminus W_n^{r+1}$$

is a holomorphic vector bundle for $r = 0,1,\cdots$.

Let $P(H^0(V, \mathcal{O}(F_x)))$ be the complex projective space of all 1-dimensional linear subspaces of $H^0(V, \mathcal{O}(F_x))$. (It is empty if $h^0(F_x) = 0$.) Then, by the theorem, the (disjoint) union

$$\mathbb{G}^0 = \cup_{x\in J(V)} P(H^0(V, \mathcal{O}(F_x)))$$

admits a complex space structure. The <u>divisor map</u>

$$\mathcal{D} : \eta \in H^0(V, \mathcal{O}(F_x)) - 0 \longmapsto (\eta) \in S^nV$$

((η) = the zero divisor of the section η.) induces a bijective map

$$\hat{\mathcal{D}} : \mathbb{G}^0 \longrightarrow S^nV.$$

We can easily show that it is a biholomorphic map (see Chapter 5). In particular,

<u>Corollary 1.3.10</u>. For $r = 0,1,\cdots$, $(G_n^r - G_n^{r+1}, \phi, W_n^r \setminus W_n^{r+1})$ is a fiber bundle with the standard fiber $\mathbb{P}^r$.

<u>Remark 1.3.11</u>. Note that Corollary 1.3.10 is not described in Gunning [28].

Now, let $\{U_i\}$ be an open covering of $W_n^r \setminus W_n^{r+1}$ and let

$$\sigma_i : \lambda^{-1}(U_i) \longrightarrow E \times U_i$$

be a local trivialization of the vector bundle

$$\lambda : \lambda^{-1}(W_n^r \setminus W_n^{r+1}) \longrightarrow W_n^r \setminus W_n^{r+1}.$$

Here, E is the standard fiber, a complex vector space. We fix a basis

$\{e_0, \cdots, e_r\}$ of E and put

$$\eta_k^i(x) = \sigma_i^{-1}(e_k, x), \quad k = 0, 1, \cdots, r,$$

for $x \in U_i$. Then $\{\eta_0^i(x), \cdots, \eta_r^i(x)\}$ form a basis of $H^0(V, \mathcal{O}(F_x))$.
Let

$$\phi^* : (G_n^r - G_n^{r+1})^* \longrightarrow W_n^r \setminus W_n^{r+1}$$

be the dual bundle to the projective bundle

$$\phi : (G_n^r - G_n^{r+1}) \longrightarrow W_n^r \setminus W_n^{r+1} .$$

We denote by

$$\sigma_i^* : \phi^{*-1}(U_i) \longrightarrow \mathbb{P}^r \times U_i$$

the local trivialization corresponding to σ_i. Here $\mathbb{P}^r$ is the dual space of the complex projective space of all 1-dimensional linear subspaces of E. We then define a meromorphic map

$$\Phi : V \times (W_n^r \setminus W_n^{r+1}) \longrightarrow (G_n^r - G_n^{r+1})^*$$

by

$$\Phi(P, x) = \sigma_i^{*-1}((\eta_0^i(x)(P) : \cdots : \eta_r^i(x)(P)), x).$$

Φ is well defined. In fact, fora point $P \in V$,

$$\hat{P}_\alpha(x) : \xi = \{\xi_\alpha\} \in \lambda^{-1}(x) = H^0(V, \mathcal{O}(F_x)) \longmapsto \xi_\alpha(P) \in \mathbb{C}$$

is an element of $\lambda^{-1}(x)^*$, the dual space of $\lambda^{-1}(x)$. Here, ξ_α is the representative of ξ with respect to an open covering $\{W_\alpha\}$ of V. For β with $P \in W_\beta$, $\hat{P}_\beta(x)$ differs from $\hat{P}_\alpha(x)$ by a non-zero scalar multiple. Hence $\hat{P}_\alpha(x)$ determines a point $\hat{P}(x)$ in $\phi^{*-1}(x)$, the dual space of $\phi^{-1}(x)$. It is then easy to see that $\Phi(P, x) = \hat{P}(x)$.

It is clear that Φ is holomorphic on $V \times (W_n^r \setminus F_n^r)$. Assume that $r \geqq 2$. Let $x \in W_n^r \setminus F_n^r$. Let G_x^* be as above the Grassmann variety of all (r-2)-dimensional linear subspaces in $\phi^{*-1}(x)$. Then, the (disjoint) union

$$\mathbb{G}^*(n,r) = \cup\, G_x^*, \quad x \in W_n^r \setminus F_n^r,$$

is a Grassmann bundle. We denote by $\mathbb{H}^*(n,r)$ the Zariski open subset of $\mathbb{G}^*(n,r)$ consisting of all elements which do not intersect with $\Phi(V \times (W_n^r \setminus F_n^r))$.

Now, by similar arguments to the proof of Theorem 1.3.7, we get

Theorem 1.3.12. (1) If $r \geqq 2$, then $(R_n^r(V) - R_n^{r+1}(V))/\mathrm{Aut}(\mathbb{P}^1)$ is biholomorphic to $\mathbb{H}^*(n,r)$.

(2) $(R_n(V) - R_n^2(V))/\mathrm{Aut}(\mathbb{P}^1)$ is biholomorphic to $W_n^1 \setminus F_n^1$.

Let $n \geqq 2g$. Then it is well known that, for every $D \in S^nV$, $|D|$ has no fixed point and of dimension n-g. Hence,

$$W_n^{n-g} = J(V), \quad W_n^{n-g+1} = \phi, \quad F_n^{n-g} = \phi.$$

Hence $\mathbb{G}^*(n,n-g)$ is a Grassmann bundle over $J(V)$ and $\mathbb{H}^*(n,n-g)$ is the Zariski open subset of $\mathbb{G}^*(n,n-g)$ consisting of all elements which do not intersect with $\Phi(V \times J(V))$, where

$$\Phi : V \times J(V) \longrightarrow (S^nV)^*$$

is the holomorphic map defined above. Thus

Corollary 1.3.13. (1) If $g \geqq 2$ and $n \geqq 2g$, Then $R_n(V)/\mathrm{Aut}(\mathbb{P}^1)$ is biholomorphic to $\mathbb{H}^*(n,n-g)$.

(2) If $g = 1$ and $n \geqq 3$, then $R_n(V)/\mathrm{Aut}(\mathbb{P}^1)$ is biholomorphic to $\mathbb{H}^*(n,n-1)$.

(3) If $g = 1$, then $R_2(V)/\mathrm{Aut}(\mathbb{P}^1)$ is biholomorphic to $J(V)$, hence to V.

Similar arguments can be done for $n = 2g-1$. Let κ be the point of $J(V)$ corresponding to the canonical bundle K_V, i.e., $K_V = F_\kappa$. κ is called the canonical point. Then it is well known (e.g., Gunning [28]) that

$$W_{2g-2}^{g-1} = \kappa, \quad W_{2g-1}^{g-1} = J(V), \quad W_{2g-1}^{g} = \phi.$$

Hence $F_{2g-1}^{g-1} = \kappa + W_1$ is biholomorphic to V. Thus $\mathbb{G}^*(2g-1,g-1)$ is a Grassmann bundle over $J(V)\setminus(\kappa+W_1)$ and $\mathbb{H}^*(2g-1,g-1)$ is the Zariski open subset of $\mathbb{G}^*(2g-1,g-1)$ consisting of all elements which do not intersect with $\Phi(V \times (J(V)\setminus(\kappa+W_1)))$.

Corollary 1.3.14. (1) If $g \geqq 3$, then $R_{2g-1}(V)/\mathrm{Aut}(\mathbb{P}^1)$ is biholomorphic to $\mathbb{H}^*(2g-1,g-1)$.

(2) If $g = 2$, then $R_3(V)/\mathrm{Aut}(\mathbb{P}^1)$ is biholomorphic to $J(V)\setminus(\kappa+W_1)$.

For later use, we state the following well known proposition. Its proof is found, e.g., in Meis [53].

Proposition 1.3.15. Let $r \geqq 0$ and $n \leqq g$. Then (1) $G_n^r - G_n^{r+1}$ is open dense in G_n^r. (2) $W_n^r \setminus W_n^{r+1}$ is open dense in W_n^r.

1.4. Families of elliptic functions.

Let V be a complex 1-torus. Then the automorphism group $\mathrm{Aut}(V)$ is a compact complex Lie group. It was shown in Namba [64] that the action

$$(b,a,f) \in \mathrm{Aut}(\mathbb{P}^1) \times \mathrm{Aut}(V) \times R_n(V) \longmapsto bfa^{-1} \in R_n(V)$$

is proper (see Chapter 3). Hence, by Holmann's theorem [31], the orbit space $E_n(V) = R_n(V)/(\mathrm{Aut}(\mathbb{P}^1) \times \mathrm{Aut}(V))$ is a complex space. This space can be considered as the moduli space of elliptic functions of order n on V. We determine its structure.

Let $\omega \in \mathbb{H}$ ($\mathbb{H}$ = the upper half plane). We consider the complex 1-torus

$$V = V_\omega = \mathbb{C}/\Delta,$$

where $\Delta = \mathbb{Z} + \mathbb{Z}\omega$ ($\mathbb{Z}$ = the additive group of integers).

For any point $Q \in V$, we denote by t_Q the translation of V by Q, i.e.,

$$t_Q : P \in V \longmapsto P + Q \in V.$$

It is an element of $\mathrm{Aut}(V)$. In fact,

$$\mathrm{Aut}_o(V) = \{t_Q \mid Q \in V\}$$

is the identity component of the complex Lie group $\mathrm{Aut}(V)$. It is isomorphic to V as complex Lie groups.

The map

$$s_{-1} : P \in V \longmapsto -P \in V$$

is also an element of $\mathrm{Aut}(V)$.

If V is neither isomorphic to $\mathbb{C}/(\mathbb{Z} + \mathbb{Z}i)$ nor to $\mathbb{C}/(\mathbb{Z} + \mathbb{Z}\zeta)$, ($i = \sqrt{-1}$, $\zeta = \frac{1+\sqrt{-3}}{2}$), then

$$\mathrm{Aut}(V) = \mathrm{Aut}_0(V) + s_{-1}\mathrm{Aut}_0(V) \quad \text{(the coset decomposition).}$$

If $V = \mathbb{C}/(\mathbb{Z} + \mathbb{Z}i)$, then the maps

$$s_{i^k} : P \in V \longmapsto i^k P \in V, \quad k = 0,1,2,3,$$

are well defined and are elements of $\mathrm{Aut}(V)$. In this case,

$$\mathrm{Aut}(V) = \sum_k s_{i^k}\mathrm{Aut}_0(V).$$

If $V = \mathbb{C}/(\mathbb{Z} + \mathbb{Z}\zeta)$, then the maps

$$s_{\zeta^k} : P \in V \longmapsto \zeta^k P \in V, \quad k = 0,1,2,3,4,5,$$

are well defined and are elements of $\mathrm{Aut}(V)$. In this case,

$$\mathrm{Aut}(V) = \sum_k s_{\zeta^k}\mathrm{Aut}_0(V), \quad \text{(see §3.5).}$$

In order to distinguish the addition of divisors on V from the addition of elements of V as a group, we temporary use the notation

$$P \dot{+} Q, \quad \text{for} \quad P,Q \in V,$$

for the addition of point divisors.

Now, we fix an integer $n \geqq 3$. Let 0 be zero of V (as a group). We take 0 as the base point. Then the Jacobi map

$$\phi : S^n V \longmapsto J(V) \cong V$$

is given by

$$\phi(P_1 \dot{+} \cdots \dot{+} P_n) = \sum_k \int_0^{P_k} dz = P_1 + \cdots + P_n,$$

where z is the standard coordinate in $\mathbb{C}$ and dz is regarded as an element of $H^0(V, \mathcal{O}(K_V))$. Hence, for a point $Q \in V$,

$$\phi^{-1}(Q) = \{P_1 \dot{+} \cdots \dot{+} P_n \in S^n V \mid \sum_k P_k = Q\}.$$

For any point $Q \in V$, we denote by F_Q the line bundle $[D]$ for $D \in \phi^{-1}(Q)$. Among them, $F_0 = [0 \dot{+} \cdots \dot{+} 0]$ (n-times) is the most important in our consideration. Let

$$\{\xi_0, \cdots, \xi_{n-1}\}$$

be a basis of $H^0(V, \mathcal{O}(F_0))$. Then the associated meromorphic map

$$\Phi_0 : P \in V \longmapsto (\xi_0(P):\cdots:\xi_{n-1}(P)) \in \mathbb{P}^{n-1}$$

gives a holomorphic imbedding of V into $\mathbb{P}^{n-1}$. We denote its image by $C = C_n$.

Let $\mathcal{S}$ be the (Zariski) open subspace of the Grassmann variety consisting of all (n-3)-dimensional linear subspaces of $\mathbb{P}^{n-1}$ which do not intersect with C. It is non-singular and of 2(n-2)-dimensional.

By Corollary 1.3.13, we know that

$$\psi : R_n(V)/\mathrm{Aut}(\mathbb{P}^1) \longrightarrow V$$

is surjective and of maximal rank at every point of $R_n(V)/\mathrm{Aut}(\mathbb{P}^1)$. Hence every fiber of ψ is a complex manifold of dimension 2(n-2). By Theorem 1.3.7, $\psi^{-1}(0)$ is biholomorphic to $\mathcal{S}$.

Now the map

$$s_n : P \in V \longmapsto nP \in V$$

is an unramified covering map of order n^2. Let

$$p : \mathcal{S} \times V \longrightarrow V$$

be the projection.

Lemma 1.4.1. There is a map

$$\tilde{s}_n : \mathcal{S} \times V \longrightarrow R_n(V)/\mathrm{Aut}(\mathbb{P}^1)$$

which makes the diagram

$$\begin{array}{ccc} \mathcal{S} \times V & \xrightarrow{\tilde{s}_n} & R_n(V)/\mathrm{Aut}(\mathbb{P}^1) \\ {\scriptstyle p}\downarrow & & \downarrow{\scriptstyle \psi} \\ V & \xrightarrow{s_n} & V \end{array}$$

commutative. Moreover, this diagram makes $\mathcal{S} \times V$ the fiber product of V and $R_n(V)/\mathrm{Aut}(\mathbb{P}^1)$ over V.

Proof. Let

$$\beta : R_n(V) \longrightarrow V$$

$$\tilde{\omega} : R_n(V) \longrightarrow R_n(V)/\mathrm{Aut}(\mathbb{P}^1)$$

be the maps defined in §1.3. Then $\mathrm{Aut}(\mathbb{P}^1)$ acts on $\beta^{-1}(0)$. $\beta^{-1}(0)/\mathrm{Aut}(\mathbb{P}^1)$ is then identified with $\psi^{-1}(0)$, which is identified with $\mathcal{S}$:

$$\mathcal{S} = \psi^{-1}(0) = \beta^{-1}(0)/\mathrm{Aut}(\mathbb{P}^1).$$

Let U be an open subset of $\mathcal{S}$ and let $u : U \longrightarrow \beta^{-1}(0)$ be a holomorphic cross section. We define a map

$$\tilde{s}_n : \mathcal{S} \times V \longrightarrow R_n(V)/\mathrm{Aut}(\mathbb{P}^1)$$

by

$$\tilde{s}_n : (S,P) \longmapsto \tilde{\omega}(u(S)t_{-P}).$$

Then it is well defined (i.e., does not depend on the choice of u) and holomorphic. It is clear that $\tilde{s}_n$ makes the above diagram commutative.

Next, let W be a complex space and let

$$\begin{array}{ccc} W & \xrightarrow{\ r\ } & R_n(V)/\mathrm{Aut}(\mathbb{P}^1) \\ {\scriptstyle q}\downarrow & & \downarrow{\scriptstyle \psi} \\ V & \xrightarrow{\ s_n\ } & V \end{array}$$

be a commutative diagram. Let U be an open subset of $R_n(V)/\mathrm{Aut}(\mathbb{P}^1)$ and let $v : U \longrightarrow R_n(V)$ be a holomorphic cross section. We define a map

$$w : W \longrightarrow \mathcal{S} \times V$$

by

$$w(x) = (\widetilde{\omega}(v(r(x))t_{q(x)}), q(x)).$$

Then it is well defined (i.e., does not depend on the choice of v) and holomorphic. It is now clear that the following diagram commutes:

$$\begin{array}{ccccc} & & W & & \\ & {\scriptstyle q}\swarrow & \downarrow{\scriptstyle w} & \searrow{\scriptstyle r} & \\ V & & & & R_n(V)/\mathrm{Aut}(\mathbb{P}^1) \\ & {\scriptstyle p}\nwarrow & & \nearrow{\scriptstyle \widetilde{s}_n} & \\ & & \mathcal{S} \times V & & \end{array}$$

<u>Q.E.D.</u>

By Lemma 1.4.1, we conclude

<u>Theorem 1.4.2</u>. (1) For $n \geqq 3$, $R_n(V)/\mathrm{Aut}(\mathbb{P}^1)$ is a fiber bundle over V with the standard fiber $\mathcal{S}$. The induced bundle of $R_n(V)/\mathrm{Aut}(\mathbb{P}^1)$ over s_n is a trivial bundle.

(2) $R_2(V)/\mathrm{Aut}(\mathbb{P}^1)$ is biholomorphic to V.

Now, we put

$$G_o = \{ t_P \mid P \in V,\ nP = 0 \}.$$

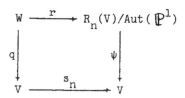

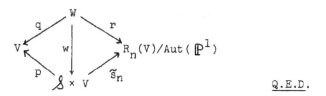

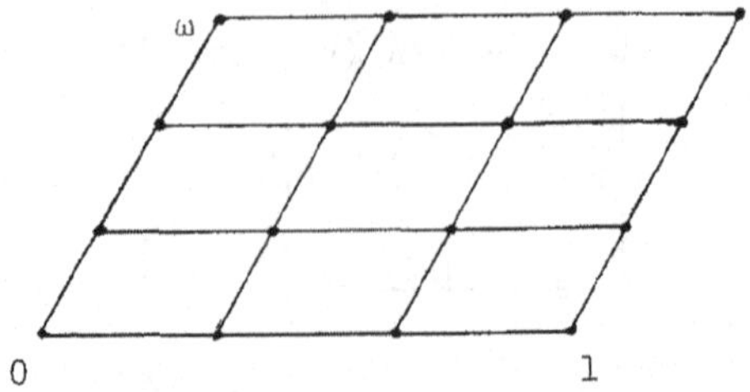

Then G_o is a finite subgroup of $\mathrm{Aut}_o(V)$ of order n^2.

We define a finite subgroup G of $\mathrm{Aut}(V)$ as follows:

(1) If V is neither biholomorphic to $\mathbb{C}/(\mathbb{Z}+\mathbb{Z}i)$ nor to $\mathbb{C}/(\mathbb{Z}+\mathbb{Z}\zeta)$, then G is the subgroup generated by G_o and s_{-1}. (ord $G = 2n^2$.)

(2) If $V = \mathbb{C}/(\mathbb{Z}+\mathbb{Z}i)$, then G is the subgroup generated by G_o and s_i. (ord $G = 4n^2$.)

(3) If $V = \mathbb{C}/(\mathbb{Z}+\mathbb{Z}\zeta)$, then G is the subgroup generated by G_o and s_ζ. (ord $G = 6n^2$.)

<u>Lemma 1.4.3</u>. Every element of G can be extended to a projective transformation of $\mathbb{P}^{n-1}$ mapping C onto C. In particular, G acts on $\mathcal{S}$.

<u>Proof</u>. We only consider the case of elements of G_o. Other elements of G can be treated in similar ways. Take $P \in V$ with $nP = 0$. For $D = P_1 \dot{+} \cdots \dot{+} P_n \in \phi^{-1}(0)$, put

$$t_P^* D = (P_1 - P) \dot{+} \cdots \dot{+} (P_n - P) \in \phi^{-1}(0).$$

Then the map $t_P^* : \phi^{-1}(0) \longrightarrow \phi^{-1}(0)$ is a projective transformation of $\phi^{-1}(0)$. In fact, for a fixed $D' = P_1' \dot{+} \cdots \dot{+} P_n' \in \phi^{-1}(0)$, put

$$L(D') = \{\text{all elliptic functions } f \text{ on } V \text{ with } (f) + D' \geqq 0\}.$$

Put $D'' = (P_1' - P) \dot{+} \cdots \dot{+} (P_n' - P) \in \phi^{-1}(0)$. Then, the map

$$t_P^* : f \in L(D') \longmapsto t_P^* f \in L(D'')$$

$((t_P^* f)(Q) = f(P+Q).)$ is a linear isomorphism. $\phi^{-1}(0)$ is regarded as the projective space of all lines through 0 of both $L(D')$ and $L(D'')$. (The canonical projection is $f \in L(D') \longmapsto (f)+D' \in \phi^{-1}(0)$.) Hence the linear isomorphism t_P^* induces a projective transformation of $\phi^{-1}(0)$, which is clearly equal to the map t_P^* defined above.

Dualizing it, we get a projective transformation t_P of $\mathbb{P}^{n-1}$. It is easy to see that the diagram

$$\begin{array}{ccc} V & \xrightarrow{\ \Phi_0\ } & C \\ {\scriptstyle t_P}\downarrow & & \downarrow{\scriptstyle t_P} \\ V & \xrightarrow{\ \Phi_0\ } & \mathbb{P}^{n-1} \end{array}$$

is commutative. Hence t_P maps C onto C. Q.E.D.

The following lemma is easy to see.

Lemma 1.4.4. If we identify $\mathcal{S}$ with $\beta^{-1}(0)/\mathrm{Aut}(\mathbb{P}^1)$, then the action of G_o on $\beta^{-1}(0)/\mathrm{Aut}(\mathbb{P}^1)$ is given as follows: Let $t_P \in G_0$ and $g \in \beta^{-1}(0)$. Then

$$t_P : g \pmod{\mathrm{Aut}(\mathbb{P}^1)} \longmapsto g t_{-P} \pmod{\mathrm{Aut}(\mathbb{P}^1)}.$$

The action of G on $\beta^{-1}(0)/\mathrm{Aut}(\mathbb{P}^1)$ is given in a similar way.

Now, we have

Theorem 1.4.5. If $n \geqq 3$, then

(1) $E_n^o(V) = R_n(V)/(\mathrm{Aut}(\mathbb{P}^1) \times \mathrm{Aut}_o(V))$ is biholomorphic to $\mathcal{S}/G_o$.
($E_2^o(V)$ is one point.)

(2) $E_n(V) = R_n(V)/(\mathrm{Aut}(\mathbb{P}^1) \times \mathrm{Aut}(V))$ is biholomorphic to $\mathcal{S}/G$.

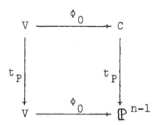

Proof. We prove (1). (2) can be proved in a similar way. As in the proof of Lemma 1.4.1, there are identifications:

$$\mathcal{S} = \psi^{-1}(0) = \beta^{-1}(0)/\mathrm{Aut}(\mathbb{P}^1).$$

Let $u : U \longrightarrow \beta^{-1}(0)$ be a holomorphic cross section as in the proof of Lemma 1.4.1, where U is an open subset of $\mathcal{S}$. We define a map $\mathcal{S} \longrightarrow E_n^o(V)$ by

$$S \longmapsto u(S) \ (\mathrm{mod}(\mathrm{Aut}(\mathbb{P}^1) \times \mathrm{Aut}_o(V))).$$

It is well defined (i.e., does not depend on the choice of u) and holomorphic. For $g, g' \in \beta^{-1}(0)$, if

$$g \equiv g' \ (\mathrm{mod}(\mathrm{Aut}(\mathbb{P}^1) \times \mathrm{Aut}_o(V))),$$

then there are $b \in \mathrm{Aut}(\mathbb{P}^1)$ and $P \in V$ such that $g' = bgt_{-P}$. If $nP \neq 0$, then $\beta(g') = nP \neq 0$, a contradiction. Hence $nP = 0$, i.e., $t_P \in G_o$. This implies that the induced holomorphic map

$$S \ (\mathrm{mod}\ G_o) \in \mathcal{S} \longmapsto u(S) \ (\mathrm{mod}(\mathrm{Aut}(\mathbb{P}^1) \times \mathrm{Aut}_o(V))) \in E_n^o(V)$$

is injective. We show that it is surjective and the inverse is also holomorphic. In fact, for any $h \in R_n(V)$, we put $Q = \beta(h)$. Locally, we can choose P with $nP = Q$, depending holomorphically on Q. Hence, locally, we can define a holomorphic map

$$h \in R_n(V) \longmapsto \tilde{\omega}(ht_P) \in \psi^{-1}(0),$$

which induces the inverse of the map defined above. Q.E.D.

Finally, we consider the case $n = 3$. The curve $C = C_3$ is, in this case, the image of the map

$$z \in \mathbb{C} \longmapsto (1 : \wp(z) : \wp'(z)) \in \mathbb{P}^2,$$

where $\wp$ is the Weierstrass $\wp$-function.

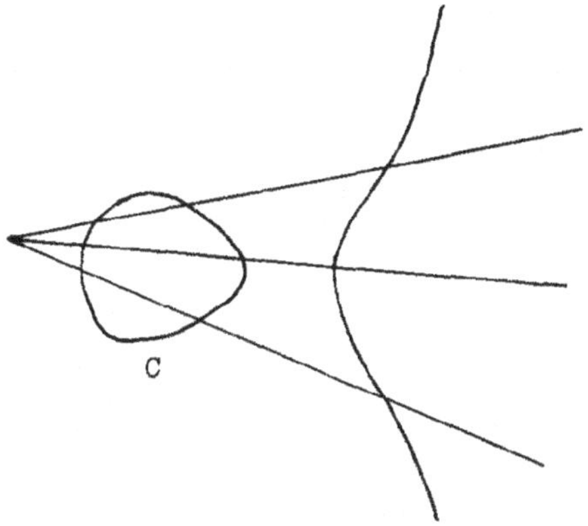

The proof of the following theorem is omitted.

Theorem 1.4.6.

(1) $E_3^o(V) \cong (\mathbb{P}^2 - C)/G_o$ has just 4 singular points. Every singular point is of the same type as the origin of

$$A = \{(x,y,z) \in \mathbb{C}^3 \mid xy = z^3\} .$$

(2) If V is neither biholomorphic to $\mathbb{C}/(\mathbb{Z}+\mathbb{Z}i)$ nor to $\mathbb{C}/(\mathbb{Z}+\mathbb{Z}\zeta)$, then $E_3(V) \cong (\mathbb{P}^2 - C)/G$ is non-singular.

(3) If V is biholomorphic to $\mathbb{C}/(\mathbb{Z}+\mathbb{Z}i)$ or to $\mathbb{C}/(\mathbb{Z}+\mathbb{Z}\zeta)$, then $E_3(V) \cong (\mathbb{P}^2 - C)/G$ has a unique singular point, which is of the same type as the origin of

$$B = \{(x,y,z) \in \mathbb{C}^3 \mid xy = z^2\} .$$

Chapter 2. $R_n(V)$ for $n \leqq g$.

2.1. $R_n(V)$ for $n \leqq g$.

We start with the following simple lemma.

Lemma 2.1.1. (e.g., Meis [53]). For a meromorphic function f on V of order $n \leqq g$, there are holomorphic 1-forms ω_1 and ω_2 on V such that $\omega_2 = f\omega_1$.

Proof. By Riemann-Roch theorem,

$$h^0(K_V \otimes [D_\infty(f)]^{-1}) = h^0(D_\infty(f))-n-1+g$$

$$\geqq 2-n-1+g = (g-n) + 1 \geqq 1.$$

Hence, there is a holomorphic 1-form ω_1 such that ω_1 vanishes on $D_\infty(f)$. Hence $\omega_2 = f\omega_1$ is also a holomorphic 1-form. Q.E.D.

Let f be a meromorphic function of order n on V. Let e_P be the ramification exponent of a point $P \in V$ of f. Then, the Riemann-Hurwitz formula holds:

$$\Sigma_P(e_P - 1) = 2n + 2g - 2,$$

where Σ is extended over all points of V.

(The general Hurwitz formula for a holomorphic map $f : V \longrightarrow V'$ of mapping order n of compact Riemann surfaces is given by

$$2 - 2g = n(2 - 2g') - \Sigma_P(e_P - 1),$$

where g and g' are the genera of V and V', respectively and e_P is the ramification exponent of a point $P \in V$ of f. One of the simplest proof of the formula is as follows:

Triangulate V' so that all branch points in V' of f are vertices of it. Pulling it back, we get a triangulation of V.

Counting the numbers of faces, sides and vertices, the relation of Euler-Poincaré characteristices is

$$\chi(V) = n\chi(V') - \Sigma_P(e_P - 1),$$

which is nothing but the above formula.)

Proposition 2.1.2. (e.g., Meis [53]). Let V be a hyperelliptic compact Riemann surface and let $\pi \in R_2(V)$. Then, for any $f \in R_n(V)$ with $n \leq g$, there is a rational function $h : \mathbb{P}^1 \longrightarrow \mathbb{P}^1$ of order $n/2$ such that $f = h\pi$ (the composition of the maps). In paticular, n must be even.

Proof. By replacing π by $b\pi$, $b \in \mathrm{Aut}(\mathbb{P}^1)$, if necessary, we may assume that $\infty \in \mathbb{P}^1$ is not a branch point of π. By the Riemann-Hurwitz formula, there are just $2g+2$ points of V at which π ramifies. Let $\{a_1, \cdots, a_{2g+2}\}$ be the set of branch points, i.e., the image of π of these points.

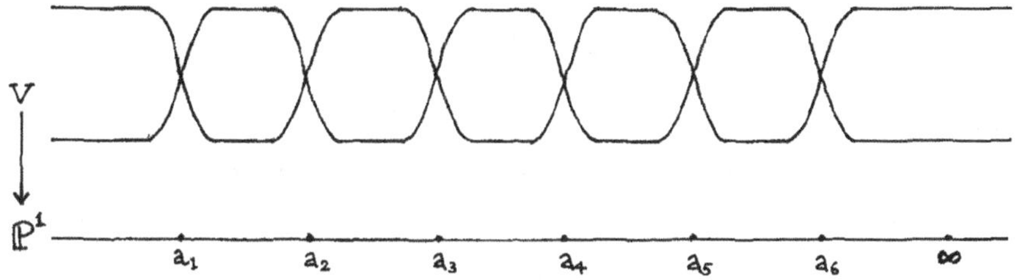

Then it is easy to see that V is a non-singular medel of the closure C in $\mathbb{P}^2$ of the curve:

$$y^2 = (x - a_1)(x - a_2) \cdots (x - a_{2g+2}).$$

The coordinate x is regarded as a meromorphic function on V and is equal to π. The coordinate y is also regarded as a meromorphic function on V. We can easily check that

$$\frac{dx}{y}, \frac{xdx}{y}, \ldots, \frac{x^{g-1}dx}{y}$$

are holomorphic 1-forms on V and form a basis of $H^0(V, \mathcal{O}(K_V))$. Hence every $\omega \in H^0(V, \mathcal{O}(K_V))$ is written as

$$\omega = p(x)\frac{dx}{y},$$

where $p(x)$ is a polynomial of x of degree at most $g-1$. Now, let $f \in R_n(V)$, $n \leqq g$. Then, by Lemma 2.1.1, there are holomorphic 1-forms

$$\omega_1 = p_1(x)\frac{dx}{y} \quad \text{and} \quad \omega_2 = p_2(x)\frac{dx}{y}$$

on V such that $\omega_2 = f\omega_1$. Hence $f = p_2(x)/p_1(x)$. Q.E.D.

Corollary 2.1.3. Let V be a hyperelliptic compact Riemann surface of genus g. Then, for $n \leqq g$,

$$R_n(V) \cong R_{n/2}(\mathbb{P}^1), \quad \text{if } n \text{ is even},$$

$$= \phi\,(\text{empty}), \quad \text{if } n \text{ is odd}.$$

Proof. By Proposition 1.1.7, the map

$$h \in R_{n/2}(\mathbb{P}^1) \longmapsto h\pi \in R_n(V)$$

is a holomorphic imbedding. Since it is surjective, it is biholomorphic. (Note that we consider only reduced complex spaces.) Q.E.D.

Proposition 2.1.4. Let V be a non-hyperelliptic compact Riemann surface of genus $g \geqq 3$. Then, $R_g(V)$ is non-empty, non-singular and of dimension $g+1$.

Proof. By Proposition 1.1.4, it suffices to prove that $R_g(V)$ is non-empty. By Theorem 1.3.12, it suffices to show that $W_g^1 \setminus F_g^1$ is non-empty. It is well known (see Gunning [28]) that $W_g^1 = \kappa - W_{g-2}$ is irreducible and of dimension $g-2$.

On the other hand, each irreducible component of W_{g-1}^1 has dimension g-4 (see Martens [52,I]). Hence $F_g^1 = W_{g-1}^1 + W_1$ has pure g-3 dimension (see Martens [52, I, p.115]). This shows that $W_g^1 \setminus F_g^1$ is non-empty. Q.E.D.

Example 2.1.5. Let V be non-hyperelliptic and have genus 3. Then $R_2(V)$ is empty and $R_3(V)/\mathrm{Aut}(\mathbb{P}^1)$ is biholomorphic to V. In fact, by Mumford [59, p.54], W_2^1 and W_3^2 are empty and W_3^1 is biholomorphic to V. The assertion then follows from Theorem 1.3.12.

The following interesting theorem was communicated by T. Akahori.

Theorem 2.1.6. (Akahori-Namba [2]). Let $f \in R_{g-1}(V)$.

(1) Assume that $[2D_\infty(f)] \neq K_V$. Then f is unobstructed.

(2) Assume that $[2D_\infty(f)] = K_V$. Then f is unobstructed if and only if $2h^0(D_\infty(f)) - 1 = g$.

Proof. (1). If $[2D_\infty(f)] \neq K_V$, then $h^1(2D_\infty(f)) = 0$. Hence, by Proposition 1.1.2, f is unobstructed and

$$\dim_f R_{g-1}(V) = 2(g-1) + 1 - g = g - 1.$$

(2). Assume that $[2D_\infty(f)] = K_V$. If f is unobstructed, then there is a g-dimensional connected non-singular open neighborhood U of f in $R_{g-1}(V)$ such that every $h \in U$ is also unobstructed and

$$h^0(2D_\infty(h)) = h^0(2D_\infty(f)) = g.$$

Hence

$$[2D_\infty(h)] = K_V = [2D_\infty(f)] \quad \text{for all} \quad h \in U.$$

Let $\phi : S^{g-1}V \longrightarrow J(V)$ be as before the Jacobi map. Then $\phi(D_\infty(h))$ depends holomorphically on $h \in U$. On the other hand,

$$x \in J(V) \longmapsto 2x \in J(V)$$

is a <u>finite</u> unramified covering map (with the covering order 2^{2g}). Note that the diagram

$$\begin{array}{ccc} D \in S^{g-1}V & \longrightarrow & 2D \in S^{2g-2}V \\ \downarrow \phi & & \downarrow \phi \\ x \in J(V) & \longrightarrow & 2x \in J(V) \end{array}$$

commutes. Thus we have

$$[D_\infty(h)] = [D_\infty(f)] \quad \text{for all} \quad h \in U.$$

This implies that $U \subset \beta^{-1}(x)$, where $x = \phi(D_\infty(f)) = \beta(f)$. ($\beta : R_{g-1}(V) \longrightarrow J(V)$ is the map defined in §1.3.) Thus U is a complex manifold of dimension

$$\dim \beta^{-1}(x) = 2h^0(D_\infty(f)) - 1.$$

Hence

$$g = 2h^0(D_\infty(f)) - 1.$$

Conversely, assume that the equality holds. This implies that $\dim \beta^{-1}(x) = g$. Hence, by Proposition 1.1.3,

$$g = \dim \beta^{-1}(x) \leqq \dim_f R_{g-1}(V) \leqq \dim T_f R_{g-1}(V) \leqq g.$$

Hence f is unobstructed. <u>Q.E.D.</u>

The equality $2h^0(D_\infty(f)) - 1 = g$ in (2) of the theorem occurs only when V is hyperelliptic and its genus g is odd. In fact, if V is non-hyperelliptic, then

$$h^0(D_\infty(f)) \leqq [g/2],$$

where [ ] is the Gauss' notation. (<u>Clifford's theorem</u>. See Gunning [28, p.60].) Thus

Corollary 2.1.7. Let V be a non-hyperelliptic comact Riemann surface of genus $g \geqq 4$. Assume that, for $f \in R_{g-1}(V)$, $[2D_\infty(f)] = K_V$. Then f is obstructed.

Corollary 2.1.8. Let V be non-hyperelliptic and have genus $g \geqq 4$. Assume that $R_{g-1}(V)$ is non-empty. Then, it is of pure $g-1$ dimension.

Proof. This follows from Corollary 2.1.7 and Proposition 1.1.3.

Q.E.D.

Problem. When is $R_{g-1}(V)$ non-empty?

Finally, we show

Theorem 2.1.9. Let n be the minimal positive number such that $R_n(V)$ is non-empty. Then W_n^2 is empty and the map

$$\psi : R_n(V)/\mathrm{Aut}(\mathbb{P}^1) \longrightarrow W_n^1,$$

defined in §1.3, is a biholomorphic map.

Proof. By Theorem 1.3.12,

$$\psi : (R_n(V) - R_n^2(V))/\mathrm{Aut}(\mathbb{P}^1) \longrightarrow W_n^1 \setminus F_n^1$$

is a biholomorphic map. Note that F_n^1 is empty, for W_{n-1}^1 is empty by the assumption on n. Hence W_n^2 $(\subset F_n^1)$ and so $R_n^2(V)$ are empty.

Q.E.D.

Example 2.1.10. Let V be non-hyperelliptic and have genus 4. Then W_3^1 consists of either one point or two points (see Mumford [59, p.55]). Accordingly, $R_3(V)/\mathrm{Aut}(\mathbb{P}^1)$ consists of one point or two points. In fact, the canonical curve $C = \Phi_{|K|}(V)$ of V is the complete intersection of a quadric surface F and a cubic surface G in $\mathbb{P}^3$ meeting transversally.

There are two caces: (1) F is singular, i.e., a quadric cone and (2) F is non-singular. There are: only one ruling on F in the case (1) and two rulings in the case (2). Accordingly, the above two cases of $R_3(V)/\mathrm{Aut}(\mathbb{P}^1)$ occur.

In the case (1), $f \in R_3(V)$ satisfies $[2D_\infty(f)] = K_V$. Hence f is obstructed by Corollary 2.1.7.

For example, a non-singular model of the closure in $\mathbb{P}^2$ of the curve:

$$y^3 = x^6 - 1$$

is in the case (1). The meromorphic function $x \in R_3(V)$ is thus obstructed and

$$R_3(V)/\mathrm{Aut}(\mathbb{P}^1) = \{ x \pmod{\mathrm{Aut}(\mathbb{P}^1)} \} .$$

A non-singular medel of the closure in $\mathbb{P}^2$ of the curve:

$$y^3 = \frac{x^3 - 1}{x^3 + 1}$$

is in the case (2). In this case,

$$R_3(V)/\mathrm{Aut}(\mathbb{P}^1) = \{ x \pmod{\mathrm{Aut}(\mathbb{P}^1)},\ \ y \pmod{\mathrm{Aut}(\mathbb{P}^1)} \} .$$

(See Meis [53].)

2.2. An example of $R_n(V)$ with singular points.

It is not known (to my knowledge) any example of Hol(V,W) having singular points, where V and W are compact complex manifolds. In this section, we present such an example.

Let $(Z_0:Z_1:Z_2)$ be the standard homogeneous coordinate system in $\mathbb{P}^2$. We put

$$x = Z_1/Z_0, \quad y = Z_2/Z_0 .$$

We consider an irreducible curve C in $\mathbb{P}^2$ defined by the equation:

$$Z_0^5 Z_2^3 = Z_1^8 - Z_0^8.$$

In the affine coordinate system (x,y), C is defined by

$$y^3 = x^8 - 1.$$

C has the unique singular point $(Z_0:Z_1:Z_2) = (0:0:1)$. Moreover C is irreducible at the point. We denote this point by ∞.

Let V be a non-singular model of C. We sometimes identify each point of C with the point of V corresponding to it.

Lemma 2.2.1. The genus of V is 7.

Proof. Apply the Riemann-Hurwitz formula (§2.1) to the meromorphic function x. Note that the ramification exponent at ∞ is 3. Q.E.D.

The meromorphic function x has the order 3. Hence, by Proposition 2.1.2, V is non-hyperelliptic.

Lemma 2.2.2. The following abelian differentials form a basis of $H^0(V, \mathcal{O}(K_V))$:

$$\frac{dx}{y},\ \frac{xdx}{y},\ \frac{dx}{y^2},\ \frac{xdx}{y^2},\ \frac{x^2dx}{y^2},\ \frac{x^3dx}{y^2},\ \frac{x^4dx}{y^2}\ .$$

Proof. It is easy to see that they have no poles on V. Moreover, they have the point ∞ as a zero of order

$$4,\ 1,\ 12,\ 9,\ 6,\ 3,\ 0,$$

respectively. Hence they are linearly independent. Since V has the genus 7, they form a basis of $H^0(V, \mathcal{O}(K_V))$. Q.E.D.

For a holomorphic 1-form ω, we denote by (ω) the zero divisor of ω. Then we have

$$(dx/y^2) = 12\infty\ .$$

Now, we show that $R_6(V)$ has singularity. The meromorphic function $f = x^2$ is an element of $R_6(V)$. Moreover,

$$2D_\infty(f) = 12\infty = (dx/y^2).$$

Hence

$$[2D_\infty(f)] = K_V.$$

Thus, by Corollary 2.1.7, f is obstructed. However, this does not necessarily imply that f is a singular point of $R_6(V)$, for we always assume that a complex space is reduced (see e.g., Example 2.1.10).

For our purpose, it is enough to prove the following proposition.

Proposition 2.2.3. $f = x^2$ is a singular point of $R_6(V)$. In fact, the tangent cone to $R_6(V)$ at f is given by

$$\{ (z_1,z_2,\cdots,z_7) \in \mathbb{C}^7 \mid z_1z_2 = 0 \}.$$

Proof. For any $b \in \mathrm{Aut}(\mathbb{P}^1)$, f and bf have biholomorphic open neighborhoods in $R_6(V)$. Hence we may compute the tangent cone at bf for a suitable $b \in \mathrm{Aut}(\mathbb{P}^1)$ instead of f. Let $b \in \mathrm{Aut}(\mathbb{P}^1)$ be such that

$$b(2) = 0 \quad \text{and} \quad b(-2) = \infty,$$

where we use the inhomogeneous coordinate $\xi = Z_1/Z_0$, $(\infty = 1/0)$, in $\mathbb{P}^1$. Then

$$D_0(bf) = f^{-1}(2) = \{(\alpha,\beta),\ (\alpha,\rho\beta),\ (\alpha,\rho^2\beta),\ (-\alpha,\beta),\ (-\alpha,\rho\beta),\ (-\alpha,\rho^2\beta)\},$$

$$D_\infty(bf) = f^{-1}(-2) = \{(i\alpha,\beta),\ (i\alpha,\rho\beta),\ (i\alpha,\rho^2\beta),\ (-i\alpha,\beta),\ (-i\alpha,\rho\beta),$$
$$(-i\alpha,\rho^2\beta)\},$$

where

$$\alpha = \sqrt{2},\ \beta = \sqrt[3]{15},\ i = \sqrt{-1},\ \rho = \frac{-1+\sqrt{-3}}{2}.$$

Note that both divisors $D_0(bf)$ and $D_\infty(bf)$ consist of 6 distinct points.

By Theorem 1.2.1, there are biholomorphic open neighborhoods S_1 of $bf \pmod{\mathbb{C}^*}$ in $R_6(V)/\mathbb{C}^*$ and S_2 of $(D_0(bf), D_\infty(bf))$ in $Q_6(V)$. We regard

$$((\alpha,\beta),\cdots,(-\alpha,\rho^2\beta)) \quad \text{and} \quad ((i\alpha,\beta),\cdots,(-i\alpha,\rho^2\beta))$$

as points of V^6 (the Cartesian product of V).

Let B_1 and B_2 be small open neighborhoods of these points in V_6, respectively. Then S_2 is given as the set of zeros in $B_1 \times B_2$ of the functions

$$f_1 = \left(\int_{(\alpha,\beta)}^{(x_1,y_1)} + \cdots + \int_{(-\alpha,\rho^2\beta)}^{(x_6,y_6)} - \int_{(i\alpha,\beta)}^{(\hat{x}_1,\hat{y}_1)} - \cdots - \int_{(-i\alpha,\rho^2\beta)}^{(\hat{x}_6,\hat{y}_6)}\right) \frac{dx}{y},$$

....

$$f_7 = \left(\int_{(\alpha,\beta)}^{(x_1,y_1)} + \cdots + \int_{(-\alpha,\rho^2\beta)}^{(x_6,y_6)} - \int_{(i\alpha,\beta)}^{(\hat{x}_1,\hat{y}_1)} - \cdots - \int_{(-i\alpha,\rho^2\beta)}^{(\hat{x}_6,\hat{y}_6)}\right) \frac{x^4dx}{y^2},$$

where $((x_1,y_1),\cdots,(x_6,y_6))\in B_1$ and $((\hat{x}_1,\hat{y}_1),\cdots,(\hat{x}_6,\hat{y}_6))\in B_2$.
We put

$$s_1 = x_1-\alpha,\cdots,\quad s_6 = x_6-(-\alpha),$$

$$t_1 = \hat{x}_1-i\alpha,\cdots,\quad t_6 = \hat{x}_6-(-i\alpha)$$

Then $(s_1,\cdots,s_6,t_1,\cdots,t_6)$ forms a local coordinate system in $B_1\times B_2$.

We expand the above functions $f_1,\cdots,f_7$ in the power series of $(s_1,\cdots,s_6,t_1,\cdots t_6)$:

$$\begin{aligned} f_1 = {} & \frac{1}{\beta}s_1+\frac{1}{\rho\beta}s_2+\frac{1}{\rho^2\beta}s_3+\frac{1}{\beta}s_4+\frac{1}{\rho\beta}s_5+\frac{1}{\rho^2\beta}s_6 \\ & -\frac{1}{\beta}t_1-\frac{1}{\rho\beta}t_2-\frac{1}{\rho^2\beta}t_3-\frac{1}{\beta}t_4-\frac{1}{\rho\beta}t_5-\frac{1}{\rho^2\beta}t_6 \\ & +*, \end{aligned}$$

$$\cdots\cdots$$

$$\begin{aligned} f_7 = {} & \frac{\alpha^4}{\beta^2}s_1+\frac{\alpha^4}{\rho^2\beta^2}s_2+\frac{\alpha^4}{\rho\beta^2}s_3+\frac{\alpha^4}{\beta^2}s_4+\frac{\alpha^4}{\rho^2\beta^2}s_5+\frac{\alpha^4}{\rho\beta^2}s_6 \\ & -\frac{\alpha^4}{\beta^2}t_1-\frac{\alpha^4}{\rho^2\beta^2}t_2-\frac{\alpha^4}{\rho\beta^2}t_3-\frac{\alpha^4}{\beta^2}t_4-\frac{\alpha^4}{\rho^2\beta^2}t_5-\frac{\alpha^4}{\rho\beta^2}t_6 \\ & +*, \end{aligned}$$

where $*$ indicates higher order terms. Note that the function

$$f_8 = f_7-\alpha^4f_3 = f_7-4f_3$$

has no linear term. It is expanded in the power series as follows:

$$\begin{aligned} f_8 = {} & c(s_1^2+\rho s_2^2+\rho^2s_3^2-s_4^2-\rho s_5^2-\rho^2s_6^2 \\ & +it_1^2+i\rho t_2^2+i\rho^2t_3^2-it_4^2-i\rho t_5^2-i\rho^2t_6^2)+*, \end{aligned}$$

where $c = 180\sqrt{2}/(3\beta^5)$.

We show that the rank of the (6×12)-matrix of the coefficients of the linear terms of the functions $f_1,\cdots,f_6$ is 6. Omitting α, $1/\beta$,

$1/\beta^2$, etc., in the expansions, it is enough to consider the following matrix:

$$A = \begin{pmatrix} 1 & \rho^2 & \rho & 1 & \rho^2 & \rho & -1 & -\rho^2 & -\rho & -1 & -\rho^2 & -\rho \\ 1 & \rho^2 & \rho & -1 & -\rho^2 & -\rho & -i & -i\rho^2 & -i\rho & i & i\rho^2 & i\rho \\ 1 & \rho & \rho^2 & 1 & \rho & \rho^2 & -1 & -\rho & -\rho^2 & -1 & -\rho & -\rho^2 \\ 1 & \rho & \rho^2 & -1 & -\rho & -\rho^2 & -i & -i\rho & -i\rho^2 & i & i\rho & i\rho^2 \\ 1 & \rho & \rho^2 & 1 & \rho & \rho2 & 1 & \rho & \rho^2 & 1 & \rho & \rho^2 \\ 1 & \rho & \rho^2 & -1 & -\rho & -\rho2 & i & i\rho & i\rho^2 & -i & -i\rho & -i\rho^2 \end{pmatrix}.$$

The matrix

$$B = \begin{pmatrix} 1 & \rho^2 & 1 & \rho^2 & -1 & -1 \\ 1 & \rho^2 & -1 & -\rho^2 & -i & i \\ 1 & \rho & 1 & \rho & -1 & -1 \\ 1 & \rho & -1 & -\rho & -i & i \\ 1 & \rho & 1 & \rho & 1 & 1 \\ 1 & \rho & -1 & -\rho & i & -i \end{pmatrix}$$

consisting of 1,2,4,5,7,10-th columns of A is non-singular. In fact,

$$B^{-1} = \begin{pmatrix} -a\rho & -a\rho & a\rho+\frac{1}{4} & a\rho+\frac{1}{4} & \frac{1}{4} & \frac{1}{4} \\ a & a & -a & -a & 0 & 0 \\ -a\rho & a\rho & a\rho+\frac{1}{4} & -a\rho-\frac{1}{4} & \frac{1}{4} & -\frac{1}{4} \\ a & -a & -a & a & 0 & 0 \\ 0 & 0 & -\frac{1}{4} & \frac{i}{4} & \frac{1}{4} & -\frac{i}{4} \\ 0 & 0 & -\frac{1}{4} & -\frac{i}{4} & \frac{1}{4} & \frac{i}{4} \end{pmatrix},$$

where $a = (\rho - \rho^2)/6$.

Hence, by the implicit function theorem, the set

$$\{f_1 = f_2 = \cdots = f_6 = 0\}$$

is given by the equations:

$$s_1 = s_3 - \frac{\rho(1+i)}{2}t_2 + \frac{\rho(1+i)}{2}t_3 - \frac{\rho(1-i)}{2}t_5 + \frac{\rho(1-i)}{2}t_6 + * ,$$

$$s_2 = s_3 + \frac{1+i}{2}t_2 - \frac{1+i}{2}t_3 + \frac{1-i}{2}t_5 - \frac{1-i}{2}t_6 + * ,$$

$$s_4 = s_6 - \frac{\rho(1-i)}{2}t_2 + \frac{\rho(1-i)}{2}t_3 - \frac{\rho(1+i)}{2}t_5 + \frac{\rho(1+i)}{2}t_6 + * ,$$

$$s_5 = s_6 + \frac{1-i}{2}t_2 - \frac{1-i}{2}t_3 + \frac{1+i}{2}t_5 - \frac{1+i}{2}t_6 + * ,$$

$$t_1 = -\rho t_2 - \rho^2 t_3 + * ,$$

$$t_4 = -\rho t_5 - \rho^2 t_6 + * .$$

For convenience, we put

$$y_1 = s_3, \quad y_2 = s_6, \quad y_3 = t_2 - t_3, \quad y_4 = t_5 - t_6,$$

$$y_5 = t_2 + t_3, \quad y_6 = t_5 + t_6.$$

Then $(y_1, \cdots, y_6)$ forms a local coordinate system in an ambient space of $Q_6(V)$ around $(D_0(bf), D_\infty(bf))$. Moreover, we get

$$s_1 = y_1 - \frac{\rho(1+i)}{2}y_3 - \frac{\rho(1-i)}{2}y_4 + * ,$$

$$s_2 = y_1 + \frac{1+i}{2}y_3 + \frac{1-i}{2}y_4 + * ,$$

$$s_4 = y_2 - \frac{\rho(1-i)}{2}y_3 + \frac{\rho(1+i)}{2}y_4 + * ,$$

$$s_5 = y_2 + \frac{1-i}{2}y_3 + \frac{1+i}{2}y_4 + * ,$$

$$t_1^2 + \rho t_2^2 + \rho^2 t_3^2 = -y_3^2 + * ,$$

$$t_4^2 + \rho t_5^2 + \rho^2 t_6^2 = -y_4^2 + * .$$

Substituting these expansions into f_8, we get the defining equation of $Q_6(V)$ in the ambient space around $(D_0(bf), D_\infty(bf))$. The result is, (omitting a constant),

$$0 = y_3^2 - y_4^2 + * .$$

It is clear that this implies the proposition. Q.E.D.

The following proposition follows easily from Proposition 2.2.3.

Proposition 2.2.4. The tangent cone to $R_6(V)/\mathrm{Aut}(\mathbb{P}^1)$ at f (mod $\mathrm{Aut}(\mathbb{P}^1)$) is given by

$$\{(z_1, z_2, z_3, z_4) \in \mathbb{C}^4 \mid z_1 z_2 = 0\}$$

Our compact Riemann surface V is a very special one. It is so called trigonal. (A compact Riemann surface of genus $g \geqq 3$ is said to be trigonal if $R_3(V)$ is non-empty.) We describe $R_n(V)$ for $n \leqq 6$.

For a polynomial $A = A(x)$ of x, we denote by $\deg A$ the degree of A. When we write a polynomial A_k, the subscript k means that

$$\deg A_k \leqq k.$$

We put

$$\mathbb{A}_2 = \left\{ \frac{A_0 y + A_2}{B_0 y + B_2} \;\middle|\; \deg (A_0 B_2 - B_0 A_2) = 2, \ A_0 B_2 - B_0 A_2 \text{ divides both } A_0^3(x^8 - 1) + A_2^3 \text{ and } B_0^3(x^8 - 1) + B_2^3 \right\},$$

$$\mathbb{A}_3 = \left\{ \frac{A_0 y + A_3}{B_0 y + B_3} \;\middle|\; \deg (A_0 B_3 - B_0 A_3) = 3, \ A_0 B_3 - B_0 A_3 \text{ divides both } A_0^3(x^8 - 1) + A_3^3 \text{ and } B_0^3(x^8 - 1) + B_3^3 \right\},$$

$$\mathbb{B} = \{ b_2(x) \mid b_2 \in R_2(\mathbb{P}^1) \},$$

where $b_2(x)$ is the composition of the functions $x = Z_1/Z_0 \in R_3(V)$ and b_2. Then some calculations show the following propositions, whose proofs are omitted.

Proposition 2.2.5.

(1) $R_3(V) = \{ b_1(x) \mid b_1 \in \mathrm{Aut}(\mathbb{P}^1) \}$.

(2) $R_4(V)$ and $R_5(V)$ are empty.

(3) $R_6(V) = \mathbb{A}_2 \cup \mathbb{A}_3 \cup \mathbb{B}$ (disjoint union).

(4) $\mathbb{A}_3$ is a non-singular open subset of $R_6(V)$.

(5) $\mathbb{A}_2$ and $\mathbb{B}$ are non-singular and of dimension 5.

(6) $\mathrm{Aut}(\mathbb{P}^1)$ acts on each of $\mathbb{A}_2$, $\mathbb{A}_3$ and $\mathbb{B}$.

(7) $\mathbb{B}/\mathrm{Aut}(\mathbb{P}^1)$ is biholomorphic to $\mathbb{P}^2 - C$, (C = a conic).

Proposition 2.2.6.

(1) $W_3^1 = \{x_o\}$, one point.

(2) $W_6^2 = \{2x_o\}$, one point.

(3) $F_6^1 = \{x_o\} + W_3$.

(4) W_6^1 has two irreducible components of dimension 3. One of them is F_6^1.

(5) The induced map $\psi : R_6(V)/\mathrm{Aut}(\mathbb{P}^1) \longrightarrow W_6^1$ maps

$$\mathbb{B}/\mathrm{Aut}(\mathbb{P}^1) \text{ to } \{2x_o\} \text{ and}$$

$$(\mathbb{A}_2 \cup \mathbb{A}_3)/\mathrm{Aut}(\mathbb{P}^1) \text{ biholomorphically onto } W_6^1 \setminus F_6^1.$$

2.3. A theorem on non-singular plane curves.

It is a difficult problem to determine positive integers $n \leqq g$ with non-empty $R_n(V)$ and to determine the structure of $R_n(V)$ for such n. We only know the following famous fact:

There exists a positive integer $n \leqq \frac{g+3}{2}$ such that $R_n(V)$ is non-empty.

We shall give a proof of this fact in Chapter 5, following the idea of Meis [53].

Concerning with this problem, we give a theorem in this section and another one in the next section. They were already announced in Namba [66].

Theorem 2.3.1. Let $V = C$ be a non-singular plane curve of degree $d \geqq 2$. Then the minimal positive number n such that $R_n(C)$ is non-empty is $d-1$. If $d \geqq 3$, then the projection

$$\pi_P : C \longrightarrow \mathbb{P}^1$$

with the center $P \in C$ gives an element of $R_{d-1}(C)/\mathrm{Aut}(\mathbb{P}^1)$ and the map

$$P \in C \longmapsto \pi_P \in R_{d-1}(C)/\mathrm{Aut}(\mathbb{P}^1)$$

is biholomorphic.

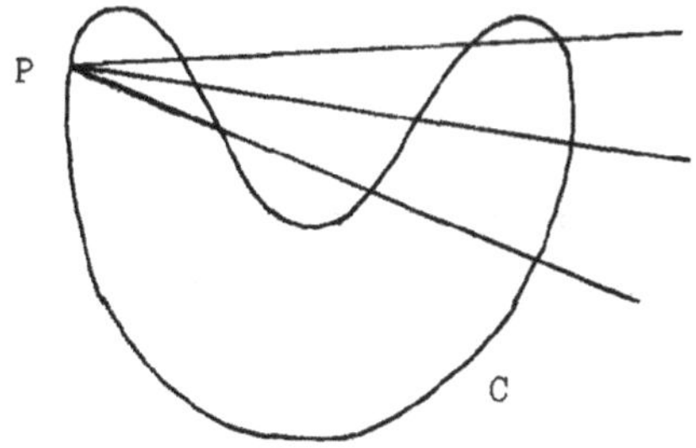

In order to prove Theorem 2.3.1, we need some preparations. For two plane curves C and C' and for a point $P \in \mathbb{P}^2$, we denote by $I(P, C \cap C')$ the intersection multiplicity of C and C' at P.

Lemma 2.3.2. Let C, F and G be plane curves. Assume that C is irreducible and is neither a component of F nor of G. Let 0 be a non-singular point of C. Then

$$\operatorname{Min}\{I(0,C \cap F),\ I(0,C \cap G)\} \leqq I(0,F \cap G).$$

Proof. We take a local coordinate system (x,y) with the origin 0 such that C coincides locally with the y-axis, i.e., $C : x = 0$. Let $f(x,y) = 0$ and $g(x,y) = 0$ be the local equations of F and G around 0, respectively. We expand the holomorphic functions $f(0,y)$ and $g(0,y)$ of y as follows:

$$f(0,y) = y^s(c_0 + c_1 y + \cdots), \quad c_0 \neq 0,$$

$$g(0,y) = y^t(d_0 + d_1 y + \cdots), \quad d_0 \neq 0.$$

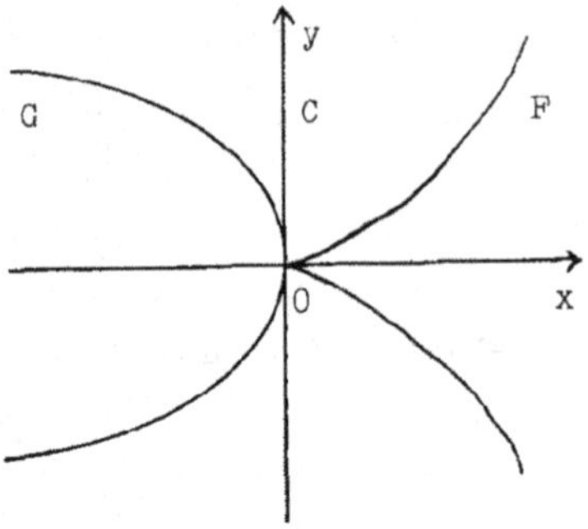

Then, it is clear that

$$I(0,C \cap F) = s,$$

$$I(0,C \cap G) = t.$$

If one of s and t is zero, then there is nothing to prove. Assume that $t \geqq s \geqq 1$. By Weierstrass division theorem (see Gunning-Rossi[29]), we may write

$$g = af + (b_1 y^{s-1} + \cdots + b_s), \tag{1}$$

where $a = a(x,y)$ is a holomorphic function around 0 and $b_1, \cdots, b_s$ are holomorphic functions of x. In (1), if we put $x = 0$, then we get

$$y^t(d_0 + d_1 y + \cdots) = a(0,y)y^s(c_0 + c_1 y + \cdots) + b_1(0)y^{s-1} + \cdots + b_s(0) \cdot$$

This implies that

$$b_1(0) = \cdots = b_s(0) = 0.$$

Hence, we may write

$$b_i(x) = xe_i(x), \quad i = 1,\cdots,s,$$

where each $e_i(x)$ is a holomorphic function of x.

We define a curve H around 0 by

$$H = \{ (x,y) \mid b_1 y^{s-1} + \cdots + b_s = 0 \}.$$

Then, locally, H has C as one of its components, for

$$b_1 y^{s-1} + \cdots + b_s = x(e_1 y^{s-1} + \cdots + e_s).$$

Now, by (1),

$$I(0,F \cap G) = I(0,F \cap H) \geqq I(0,F \cap C) = s. \qquad \text{Q.E.D.}$$

Proposition 2.3.3. Let C be an irreducible plane curve of degree $d \geqq 3$. Let V be a non-singular model of C. Let $\mathrm{Reg}(C)$ be the open subset of C of all non-singular points. For a point $P \in \mathrm{Reg}(C)$, let π_P be the projection of C with the center P. Then the map

$$\pi : P \in \mathrm{Reg}(C) \longmapsto \pi_P \in R_{d-1}(V)/\mathrm{Aut}(\mathbb{P}^1)$$

is a holomorphic imbedding.

Proof. First of all, we show that π is an injection. Assume that

$$\pi_P = \pi_Q \in R_{d-1}(V)/\mathrm{Aut}(\mathbb{P}^1),$$

for two distinct points P and Q in $\mathrm{Reg}(C)$. This means that, if we regard π_P and π_Q as meromorphic functions by choosing a fixed coordinate system in $\mathbb{P}^1$, then there is $b \in \mathrm{Aut}(\mathbb{P}^1)$ such that $\pi_Q = b\pi_P$. Let R be a point of $\mathrm{Reg}(C)$ such that P, Q and R are not colinear. By the assumption $d \geqq 3$, we may assume that there is a point $S \in C$ with $S \neq P$ and $S \neq R$ such that P, R, and S are colinear. Then $\pi_P(R) = \pi_P(S)$. Hence

$$\pi_Q(R) = b\pi_P(R) = b\pi_P(S) = \pi_Q(S).$$

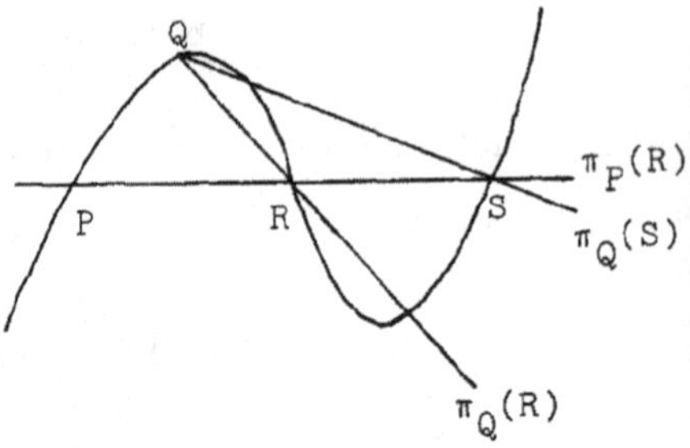

This means that Q, R and S are colinear. Hence P, Q and R are colinear, a contradiction. Hence π is injective.

Now, let $(Z_0:Z_1:Z_2)$ be a homogeneous coordinate system in $\mathbb{P}^2$. We put

$$x = Z_1/Z_0 \quad \text{and} \quad y = Z_2/Z_0 .$$

Then x and y are regarded as meromorphic functions on V through the birational map $V \longrightarrow C$. We denote them by g_1 and g_2, respectively.

We may assume that $(x,y) = (0,0)$ is a non-singular point of C. We denote by 0 the point of V corresponding to $(0,0)$. Let U be a connected, non-singular open neighborhood of 0 in V and t be a coordinate in U such that $t(0) = 0$. Then g_1 and g_2 are locally regarded as holomorphic functions of t. The curve C is, then, locally defined by

$$C : t \longmapsto (g_1(t), g_2(t)).$$

Hence

(1) $$\left|\frac{dg_1}{dt}(0)\right| + \left|\frac{dg_2}{dt}(0)\right| \neq 0.$$

Now, for a point $P = (g_1(t), g_2(t))$ on C near $(0,0)$, π_P is represented by the meromorphic function

$$f_t(Q) = \frac{g_2(Q)-g_2(t)}{g_1(Q)-g_1(t)} \quad \text{for} \quad Q \in V.$$

Now, for a point $P = (g_1(t), g_2(t))$ on C near $(0,0)$, π_P is represented by the meromorphic function

$$f_t(Q) = \frac{g_2(Q)-g_2(t)}{g_1(Q)-g_1(t)} \quad \text{for} \quad Q \in V.$$

Taking U sufficiently small, we may assume that the map

$$\mathfrak{F} : (Q,t) \in V \times U \longmapsto f_t(Q) \in \mathbb{P}^1$$

is a holomorphic map. In fact, by (1), we may assume, say,

$$\frac{dg_1}{dt}(t) \neq 0 \quad \text{for all} \quad t \in U.$$

If $g_1(Q) = g_1(t)$ and $g_2(Q) = g_2(t)$, then $Q = t$. In this case, $f_t(t)$ is defined by

$$f_t(t) = \frac{(\frac{dg_2}{dt})(t)}{(\frac{dg_1}{dt})(t)} .$$

$\mathfrak{F}$ is thus continuously continuated to the subset

$$\{ (Q,t) \mid Q = t \}.$$

Assume that $g_1(Q) = g_2(Q) = \infty$. The image point of Q by the map $V \longrightarrow C$ has the homogeneous coordinate $(Z_0^o:Z_1^o:Z_2^o)$ with $Z_0^o = 0$. Assume that $Z_2^o \neq 0$. Put

$$h_1 = Z_1/Z_2 = g_1/g_2 \quad \text{and} \quad h_2 = Z_0/Z_2 = 1/g_2.$$

Then h_1 and h_2 are meromorphic functions on V and $h_2(Q) = 0$. Thus, if W is a small open neighborhood of Q in V, then h_2 is a holomorphic function on W. Note that

$$f_t(R) = \frac{1 - g_2(t)h_2(R)}{h_1(R) - g_1(t)h_2(R)} \quad \text{for} \quad (R,t) \in (W - \{0\}) \times U.$$

This is continuously continuated to the subset

$$\{ (R,t) \mid R = Q \}$$

by defining

$$f_t(R) = \frac{1}{h_1(Q)}, \quad \text{if} \quad h_1(Q) \neq 0,$$

$$= \infty, \quad \text{if} \quad h_1(Q) = 0.$$

Thus $\mathfrak{F}$ is a holomorphic map. Hence $\{f_t\}_{t\in U}$ is a family of meromorphic functions on V with the parameter space U (see §1.1).

We put

$$V' = \{ P \in V \mid g_1(P) \neq 0,\infty \text{ and } g_2(P) \neq 0,\infty \}.$$

Then, for any point $P \in V'$,

$$(\frac{\partial \mathfrak{F}}{\partial t})(P,0) = \frac{(dg_1/dt)(0)\cdot g_2(P) - (dg_2/dt)(0)\cdot g_1(p)}{g_1(P)^2} .$$

Note that $(\frac{\partial \mathfrak{F}}{\partial t})_{t=0} = \sigma_0(\frac{\partial}{\partial t})$ is an element of $H^0(V, \mathcal{O}([2D_\infty(f_0)]))$ (see §1.1). If it is zero, then

$$\frac{dg_1}{dt}(0)\cdot g_2(P) = \frac{dg_2}{dt}(0)\cdot g_1(P) \quad \text{for all} \quad P \in V' .$$

This occurs only when

$$\frac{dg_1}{dt}(0) = \frac{dg_2}{dt}(0) = 0 .$$

This contradicts to (1) above. Hence the linear map

$$\sigma_0 : T_0U \longrightarrow H^0(V, \mathcal{O}([2D_\infty(f_0)]))$$

is injective. On the other hand, the linear map

$$\sigma_{f_0} : T_{f_0}\mathrm{Hol}(V, \mathbb{P}^1) \longrightarrow H^0(V, \mathcal{O}([2D_\infty(f_0)]))$$

(see §1,1) satisfies

$$\sigma_0 = \sigma_{f_0}(di)_0,$$

where $(di)_0$ is the differential at 0 of the holomorphic map

$$i : t \in U \longmapsto f_t \in R_{d-1}(V) .$$

Hence, $(di)_0$ is an injective linear map. Note that i is injective, for $\tilde{\omega} i = \pi$. ($\tilde{\omega} : R_{d-1}(V) \longrightarrow R_{d-1}(V)/\mathrm{Aut}(\mathbb{P}^1)$ is the projection). Hence i is a holomorphic imbedding, provided U is sufficiently small.

In order to prove that the map

$$\pi : t \in U \longmapsto f_t \pmod{\mathrm{Aut}(\mathbb{P}^1)} \in R_{d-1}(V)/\mathrm{Aut}(\mathbb{P}^1)$$

is a holomorphic imbedding at 0, it is enough to show that

$$(di)_0(T_0U) \cap (dj)_e(T_e\mathrm{Aut}(\mathbb{P}^1)) = \{0\},$$

where e is the identity of $\mathrm{Aut}(\mathbb{P}^1)$ and

$$j : b \in \mathrm{Aut}(\mathbb{P}^1) \longmapsto bf_0 \in R_{d-1}(V) .$$

By Lemma 1.1.8, it suffices to prove that, if $X \in H^0(\mathbb{P}^1, \mathcal{O}(T\mathbb{P}^1))$ and $s \in \mathbb{C}$ satisfy

$$f_0^*X = \sigma_0(s(\tfrac{d}{dt})_0) ,$$

then $X = 0$ and $s = 0$. We may write

$$X = (a_0 + a_1\xi + a_2\xi^2)\frac{d}{d\xi} ,$$

where $a_0, a_1, a_2 \in \mathbb{C}$ and $\xi = Z_1/Z_0$ is an inhomogeneous coordinate in $\mathbb{P}^1$. Then the above equality implies

$$a_0 + a_1(g_2/g_1) + a_2(g_2/g_1)^2$$
$$= s((dg_1/dt)(0)\cdot g_2 - (dg_2/dt)(0)\cdot g_1)/g_1^2$$

on V'. Hence, as meromorphic functions on V,

$$a_0 g_1^2 + a_1 g_1 g_2 + a_2 g_2^2 = s((dg_1/dt)(0)\cdot g_2 - (dg_2/dt)(0)\cdot g_1) .$$

But C is not a conic. Hence, we get

$$a_0 = a_1 = a_2 = s = 0 .$$ Q.E.D.

Now, we are ready to prove Theorem 2.3.1.

Proof of Theorem 2.3.1. Let C be a non-singular plane curve of degree d. We first assume that $d \geqq 4$. Let $(Z_0:Z_1:Z_2)$ be as before a homogeneous coordinate system in $\mathbb{P}^2$. Let

$$F(Z_0:Z_1:Z_2) = 0$$

be the defining equation of the curve C. Then, by Lefschetz [51, p.159], we may assume that any holomorphic 1-form ω on C can be written as

$$\omega = G \frac{\widetilde{D}}{\Sigma c_i F_i} ,$$

where $G = G(Z_0,Z_1,Z_2)$ is a homogeneous polynomial of degree d-3,

$$\widetilde{D} = \begin{vmatrix} Z_0 & Z_1 & Z_2 \\ dZ_0 & dZ_1 & dZ_2 \\ c_0 & c_1 & c_2 \end{vmatrix}$$

for some constants c_0, c_1, c_2 and

$$F_i = \partial F/\partial Z_i, \quad i = 0,1,2.$$

Hence, by Lemma 2.1.1, every meromorphic function f of order

$$n \leqq g = \frac{(d-1)(d-2)}{2}$$

can be written as

$$f = G_1/G_2 ,$$

where G_1 and G_2 are homogeneous polynomials of degree $d-3$. Let $G = (G_1, G_2)$ be the G.C.M. of G_1 and G_2 and put

$$H_1 = G_1/G \quad \text{and} \quad H_2 = G_2/G .$$

Then

$$f = H_1/H_2 .$$

Let k be the degree of G. For convenience, we use the same letter H_i $(i = 1, 2)$, for the curve $\{H_i = 0\}$. The intersection divisors $C \cap H_1$ and $C \cap H_2$ on C can be written as

$$C \cap H_1 = a_1P_1 + \cdots + a_sP_s ,$$

$$C \cap H_2 = b_1P_1 + \cdots + b_sP_s ,$$

where a_i, b_i $(i = 1,\cdots,s)$ are non-negative integers such that

$$\Sigma a_i = \Sigma b_i = d(d-3-k) .$$

We put

$$c_i = \min\{a_i, b_i\}, \quad i = 1,\cdots,s,$$

$$D = c_1P_1 + \cdots + c_sP_s .$$

Then it is clear that the pencil

$$\{s_1H_1 + s_2H_2\}_{(s_1:s_2) \in \mathbb{P}^1}$$

has the fixed part D. Thus

$$\deg f = d(d-3-k) - \Sigma c_i .$$

By Lemma 2.3.2,

$$\Sigma c_i \leqq \Sigma I(P_i, H_1 \cap H_2) \leqq (d-3-k)^2 .$$

Hence

$$\deg f \geqq d(d-3-k) - (d-3-k)^2 = (d-3-k)(3+k).$$

Note that $0 \leqq k \leqq d-3$. The function $(d-3-k)(3+k)$ of k takes its minimal positive value $d-1$ at $k = d-4$. Hence, if f is not a constant function, then

$$\deg f \geqq d-1 .$$

The equality occurs if and only if H_1 and H_2 are lines and $P = H_1 \cap H_2$ lies on C. f is, in this case, the projection π_P with the center P. Hence the map

$$\pi : P \in C \longmapsto \pi_P \in R_{d-1}(C)/\mathrm{Aut}(\mathbb{P}^1)$$

is surjective. Hence, by Proposition 2.3.3, π is a biholomorphic map. This proves Theorem 2.3.1 for $d \geqq 4$.

Theorem 2.3.1 for $d = 2$ is trivial.

Finally, we prove Theorem 2.3.1 for $d = 3$. In this case, $g = 1$. Hence the minimal positive number n such that $R_n(C)$ is non-empty is $2 = d-1$. By Proposition 2.3.3,

$$\pi : P \in C \longmapsto \pi_P \in R_2(C)/\mathrm{Aut}(\mathbb{P}^1)$$

is a holomorphic imbedding. On the other hand, by Corollary 1.3.13, $R_2(C)/\mathrm{Aut}(\mathbb{P}^1)$ is biholomorphic to C. Hence π must be a biholomorphic map. Q.E.D.

Remark 2.3.4. If C is a conic, then $R_1(C)/\mathrm{Aut}(\mathbb{P}^1)$ is one point.

Remark 2.3.5. If V is a non-hyperelliptic compact Riemann

surface of genus 3, then $R_3(V)/\mathrm{Aut}(\mathbb{P}^1)$ is biholomorphic to V (Example 2.1.5). This assertion again follows from Theorem 2.3.1. In fact, the canonical curve of V is a non-singular plane curve of degree $2g-2 = 4$.

The above proof of Theorem 2.3.1 shows also

Proposition 2.3.6. Let C be a non-singular plane curve of degree $d \geqq 5$. Then

$$R_d(C)/\mathrm{Aut}(\mathbb{P}^1) \cong \mathbb{P}^2 - C .$$

In fact, every element of $R_d(C)/\mathrm{Aut}(\mathbb{P}^1)$ is obtained as the projection $\pi_P : C \longrightarrow \mathbb{P}^1$ with the center $P \in \mathbb{P}^2 - C$.

Problem. Can the proof of Theorem 2.3.1 be applied to a plane curve with only nodes?

2.4. A theorem on the non-existence of functions.

Lemma 2.4.1. Let f and h be meromorphic functions of order m and n, respectively, on a compact Riemann surface V of genus g. Let k and j be positive integers such that

$$(g-1) + kj \geqq (m-1)k + (n-1)j .$$

Assume that there is no polynomial $F(x,y)$ $(\not\equiv 0)$ with $\deg_x F \leqq k$ and $\deg_y F \leqq j$ ($\deg_x F$ = the degree of F with respect to x, etc.), such that $F(f,h) = 0$ on V. Then

$$h^1(kD_\infty(f) + jD_\infty(h)) \geqq 1 .$$

Proof. By the assumption, the following $(k+1)(j+1)$ elements of

$L(kD_\infty(f)+jD_\infty(h))$ are linearly independent:

$$1,\ f,\ \cdots,\ f^k,$$

$$h,\ fh,\ \cdots,\ f^k h,$$

$$\cdot\ \cdot\ \cdot\ \cdot$$

$$h^j,\ fh^j,\ \cdots,\ f^k h^j\ .$$

Hence

$$(k+1)(j+1) - h^1(kD_\infty(f)+jD_\infty(h))$$
$$\leqq h^0(kD_\infty(f)+jD_\infty(h)) - h^1(kD_\infty(f)+jD_\infty(h))$$
$$= mk+nj+1-g\ .$$

Hence

$$h^1(kD_\infty(f)+jD_\infty(h)) \geqq (k+1)(j+1) - mk - nj - 1 + g \geqq 1\ .$$ Q.E.D.

Corollary 2.4.2. Let f and h be meromorphic functions of order m and n, respectively, on a compact Riemann surface V of genus g. Let k and j be positive integers such that

(1) $$(g-1) + kj \geqq (m-1)k + (n-1)j\ ,$$

(2) $$mk+nj \geqq 2g-1\ .$$

Then there is a polynomial $F(x,y)$ with $\deg_x F \leqq k$ and $\deg_y F \leqq j$ such that $F(f,h) = 0$ on V.

Proof. If there is not such a polynomial, then, by the lemma, $h^1(kD_\infty(f)+jD_\infty(h)) \geqq 1$. But

$$\deg(kD_\infty(f)+jD_\infty(h)) = mk+nj \geqq 2g-1\ ,$$

a contradiction. Q.E.D.

Now, we prove

<u>Theorem 2.4.3</u>. Let V be a compact Riemann surface of genus g. Let m and n be positive integers such that

(1) $$(m,n) = 1 \quad \text{(relatively prime)},$$

(2) $$(m-1)(n-1) \leqq g-1 .$$

Then, at least one of $R_m(V)$ and $R_n(V)$ is empty.

<u>Proof</u>. Let $f \in R_m(V)$ and $h \in R_n(V)$. In Corollary 2.4.2, we put $j = m-1$. Then the conditions in Corollary 2.4.2 become

$$g-1 \geqq (m-1)(n-1) ,$$

$$mk \geqq 2g-1-mn+n .$$

Hence, if we take k sufficiently large, then there is a polynomial $F(x,y)$ with $\deg_y F \leqq m-1$ such that $F(f,h) = 0$ on V. We may assume that $F(x,y)$ is irreducible. Let C be the closure in $\mathbb{P}^2$ of the affine curve $F(x,y) = 0$. Then, the correspondence

$$u : P \in V \longmapsto (1:f(P):h(P)) \in \mathbb{P}^2$$

is a holomorphic map of V onto C. Let e be the mapping order of u. The coordinates x and y can be regarded as meromorphic functions on a non-singular model of C and the compositions of them with u are f and h, respectively. Now, we have

$$m = \operatorname{ord}(f) = e \cdot \operatorname{ord}(x) ,$$

$$n = \operatorname{ord}(h) = e \cdot \operatorname{ord}(y) .$$

Hence, by the assumption (1), $e = 1$. But, then,

$$m = \operatorname{ord}(x) = \deg_y F \leqq m-1 ,$$

a contradiction. Q.E.D.

The following corollary is a generalization of Proposition 2.1.2 and its corollary.

Corollary 2.4.4. Let V be a compact Riemann surface of genus g. Let p be a prime number with non-empty $R_p(V)$. Take $f \in R_p(V)$. Let n be a positive integer suth that $(p-1)(n-1) \leqq g-1$. Then

(1) If $n \not\equiv 0 \pmod{p}$, then $R_n(V)$ is empty.

(2) If $n \equiv 0 \pmod{p}$, then every element h of $R_n(V)$ can be written as $h = r \cdot f$ (the composition of the maps), where r is a rational function: $\mathbb{P}^1 \longrightarrow \mathbb{P}^1$ of order n/p. Moreover, the correspondence $r \in R_{n/p}(\mathbb{P}^1) \longmapsto h = r \cdot f \in R_n(V)$ is a biholomorphic map.

Proof. (1) follows from the theorem. (2) follows from the proof of the theorem. In fact, letting $m = p$, the integer e in the proof of the theorem must be p. Hence

$$\deg_y F = \operatorname{ord}(x) = 1, \quad \text{i.e.,}$$

$$F(x,y) = a_1(x)y + a_0(x) ,$$

where $a_1(x)$ and $a_0(x)$ are polynomials of x. Thus

$$h = -\frac{a_0(f)}{a_1(f)} .$$

By Proposition 1.1.7, the map $r \in R_{n/p}(\mathbb{P}^1) \longmapsto r \cdot f \in R_n(V)$ is a holomorphic imbedding. Since it is surjective, it is biholomorphic. (Note that a complex space is assumed to be reduced.) Q.E.D.

Corollary 2.4.5. Let V be a compact Riemann surface of genus g. Let p be a prime number and assume that $R_p(V)$ is non-empty. Then

(1) If $(p-1)(p-2) \leqq g-1$, then $R_n(V)$ is empty for all $n \leqq p-1$.

(2) If $(p-1)^2 \leqq g-1$, then $R_p(V)/\mathrm{Aut}(\mathbb{P}^1) \cong W_p^1$ is one point.

<u>Remark 2.4.6</u>. A compact Riemann surface V of genus $g \geqq 3$ is said to be <u>trigonal</u> if $R_3(V)$ is non-empty. If V is trigonal and $g \geqq 5$, then, by Corollary 2.4.5, $R_3(V)/\mathrm{Aut}(\mathbb{P}^1)$ is one point. But, if V has genus 4 and is trigonal (i.e., non-hyperelliptic), then $R_3(V)/\mathrm{Aut}(\mathbb{P}^1)$ is either one point or two points (see Example 2.1.10).

Another corollary to (the proof of) Theorem 2.4.3 is

<u>Corollary 2.4.7</u>. Let V be a non-hyperelliptic compact Riemann surface of genus $g \geqq 10$. Assume that $R_4(V)$ is non-empty. Then, either (1) $R_4(V)/\mathrm{Aut}(\mathbb{P}^1) \cong W_4^1$ is one point or (2) there is a (ramified) double covering $u : V \longrightarrow C$ of V onto a non-singular plane cubic C such that every element of $R_4(V)/\mathrm{Aut}(\mathbb{P}^1)$ is obtained as the composition $\pi_P \cdot u$ of u and the projection $\pi_P : C \longrightarrow \mathbb{P}^1$ with the center $P \in C$. Moreover, $P \in C \longmapsto \pi_P \cdot u \in R_4(V)/\mathrm{Aut}(\mathbb{P}^1) \cong W_4^1$ is a biholomorphic map.

<u>Proof</u>. Take $f, h \in R_4(V)$. In Corollary 2.4.2, we put $m = n = 4$ and $j = 3$. Then, the conditions (1) and (2) of Corollary 2.4.2 become

$$g \geqq 10 \quad \text{and} \quad 4k \geqq 2g - 13 .$$

Hence, there is an irreducible polynomial $F(x,y)$ with $\deg_y F \leqq 3$ such that $F(f,h) = 0$ on V. Let C, u and e be as in the proof of Theorem 2.4.3. Then 4 is divisible by $e \geqq 2$. Hence, e is either 4 or 2.

If $e = 4$, then h is a rational function of order 1 of f, i.e., $h \equiv f \pmod{\mathrm{Aut}(\mathbb{P}^1)}$.

If $e = 2$, then

$$\deg_y F = \mathrm{ord}(x) = 2 ,$$

$$\deg_x F = \operatorname{ord}(y) = 4/e = 2 .$$

Hence

$$2 \leqq \deg C \leqq 4 .$$

We show that $\deg C \geqq 3$.

Let W be a non-singular model of C with a birational holomorphic map

$$v : W \longrightarrow C .$$

Then there is a (ramified) double covering $\hat{u} : V \longrightarrow W$ such that the diagram

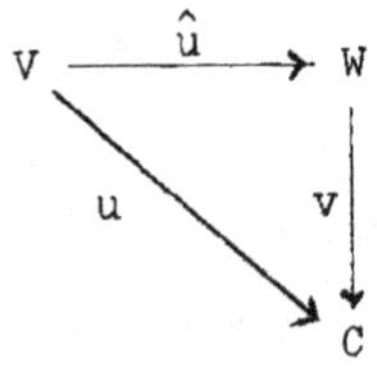

commutes. Assume that $\deg C = 2$. Then C is a conic and W is biholomorphic to $\mathbb{P}^1$. Since u is a (ramified) double covering, V must be hyperelliptic, a contradiction. Hence $\deg C \geqq 3$.

Assume that $\deg C = 3$. In this case, C is non-singular. In fact, if C is singular, then W is biholomorphic to $\mathbb{P}^1$, so that V must be hyperelliptic, a contradiction.

Assume now that $\deg C = 4$. The coordinates x and y can be regarded as meromorphic functions on W. In fact, they are identified with the compositions of v and the projections of C with the center $(Z_0:Z_1:Z_2) = (0:1:0)$ and $(Z_0:Z_1:Z_2) = (0:0:1)$, respectively. ($x = Z_1/Z_0$ and $y = Z_2/Z_0$.) Since $\operatorname{ord}(x) = \operatorname{ord}(y) = 2$, $(0:1:0)$ and $(0:0:1)$ are singular points of C with the multiplicity 2. By Fulton [21, p.201],

$$\text{the genus of } W \leqq \frac{(4-1)(4-2)}{2} - \frac{2(2-1)}{2} - \frac{2(2-1)}{2} = 1.$$

Hence the genus of W is 1, for V is not hyperelliptic. Thus there is a non-singular plane cubic C' and a biholomorphic map

$$w : W \longrightarrow C' .$$

Note that f and h are the compositions of u and meromorphic functions on C'.

Using $w \cdot \hat{u}$, C' and Theorem 2.3.1 in this case, we conclude that "there is a (ramified) double covering

$$u : V \longrightarrow C$$

of V onto a non-singular plane cubic such that both f and h are compositions of u and projections of C with centers on C."

By Proposition 1.1.9, we have the holomorphic imbedding:

$$j_u : R_2(C)/\mathrm{Aut}(\mathbb{P}^1) \longrightarrow R_4(V)/\mathrm{Aut}(\mathbb{P}^1).$$

Note that $R_2(C)/\mathrm{Aut}(\mathbb{P}^1)$ is biholomorphic to C (Theorem 2.3.1). On the other hand, since $g \geqq 10$, V is not trigonal by Theorem 2.4.3. Hence 4 is the minimal integer n such that $R_n(V)$ is non-empty. By Theorem 2.1.9, $R_4(V)/\mathrm{Aut}(\mathbb{P}^1)$ is biholomorphic to $W_4^1 = W_4^1(V)$. Hence, we have a holomorphic imbedding

$$C \longrightarrow W_4^1(V) .$$

We show that it is biholomorphic. In fact, the above argument shows that every two distinct points in $W_4^1(V)$ are connected by a non-singular elliptic curve on it. But $\dim W_4^1(V) \leqq 1$, for, otherwise,

$$g - 4 = \dim W_{g-1}^1 \geqq \dim (W_4^1 + W_{g-5}) \geqq 2 + (g-5) = g-3 ,$$

a contradiction. Hence $W_4^1(V)$ itself must be (biholomorphic to) the elliptic curve C. <u>Q.E.D.</u>

2.5. Remarks on projections of canonical curves.

Let V be as before a compact Riemann surface of genus g. We denote by $|K_V|$ the complete linear system determined by the canonical bundle K_V of V. Let f be a meromorphic function of order $n \leqq g$ on V. Then, by Lemma 2.1.1, there are holomorphic 1-forms ω_1 and ω_2 on V such that $\omega_2 = f\omega_1$. If we regard ω_1 and ω_2 as hyperplanes of $\mathbb{P}^{g-1}$, the dual space of $|K_V|$, then $\omega_2 = f\omega_1$ means that $f \pmod{\mathrm{Aut}(\mathbb{P}^1)}$ is equal to the projection π_S of the canonical curve $C_V = \Phi_{|K_V|}(V)$ of V with the center S, the $(g-3)$-dimensional linear subspace of $\mathbb{P}^{g-1}$ which is the intersection of ω_1 and ω_2.

Hence, in order to look for meromorphic functions of order $\leqq g$, it suffices to examine various projections π_S of the canonical curve.

The canonical curve C_V is not contained in any hyperplane in $\mathbb{P}^{g-1}$. Hence we can find $g-2$ points on C_V which span a $(g-3)$-dimensional linear subspace S in $\mathbb{P}^{g-1}$. Then, the order of the projection π_S is at most $(2g-2) - (g-2) = g$. If S meets no other points of C_V, then $\mathrm{ord}(\pi_S) = g$, provided V is non-hyperelliptic. If S meets other points of C_V, then $\mathrm{ord}(\pi_S) < g$ and S must be in a special position with respect to C_V. For example, if V is non-hyperelliptic and has genus 4, then C_V is the intersection of a quadric Q and a cubic in $\mathbb{P}^3$ meeting transversally (see Example 2.1.10). If a line S meets C_V in 2 points, then $\mathrm{ord}(\pi_S) = 4$. If S is contained in Q, then S meets C_V in 3 points. In this case, $\mathrm{ord}(\pi_S) = 3$.

This example indicates that quadrics containing both C_V and $(g-3)$-dimensional linear subspaces are important in our study. We explain the circumstance more precisely.

The choice of ω_1 and ω_2 above are not necessarily unique. If we take another pair ω_1' and ω_2' such that $\omega_2' = f\omega_1'$, then

$$\omega_1(P)\omega_2'(P) = \omega_2(P)\omega_1'(P), \quad \text{for all} \quad P \in V .$$

This means that

$$\omega_1\omega_2' - \omega_2\omega_1' = 0 \in H^0(V, \mathcal{O}(K_V^{\otimes 2})) .$$

If we denote by $\omega_1 \circ \omega_2$ the symmetric product of ω_1 and ω_2, i.e., $\omega_1 \circ \omega_2 \in H^0(V, \mathcal{O}(K_V)) \overset{s}{\otimes} H^0(V, \mathcal{O}(K_V))$ (symmetric product), this means that $\omega_1 \circ \omega_2' - \omega_2 \circ \omega_1'$ belongs to the kernel of the linear map

$$\omega_1 \circ \omega_2 \in H^0(V, \mathcal{O}(K_V)) \overset{s}{\otimes} H^0(V, \mathcal{O}(K_V)) \longmapsto \omega_1\omega_2 \in H^0(V, \mathcal{O}(K_V^{\otimes 2})) .$$

Geometrically, $\omega_1\omega_2' - \omega_2\omega_1' = 0$ defines a quadric Q in $\mathbb{P}^{g-1}$ which contains the canonical curve C_V. The quadric Q is a special one. It has the rank 3 or 4. The importance of such quadrics was already recognized by Martens [52,I] and Andreotti-Mayer [3].

We first generalize our consideration to a more general case.

By a <u>projective manifold</u>, we mean a compact complex manifold imbedded in a complex projective space. Let V be a projective manifold. For a holomorphic line bundle F on V, let

$$|F| = \{D \mid D \text{ is an effective divisor on } V \text{ and } [D] = F\}$$

be the <u>complete linear system determined by</u> F. It is canonically identified with the projective space

$$P(H^0(V, \mathcal{O}(F))) = \text{The set of all 1-dimensional linear subspaces of } H^0(V, \mathcal{O}(F)) .$$

Let $\mathbb{P}^N$ be the dual space of $|F|$. $|F|$ is regarded as the set of all hyperplanes of $\mathbb{P}^N$. In order to fix notations, for $\xi \in H^0(V, \mathcal{O}(F))$, we denote by (ξ) and $\langle\xi\rangle$, the zero divisor of ξ and the hyperplane of $\mathbb{P}^N$ determined by ξ, respectively.

Let $\{\phi_0, \cdots, \phi_N\}$ be a basis of $H^0(V, \mathcal{O}(F))$ and let

$$\Phi_F : P \in V \longmapsto (\phi_0(P) : \cdots : \phi_N(P)) \in \mathbb{P}^N$$

be the <u>meromorphic map associated with</u> $|F|$. The image $\Phi_F(V)$ is a closed subvariety of $\mathbb{P}^N$ which is not contained in any hyperplane in $\mathbb{P}^N$.

Now, let S be a (N-2)-dimensional linear subspace of $\mathbb{P}^N$. The composition of Φ_F and π_S, the projection of $\Phi_F(V)$ with the center S, defines a meromorphic (i.e., rational) function (mod Aut($\mathbb{P}^1$)) on V. We denote this rational function (mod Aut($\mathbb{P}^1$)) by π_S again. The subspace S is uniquely determined by a pencil

$$(\lambda_1\xi_0 - \lambda_0\xi_1)_{(\lambda_0:\lambda_1)} \in \mathbb{P}^1, \quad (\xi_0, \xi_1 \in H^0(V, \mathcal{O}(F))),$$

on V. In other words, S corresponds to a unique line in $|F|$.

Let S' be another (N-2)-dimensional linear subspace of $\mathbb{P}^N$ determined by a pencil $(\lambda_1\eta_0 - \lambda_0\eta_1)_{(\lambda_0:\lambda_1)} \in \mathbb{P}^1$, $(\eta_0, \eta_1 \in H^0(V, \mathcal{O}(F)))$ and assume that $\pi_S = \pi_{S'}$. The equality, of course, means that they are equal modulo Aut($\mathbb{P}^1$). But, if we choose η_0 and η_1 suitably, then we may assume that they are equal as rational functions on V, i.e.,

$$\xi_0(P)\eta_1(P) = \xi_1(P)\eta_0(P) \quad \text{for} \quad P \in V .$$

This means as above that

$$\xi_0\eta_1 - \xi_1\eta_0 = 0 \in H^0(V, \mathcal{O}(F^{\otimes 2}))$$

This equation defines a quadric Q in $\mathbb{P}^N$ which contains $\Phi_F(V)$.

Note that (ξ_1) and (η_1) can not be equal. In fact, if $(\xi_1) = (\eta_1)$, then there is a non-zero constant c such that $\eta_1 = c\xi_1$. Then, $\xi_1(c\xi_0 - \eta_0) = 0$ on $\Phi_F(V)$. Hence $\xi_1 = 0$ or $c\xi_0 - \eta_0 = 0$ on $\Phi_F(V)$. But $\Phi_F(V)$ is not contained in any hyperplane, a contradiction. Hence $(\xi_1) \neq (\eta_1)$. Similarly, $(\xi_0) \neq (\eta_0)$.

On the other hand, it may happen, say, $(\xi_0) = (\eta_1)$. In this case, if we put $\eta_1 = c\xi_0$, ($c \neq 0$, a constant), then Q is given by the equation

$$c\xi_0^2 - \xi_1\eta_0 = 0 .$$

This case is nothing but the case when Q has the rank 3. The vertex $V(Q)$ of Q is, in this case, given by

$$V(Q) = \{\xi_0 = \xi_1 = \eta_0 = 0\} .$$

It is a linear subspace of dimension $N-3$. Let $\pi_{V(Q)}$ be the projection with the center $V(Q)$. Then $\pi_{V(Q)}$ maps Q onto a conic $\hat{Q}$ in $\mathbb{P}^2$. Hence Q is a quadric cone over $\hat{Q}$ with the vertex $V(Q)$. For every point $x \in \hat{Q}$, the inverse image $S_x = \pi_{V(Q)}{}^{-1}(x)$ is a $(N-2)$-dimensional linear subspace of $\mathbb{P}^N$ contained in Q. The one parameter family $\{S_x\}_{x\in\hat{Q}}$ of linear subspaces determines Q and

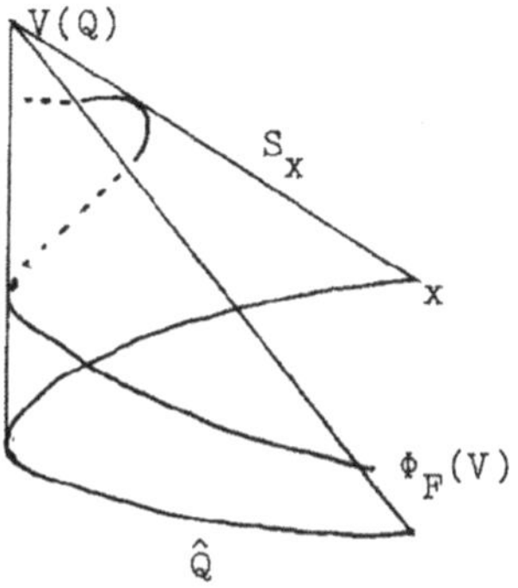

$$\pi_{S_x} \equiv \pi_{S_y} \quad (\text{mod Aut}(\mathbb{P}^1))$$

for all x and y in $\hat{Q}$.

If Q has the rank 4, then the vertex $V(Q)$ is given by

$$V(Q) = \{\xi_0 = \xi_1 = \eta_0 = \eta_1 = 0\}$$

and is a linear subspace of $\mathbb{P}^N$ of dimension $N-4$. The projection $\pi_{V(Q)}$ maps Q onto a non-singular quadric $\hat{Q}$ in $\mathbb{P}^3$. Hence, Q is a quadric cone over $\hat{Q}$ with the vertex $V(Q)$, $\hat{Q}$ has two different 1-parameter families $\{l_\lambda\}_{\lambda\in\mathbb{P}^1}$ and $\{l'_\lambda\}_{\lambda\in\mathbb{P}^1}$ of lines on it. Hence Q has two different 1-parameter families of $(N-2)$-dimensional linear subspaces on it.

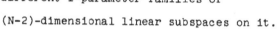

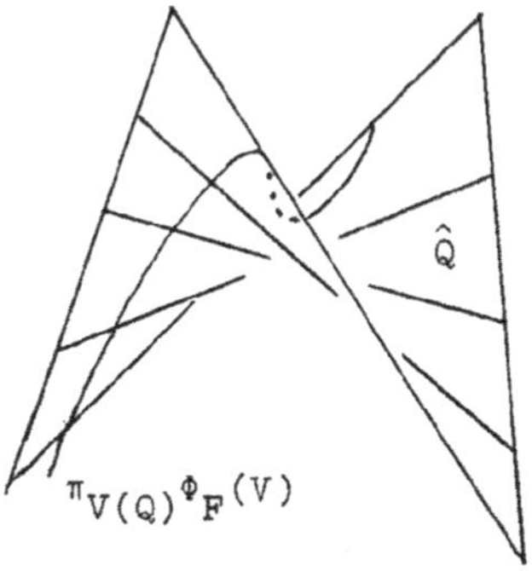

In fact, they are $\{S_{(\lambda_0:\lambda_1)}\}_{(\lambda_0:\lambda_0)\in\mathbb{P}^1}$ and $\{T_{(\lambda_0:\lambda_1)}\}_{(\lambda_0:\lambda_1)\in\mathbb{P}^1}$, where

$$S_{(\lambda_0:\lambda_1)} = \langle \lambda_1\xi_0 - \lambda_0\eta_0\rangle \cap \langle \lambda_1\xi_1 - \lambda_0\eta_1\rangle ,$$

$$T_{(\lambda_0:\lambda_1)} = \langle \lambda_1\xi_0 - \lambda_0\xi_1\rangle \cap \langle \lambda_1\eta_0 - \lambda_0\eta_1\rangle .$$

Note that $S = S_{(0:1)}$ and $S' = S_{(1:0)}$. Note also that, for any λ, $\lambda' \in \mathbb{P}^1$,

$$\pi_{S_\lambda} \equiv \pi_{S_{\lambda'}}, \quad \pi_{T_\lambda} \equiv \pi_{T_{\lambda'}}, \quad \pi_{S_\lambda} \not\equiv \pi_{T_{\lambda'}} \quad (\text{mod } \mathrm{Aut}(\mathbb{P}^1)) .$$

We sometimes write $|F-[D]|$ instead of $|F\otimes[D]^{-1}|$ for a divisor D on V. Note that

$$|F-[D]| = \{D' \mid D' \text{ is an effective divisor on } V \text{ such that } D+D' \in |F|\} .$$

We put

$$|F-D| = \{(\xi) \in |F| \;\Big|\; (\xi) \geqq D\} .$$

Then, $|F-[D]|$ and $|F-D|$ are canonically biholomorphic.

<u>Lemma 2.5.1</u>. Let f be a rational function on a projective manifold V. We denote by $D_\infty(f)$ the polar divisor of f. Then there is a bijection

$$|F-D_\infty(f)| \longrightarrow \{S \mid S \text{ is a } (N-2)\text{-dimensional linear subspace of } \mathbb{P}^N \text{ such that } \pi_S = f \;\; (\text{mod } \mathrm{Aut}(\mathbb{P}^1))\} .$$

<u>Proof</u>. For $(\xi) \in |F-D_\infty(f)|$, put $\xi' = f\xi$. Then $(\xi') \in |F|$. Now, it is easy to see that the map

$$(\xi) \longmapsto S = \langle \xi\rangle \cap \langle \xi'\rangle$$

is a bijection. Q.E.D.

We denote by $G^1(|F-D|)$ the Grassmann variety of all lines in $|F-D|$. Let f be a rational function on V. Put

$$\mathbb{Q}_f = \{Q \mid Q \text{ is a quadric of rank } 3 \text{ or } 4 \text{ in } \mathbb{P}^N \text{ containing both } \Phi_F(V) \text{ and a } (n-2)\text{-dimensional linear subspace } S \text{ such that } \pi_S = f \ (\mathrm{mod}\ \mathrm{Aut}(\mathbb{P}^1))\} .$$

Proposition 2.5.2. Let f be a rational function on a projective manifold V. Then there is a bijection between $G^1(|F-D_\infty(f)|)$ and $\mathbb{Q}_f$.

Proof. For $l \in G^1(|F-D_\infty(f)|)$, we take two distinct points (ξ_0) and (η_0) on l. We may write

$$(\xi_0) = D_\infty(f) + D, \text{ and } (\eta_0) = D_\infty(f) + D' ,$$

where D and D' are effective divisors on V. Put

$$(\xi_1) = D_0(f) + D \quad \text{and} \quad (\eta_1) = D_0(f) + D' .$$

Then, (up to constant),

$$f = \frac{\xi_1}{\xi_0} = \frac{\eta_1}{\eta_0} .$$

Hence

$$\xi_0\eta_1 - \xi_1\eta_0 = 0 \quad \text{on } V.$$

Let Q be the quadric in $\mathbb{P}^N$ defined by this equation. Then $Q \in \mathbb{Q}_f$. We consider the map $l \longmapsto Q$. Then it is easy to see that it is well defined and bijective. Q.E.D.

<u>Proposition 2.5.3. (Andreotti-Mayer [3])</u>. Let f be a rational function on a projective manifold V. Let $\{\xi_1,\cdots,\xi_r\}$ be a basis of

$$H^0(V, \mathcal{O}(F-D_\infty(f))) = \{\xi \in H^0(V, \mathcal{O}(F)) \mid (\xi) \geqq D_\infty(f)\}.$$

Then, the $\frac{r(r-1)}{2}$ quadrics $Q_{\alpha\beta} \in \mathbb{Q}_f$, $1 \leqq \alpha < \beta \leqq r$, corresponding to lines $l_{\alpha\beta} = \overline{(\xi_\alpha)(\xi_\beta)} \in G^1(|F-D_\infty(f)|)$, are linearly independent.

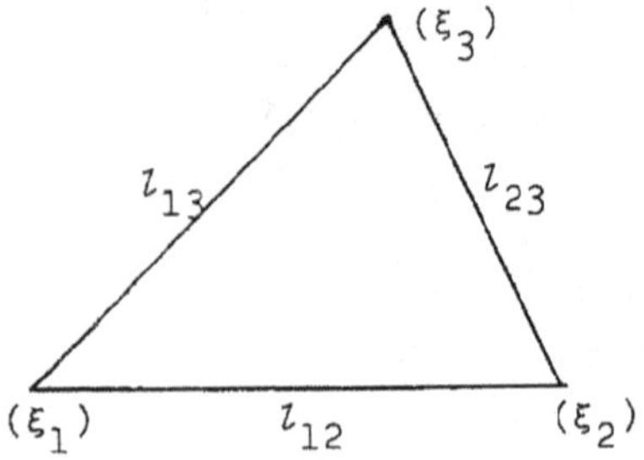

<u>Proof</u>. Put $\eta_\alpha = f\xi_\alpha$, $1 \leqq \alpha \leqq r$. Then $\eta_\alpha \in H^0(V, \mathcal{O}(F))$ and $Q_{\alpha\beta}$ is defined by the equation:

$$\omega_{\alpha\beta} = \xi_\alpha\eta_\beta - \xi_\beta\eta_\alpha = 0 .$$

Put $\omega_{\beta\alpha} = -\omega_{\alpha\beta}$. Assume that, for a $(r\times r)$-skew symmetric matrix $C = (c_{\alpha\beta})$, $\sum_{\alpha,\beta} c_{\alpha\beta}\omega_{\alpha\beta} = 0$. Take $\zeta_1,\cdots,\zeta_q \in H^0(V, \mathcal{O}(F))$ so that $\{\xi_1,\cdots,\xi_r,\zeta_1,\cdots,\zeta_q\}$ form a basis of $H^0(V, \mathcal{O}(F))$. Then $\eta_\alpha = f\xi_\alpha$ are written as

$$\eta_\alpha = \sum_\beta a_{\alpha\beta}\xi_\beta + \sum_\gamma b_{\alpha\gamma}\zeta_\gamma$$

Then

(1) $$\xi_\alpha\eta_\beta = \sum_\gamma a_{\beta\gamma}\xi_\alpha\xi_\gamma + \sum_\delta b_{\beta\delta}\xi_\alpha\zeta_\delta ,$$

(2) $$\xi_\beta\eta_\alpha = \sum_\gamma a_{\alpha\gamma}\xi_\beta\xi_\gamma + \sum_\delta b_{\alpha\delta}\xi_\beta\zeta_\delta .$$

Taking (1)-(2), we get

$$\omega_{\alpha\beta} = \sum_{\gamma} (a_{\beta\gamma}\xi_{\alpha} - a_{\alpha\gamma}\xi_{\beta})\xi_{\gamma} + \sum_{\delta} (b_{\beta\delta}\xi_{\alpha} - b_{\alpha\delta}\xi_{\beta})\zeta_{\delta} .$$

Hence

$$0 = \sum_{\alpha,\beta} c_{\alpha\beta}\omega_{\alpha\beta}$$

$$= \sum_{\alpha,\beta,\gamma} c_{\alpha\beta}(a_{\beta\gamma}\xi_{\alpha} - a_{\alpha\gamma}\xi_{\beta})\xi_{\gamma} + \sum_{\alpha,\beta,\delta} c_{\alpha\beta}(b_{\beta\delta}\xi_{\alpha} - b_{\alpha\delta}\xi_{\beta})\zeta_{\delta}$$

$$= 2\sum_{\alpha,\beta,\gamma} c_{\alpha\beta}a_{\beta\gamma}\xi_{\alpha}\xi_{\gamma} + 2\sum_{\alpha,\beta,\delta} c_{\alpha\beta}b_{\beta\delta}\xi_{\alpha}\zeta_{\delta} .$$

Note that $\xi_{\alpha}\xi_{\gamma}$ $(\alpha \leqq \gamma)$, $\xi_{\alpha}\zeta_{\delta}$ are linearly independent in $H^0(V, \mathcal{O}(F)) \overset{s}{\otimes} H^0(V, \mathcal{O}(F))$ (symmetric product). Hence

$$\sum_{\beta} c_{\alpha\beta}a_{\beta\gamma} + \sum_{\beta} c_{\gamma\beta}a_{\beta\alpha} = 0 \quad \text{and} \quad \sum_{\beta} c_{\alpha\beta}b_{\beta\delta} = 0 .$$

That is to say,

$$CA = -{}^tA{}^tC = {}^tAC \quad \text{and} \quad CB = 0 ,$$

where $A = (a_{\alpha\beta})$ and $B = (b_{\alpha\delta})$. (${}^tA$ is the transpose of A.)

Now, put $\pi_{\alpha} = \sum_{\beta} c_{\alpha\beta}\xi_{\beta} \in H^0(V, \mathcal{O}(F-D_{\infty}(f)))$. Let

$$\xi = {}^t(\xi_1,\cdots,\xi_r), \quad \zeta = {}^t(\zeta_1,\cdots,\zeta_q) \quad \text{and} \quad \pi = {}^t(\pi_1,\cdots,\pi_r)$$

be column vectors. Then

$$f\pi = {}^t(f\pi_1,\cdots,f\pi_r) = f(C\xi) = C(f\xi)$$

$$= C(A\xi+B\zeta) = CA\xi \in H^0(V, \mathcal{O}(F-D_{\infty}(f)))^r .$$

$$f^2\pi = f(f\pi) = f(CA\xi) = CA(f\xi) = CA(A\xi + B\zeta) = CA^2\xi + CAB\zeta$$

$$= CA^2\xi + {}^tACB\zeta = CA^2\xi \in H^0(V, \mathcal{O}(F-D_{\infty}(f)))^r .$$

In a similar way, we get

$$f^m\pi \in H^0(V, \mathcal{O}(F-D_{\infty}(f)))^r \quad \text{for all} \quad m \geqq 1 .$$

Take m sufficiently large. Then, this is impossible unless $\pi = 0$, i.e., $C = 0$. Q.E.D.

Let f and h be rational functions on V such that $f \not\equiv h$ $(\mathrm{mod}\ \mathrm{Aut}(\mathbb{P}^1))$. Put

$$\mathbb{Q}_{f,h} = \{Q \mid Q \text{ is a quadric in } \mathbb{P}^N \text{ with rank } 4 \text{ containing both } \Phi_F(V) \text{ and } (N-2)\text{-dimensional linear subspaces } S \text{ and } T \text{ such that } \pi_S = f\ (\mathrm{mod}\ \mathrm{Aut}(\mathbb{P}^1)),\ \pi_T = h\ (\mathrm{mod}\ \mathrm{Aut}(\mathbb{P}^1))\}.$$

$$\mathbb{Q}_{f,f} = \{Q \mid Q \text{ is a quadric in } \mathbb{P}^N \text{ with rank } 3 \text{ containing both } \Phi_F(V) \text{ and } (N-2)\text{-dimensional linear subspace } S \text{ such that } \pi_S = f\ (\mathrm{mod}\ \mathrm{Aut}(\mathbb{P}^1))\}.$$

Then

Proposition 2.5.4. Let f and h be rational functions on a projective manifold V such that $f \not\equiv h$ $(\mathrm{mod}\ \mathrm{Aut}(\mathbb{P}^1))$. Then (1) there is a bijection between $|F - (D_\infty(f)+D_\infty(h))|$ and $\mathbb{Q}_{f,h}$ and (2) there is a bijection between $|F - 2D_\infty(f)|$ and $\mathbb{Q}_{f,f}$.

Proof. (1). For $(\xi_0) \in |F - (D_\infty(f)+D_\infty(h))|$, we write

$$(\xi_0) = D_\infty(f) + D_\infty(h) + D,$$

where D is an effective divisor (or 0). We put

$$(\xi_1) = D_0(f) + D_\infty(h) + D,$$

$$(\eta_0) = D_\infty(f) + D_0(h) + D,$$

$$(\eta_1) = D_0(f) + D_0(h) + D.$$

Then, (up to constants),

$$f = \frac{\xi_1}{\xi_0} = \frac{\eta_1}{\eta_0} \quad \text{and} \quad h = \frac{\eta_0}{\xi_0} = \frac{\eta_1}{\xi_1} .$$

Hence

$$\xi_0\eta_1 - \xi_1\eta_0 = 0 \quad \text{on} \quad V .$$

Let Q be the quadric in $\mathbb{P}^N$ defined by this equation. Then $Q \in \mathbb{Q}_{f,h}$. We consider the map $(\xi_0) \longmapsto Q$. Then it is easy to see that it is well defined and bijective.

(2) is proved in a similar way. Q.E.D.

The set of all quadrics in $\mathbb{P}^N$ containing $\Phi_F(V)$ forms a complex projective space. Hence

$$\mathbb{Q}_F = \{Q \mid Q \text{ is a quadric in } \mathbb{P}^N \text{ of rank } 3 \text{ or } 4 \text{ such that } Q \supset \Phi_F(V)\} .$$

is a (possibly reducible) projective variety. Then, we have easily

Proposition 2.5.5. Let f and h be rational functions on V. Then $\mathbb{Q}_f$ and $\mathbb{Q}_{f,h}$ are closed complex subspaces of $\mathbb{Q}_F$ such that

(1) there is a bijective holomorphic map $G^1(|F-D_\infty(f)|) \longrightarrow \mathbb{Q}_f$,

(2) $|F-(D_\infty(f) + D_\infty(h))| \cong \mathbb{Q}_{f,h}$.

The quadrics in $\mathbb{Q}_F$ are constructed as follows: Let D be a divisor on V such that

$$\dim |D| \geqq 1 \quad \text{and} \quad \dim |F-[D]| \geqq 1 .$$

Let l and l' be lines in $|D|$ and $|F-[D]|$, respectively. They are linear pencils on V and are written as:

$$l = \{D_\lambda(f_l)+\hat{D}\}_{\lambda\in \mathbb{P}^1} \quad \text{and} \quad l' = \{D_\lambda(f_{l'})+\hat{D}'\}_{\lambda\in \mathbb{P}^1},$$

where $\hat{D}$ and $\hat{D}'$ are the fixed parts of l and l', respectively and f_l and $f_{l'}$ are rational functions determined by l and l' uniquely up to mod $\mathrm{Aut}(\mathbb{P}^1)$. Put

$$(\xi_0) = D_\infty(f_l) + D_\infty(f_{l'}) + \hat{D} + \hat{D}' ,$$

$$(\xi_1) = D_0(f_l) + D_\infty(f_{l'}) + \hat{D} + \hat{D}' ,$$

$$(\eta_0) = D_\infty(f_l) + D_0(f_{l'}) + \hat{D} + \hat{D}' ,$$

$$(\eta_1) = D_0(f_l) + D_0(f_{l'}) + \hat{D} + \hat{D}' .$$

(They are elements of $|F|$.) Then, as before, the equation

$$\xi_0\eta_1 - \xi_1\eta_0 = 0$$

defines a quadric in $\mathbb{P}^N$ containing $\Phi_F(V)$. It has rank 3 or 4. We denote it by $Q_{l,l'}$ and call it the quadric determined by l and l'.

Note that every quadric in $\mathbb{Q}_F$ can be constructed in this way.

Now, we return to the case of canonical curves.

In the rest of this section, we assume that V is a non-hyperelliptic compact Riemann surface of genus $g \geqq 4$ and consider the case $F = K_V$. The following theorem is classical.

Theorem 2.5.6. (Max Nöther-Enriques-Petri).

(1) The canonical map

$$J : S^*H^0(V, \mathcal{O}(K_V)) \longrightarrow \bigoplus_{n \geqq 0} H^0(V, \mathcal{O}(K_V^{\otimes n}))$$

is surjective, where $S^*H^0(V, \mathcal{O}(K_V))$ is the symmetric algebra of $H^0(V, \mathcal{O}(K_V))$.

(2) The kernel of J is generated by its elements of degree 2 and of degree 3.

(3) The kernel of J is generated by its elements of degree 2 except

in the following cases:

(i) V is biholomorphic to a non-singular plane quintic $(g = 6)$.

(ii) V is trigonal.

The proof of Theorem 2.5.6 is found in Petri [69]. Modern proofs were given by Saint-Donat [76] and Šokurov [79].

Let C_V be a canonical curve of V. Then, by (1) of Theorem 2.5.6, the set of all quadrics in $\mathbb{P}^{g-1}$ containing C_V forms a complex projective space $\Sigma = \mathbb{P}^m$, where

$$m = \frac{(g-2)(g-3)}{2} - 1 .$$

Moreover, Theorem 2.5.6 says that if V is neither trigonal nor a non-singular plane quintic, then the intersection of all such quadrics is just C_V. If V is trigonal, then the intersection of all such quadrics is a rational ruled surface (i.e., Hirzebruch surface in Example 0.1.10) $M^{(n)}$ $(= \mathbb{P}(\mathcal{O}_{\mathbb{P}^1} \oplus \mathcal{O}_{\mathbb{P}^1}(n)))$ containing C_V, where n satisfies

$$0 \leqq n \leqq \min\{\frac{g+2}{3}, g-4\} \quad \text{and} \quad n \equiv g \pmod 2 ,$$

(see Šokurov [79]). If V is a non-singular plane quintic, then the intersection is the Veronese surface in $\mathbb{P}^5$, i.e., the image of the holomorphic imbedding:

$$(z_0:z_1:z_2) \in \mathbb{P}^2 \longmapsto (z_0^2:z_0z_1:z_0z_2:z_1^2:z_1z_2:z_2^2) \in \mathbb{P}^5 .$$

Petri showed (see Saint-Donat [76] or Mumford [59]) that, if $(X_1:\cdots X_g)$ is a suitable homogeneous coordinate system in $\mathbb{P}^{g-1}$, then the set of all quadrics in $\mathbb{P}^{g-1}$ containing C_V is given by $\{Q_\lambda\}_{\lambda\in\Sigma}$ where

$$Q_\lambda : \sum_{i,j\geqq 3} \lambda_{ij} X_i X_j - \sum_{i,j,k\geqq 3} \lambda_{ij} (\alpha^1_{ijk} X_1 X_k + \alpha^2_{ijk} X_2 X_k)$$

$$- \sum_{i,j\geqq 3} \lambda_{ij} \nu_{ij} X_1 X_2 = 0 .$$

(α^1_{ijk}, α^2_{ijk} and ν_{ij} are constants symmetric in i and j.) Here, we use the homogeneous coordinate system in $\Sigma = \mathbb{P}^m$:

$$\lambda = (\lambda_{ij}), \quad 3 \leqq i, j \leqq g, \quad (\lambda_{ij} = \lambda_{ji} \text{ and } \lambda_{ii} = 0).$$

We put

$$A_\lambda = \begin{pmatrix} 0 & -\sum_{i,j} \lambda_{ij}\nu_{ij} & -\sum_{i,j} \lambda_{ij}\alpha^1_{ij3} & \cdots & -\sum_{i,j} \lambda_{ij}\alpha^1_{ijg} \\ -\sum_{i,j} \lambda_{ij}\nu_{ij} & 0 & -\sum_{i,j} \lambda_{ij}\alpha^2_{ij3} & \cdots & -\sum_{i,j} \lambda_{ij}\alpha^2_{ijg} \\ -\sum_{i,j} \lambda_{ij}\alpha^1_{ij3} & -\sum_{i,j} \lambda_{ij}\alpha^2_{ij3} & & & \\ \cdots & & & \lambda & \\ -\sum_{i,j} \lambda_{ij}\alpha^1_{ijg} & -\sum_{i,j} \lambda_{ij}\alpha^2_{ijg} & & & \end{pmatrix}.$$

Then the rank of Q_λ is nothing but the rank of the matrix A_λ. As before, we put

$$\mathbb{Q}_K = \{Q_\lambda \mid Q_\lambda \text{ has the rank} \leqq 4\}$$

$$= \{Q_\lambda \mid \text{rank } A_\lambda \leqq 4\}.$$

($K = K_V$.) Then, it is regarded as a closed complex subspace of $\Sigma = \mathbb{P}^m$.

Proposition 2.5.7. (Martens [52,I], Andreotti-Mayer [3]). If $g \geqq 4$, then $\mathbb{Q}_K$ is non-empty and every component has dimension $\geqq g-4$.

Proof. Let $(X_1:\cdots:X_g)$ be a homogeneous coordinate system in $\mathbb{P}^{g-1}$. For a quadric

$$Q = \{\Sigma\, a_{ij}X_iX_j = 0\}, \quad (a_{ij} = a_{ji}),$$

we associate a point $A = (a_{ij}) \in \mathbb{P}^\rho$, where $\rho = \frac{g(g+1)}{2} - 1$.

Then, the set of all quadrics containing the canonical curve C_V is identified with a m-dimensional linear subspace Σ, where $m = \frac{(g-2)(g-3)}{2} - 1$. Put

$$\mathcal{Y} = \{ A = (a_{\alpha\beta}) \in \mathbb{P}^\rho \mid \text{rank } A \leqq 4 \}.$$

Then $\mathcal{Y}$ is an irreducible subvariety of $\mathbb{P}^\rho$ of dimension $4g-7$. Since $\mathbb{Q}_K = \Sigma \cap \mathcal{Y}$, $\mathbb{Q}_K$ is non-empty and every component has dimension $\geqq m + (4g-7) - \rho = g-4$, provided $g \geqq 4$. Q.E.D.

By Proposition 2.5.5, for meromorphic functions f and h on V, $\mathbb{Q}_f$ and $\mathbb{Q}_{f,h}$ are closed complex subspaces of $\mathbb{Q}_K$.

Proposition 2.5.8. Let f and h be meromorphic functions on V of order l and n, respectively. Then

(1) If $l \leqq g-1$, then $\mathbb{Q}_f$ is non-empty.

(2) If $2l \leqq g+1$, then $\mathbb{Q}_{f,f}$ is non-empty.

(3) If $l \neq n$ and $l+n \leqq g+2$, then $\mathbb{Q}_{f,h}$ is non-empty.

(4) If $l = n$, $2l \leqq g+2$ and $f \not\equiv h \pmod{\mathrm{Aut}(\mathbb{P}^1)}$, then $\mathbb{Q}_{f,h}$ is non-empty.

Proof. (1). By Riemann-Roch theorem,

$$h^0(K_V - D_\infty(f)) = h^0(D_\infty(f)) - l - 1 + g \geqq 2 - l - 1 + g \geqq 2 .$$

Hence, by Proposition 2.5.2, $\mathbb{Q}_f$ is non-empty.

(2). Note that $h^0(2D_\infty(f)) \geqq 3$, for 1, f, and f^2 are linearly independent in $L(2D_\infty(f))$. If $2l \leqq g+1$, then

$$h^0(K_V - 2D_\infty(f)) = h^0(2D_\infty(f)) - 2l - 1 + g \geqq 3 - 2l - 1 + g \geqq 1 .$$

Hence, by (2) of Proposition 2.5.4, $\mathbb{Q}_{f,f}$ is non-empty.

(3) and (4) follow from (1) of Proposition 2.5.4 and Lemma 2.4.1.

Q.E.D.

Proposition 2.5.9. For $Q \in \mathbb{Q}_K$, denote by D_o the intersection divisor on V of C_V and $V(Q)$, the vertex of Q. Let f and h be the meromorphic functions (mod $\mathrm{Aut}(\mathbb{P}^1)$) determined by two 1-parameter families of $(g-3)$-dimensional linear subspaces on Q. (If Q has the rank 3, then we put $f = h$.) Then

$$\mathrm{ord}(f) + \mathrm{ord}(h) + \deg D_o = 2g-2 .$$

In particular, at least one of f and h has order $\leqq g-1$.

Proof. We assume that Q has the rank 4. (The argument is similar, when Q has the rank 3.) Then $\dim V(Q) = g-5$. Let $\pi : C_V \longrightarrow \mathbb{P}^3$ be the projection with the center $V(Q)$. Let C be its image. $\pi(Q) = \hat{Q}$ is a non-singular quadric containing C. $\hat{Q}$ has two different 1-parameter families $\{l_\alpha\}_{\alpha \in \mathbb{P}^1}$ and $\{l'_\beta\}_{\beta \in \mathbb{P}^1}$ of lines on it. Put $S_\alpha = \pi^{-1}(l_\alpha)$ and $T_\beta = \pi^{-1}(l'_\beta)$. Any l_α and l'_β meet just one point, so they span a plane $H_{\alpha\beta}$ in $\mathbb{P}^3$. Note that

$$C \cap H_{\alpha\beta} = (C \cap \hat{Q}) \cap H_{\alpha\beta} = (C \cap l_\alpha) \cup (C \cap l'_\beta) .$$

We choose α and $\beta \in \mathbb{P}^1$ so that $l_\alpha \cap l'_\beta \notin C$. Then

$$C_V \cap \pi^{-1}(H_{\alpha\beta}) = (C_V \cap S_\alpha) \cup (C_V \cap T_\beta)$$

and

$$(C_V \cap S_\alpha) \cap (C_V \cap T_\beta) = D_o .$$

This proves the proposition. Q.E.D.

As noted above, every quadric $Q \in \mathbb{Q}_K$ is obtained by lines (i.e., linear pencils) $l \subset |D|$ and $l' \subset |K-[D]|$:

$$Q = Q_{l,l'}$$

for some divisor D with $\dim |D| \geqq 1$ and $\dim |K-[D]| \geqq 1$.

Put

$$\mathbb{Q}'_K = \{ Q_{l,l'} \in \mathbb{Q}_K \mid l \text{ and } l' \text{ are \underline{complete} linear pencils of degree } g-1 \}.$$

Note that, in this definition, l and l' must satisfy

$$[D + D'] = K \quad \text{for} \quad D \in l \quad \text{and} \quad D' \in l'.$$

Problem. (Mumford [59, p.89]). Do the elements of $\mathbb{Q}'_K$ span $\Sigma = \mathbb{P}^m$?

Andreotti-Mayer [3] says that it is affirmative for any trigonal V (see Theorem 2.5.16 below). In the next section, we give an affirmative answer for $g = 5$.

The theta-divisor (ϑ) of $J(V)$ is equal to W_{g-1} up to translations $t_y : x \in J(V) \longmapsto x+y \in J(V)$ for $y \in J(V)$ with $2y = \kappa$ (see, e.g., Martens [52, II, p.96]). The set of points of (ϑ) with multiplicity 2 is just $W^1_{g-1} \setminus W^2_{g-1}$.

Every quadric $Q_{l,l'} \in \mathbb{Q}'_K$ is then the translation to the origin 0 of $J(V)$, of the projectivized tangent cone to $(\vartheta) = W_{g-1}$ at some point $x \in W^1_{g-1} \setminus W^2_{g-1}$ and conversely. (See Chapter 5, Corollary 5.1.22. Note that $T_0J(V)$ is canonically isomorphic to $H^1(V, \mathcal{O})$, which is dual to $H^0(V, \mathcal{O}(K))$.) Thus, if the problem is answered affirmatively, then we get a proof of Torelli's theorem for a compact Riemann surface V, provided V is neither hyperelliptic, trigonal nor non-singular plane quintic, for V is then the intersection of such $Q_{l,l'}$. See Mumford [59, p.89] for detail.

As the first step for solving the problem, we show

Proposition 2.5.10. The elements of $\mathbb{Q}_K$ and $\mathbb{Q}'_K$ span the same linear subspace of $\Sigma = \mathbb{P}^m$.

The idea of the proof of the proposition given below is essentially due to Andreotti-Meyer [3]. We first need some lemmas.

Lemma 2.5.11. For a divisor D on V, assume that $\dim|D| = r \geqq 1$. Then, there exist independent divisors $D_0, D_1, \cdots, D_r \in |D|$ such that the linear pencils

$$l_{\alpha\beta} = \overline{D_\alpha D_\beta} \quad \text{(the line connecting } D_\alpha \text{ and } D_\beta\text{)},$$

$0 \leqq \alpha < \beta \leqq r$, determine meromorphic functions $f_{\alpha\beta}$ (mod $\mathrm{Aut}(\mathbb{P}^1)$) such that

$$\dim|D_\infty(f_{\alpha\beta})| = 1 \quad \text{for} \quad 0 \leqq \alpha < \beta \leqq r.$$

Proof. We may assume that $r \geqq 2$ and $|D|$ is fixed point free. Thus

$$\Phi_{|D|} : V \longrightarrow \mathbb{P}^r$$

is a holomorphic map. We denote by C its image. Take independent $(r+1)$-points $Q_0, Q_1, \cdots, Q_r$ on C and put

(1) H_α = the hyperplane in $\mathbb{P}^r$ spanned by $Q_0, \cdots, \check{Q}_\alpha, \cdots, Q_r$. ($\check{Q}_\alpha$ means that we exclude Q_α).

(2) $D_\alpha = \Phi_{|D|}^{-1}(H_\alpha \cap C) \in |D|$.

(3) $S_{\alpha\beta}$ = the $(r-2)$-dimensional linear subspace of $\mathbb{P}^r$ spanned by

$$Q_0, \cdots, \check{Q}_\alpha, \cdots, \check{Q}_\beta, \cdots, Q_r.$$

(4) $\pi_{\alpha\beta} = \pi_{S_{\alpha\beta}} : \tilde{C} \longrightarrow \mathbb{P}^1$, the projection with the center $S_{\alpha\beta}$.

($\tilde{C}$ is a non-singular model of C.)

(5) $f_{\alpha\beta} = \pi_{\alpha\beta} \circ \Phi_{|D|} : V \longrightarrow \mathbb{P}^1$.

(6) $l_{\alpha\beta} = \overline{D_\alpha D_\beta}$.

Then, it is clear that $D_0, \cdots, D_r$ are independent in $|D|$ and the linear pencil $l_{\alpha\beta}$ determines the meromorphic function $f_{\alpha\beta}$.

We show that

$$\dim |D_\infty(f_{\alpha\beta})| = 1 \quad \text{for} \quad 0 \leqq \alpha < \beta \leqq r .$$

We may consider the case: $\alpha = 0$ and $\beta = 1$. Put

$$E_\alpha = \Phi_{|D|}^{-1}(Q_\alpha) \quad \text{for} \quad 0 \leqq \alpha \leqq r .$$

Then E_α is an effective divisor on V with

$$\deg E_\alpha = \text{the mapping order of } \Phi_{|D|} .$$

Note that

$$|D| - (E_2+\cdots+E_r) = \{ D' \in |D| \mid D' \geqq E_2+\cdots+E_r \}$$

is canonically biholomorphic to the complete linear system $|D-(E_2+\cdots+E_r)|$. On the other hand $|D|-(E_2+\cdots+E_r)$ corresponds to the pencil of hyperplanes containing S_{01}. Hence

$$\dim |D - (E_2+\cdots+E_r)| = 1 .$$

$|D-(E_2+\cdots+E_r)|$ may have a fixed point. Its variable part is nothing but $|D_\infty(f_{01})|$. Thus $\dim |D_\infty(f_{01})| = 1$. <u>Q.E.D.</u>

As in Proposition 2.5.9, a quadric $Q \in \mathbb{Q}_K$ determines meromorphic functions f and h (mod$\mathrm{Aut}(\mathbb{P}^1)$).

<u>Lemma 2.5.12</u>. Put

$$\mathbb{Q}_K'' = \{ Q \in \mathbb{Q}_K \mid Q \text{ determines meromorphic functions } f \text{ and } h \pmod{\mathrm{Aut}(\mathbb{P}^1)} \text{ such that } \dim |D_\infty(f)| = 1 \text{ and } \dim |D_\infty(h)| = 1 \}.$$

Then, the elements of $\mathbb{Q}_K''$ and $\mathbb{Q}_K$ span the same linear subspace of $\Sigma = \mathbb{P}^m$.

Proof. It is enough to show that every element of $\mathbb{Q}_K$ is written as a linear combination of the elements of $\mathbb{Q}''_K$. Any quadric in $\mathbb{Q}_K$ is written as $Q_{l,l'}$, where l and l' are lines in $|D|$ and $|K-[D]|$, respectively, for some D. Put

$$\dim |D| = r \geqq 1 \quad \text{and} \quad \dim |K - [D]| = s \geqq 1 \ .$$

Take independent $D_0 = (\xi_0), \cdots, D_r = (\xi_r) \in |D|$ satisfying Lemma 5.2.11. ($\xi_\alpha \in H^0(V, \mathcal{O}([D]))$, $0 \leqq \alpha \leqq r$.) Take distinct points (ζ_0) and (ζ_1) on l. Then ζ_0 and ζ_1 are written as

$$\zeta_0 = \Sigma\, a^\alpha \xi_\alpha \quad \text{and} \quad \zeta_1 = \Sigma\, b^\alpha \xi_\alpha \ .$$

Also take independent $E_0 = (\eta_0), \cdots, E_s = (\eta_s) \in |K - [D]|$ satisfying Lemma 5.2.11. ($\eta_\gamma \in H^0(V, \mathcal{O}(K - [D]))$, $0 \leqq \gamma \leqq s$.) Take distinct points (λ_0) and (λ_1) on l'. Then λ_0 and λ_1 are written as

$$\lambda_0 = \Sigma\, c^\gamma \eta_\gamma \quad \text{and} \quad \lambda_1 = \Sigma\, d^\gamma \eta_\gamma \ .$$

Now, $Q_{l,l'}$ is defined by the equation:

$$(\zeta_0\lambda_0)\cdot(\zeta_1\lambda_1) - (\zeta_0\lambda_1)\cdot(\zeta_1\lambda_0) = 0 \ .$$

But

$$\begin{aligned}
&(\zeta_0\lambda_0)\cdot(\zeta_1\lambda_1) - (\zeta_0\lambda_1)\cdot(\zeta_1\lambda_0) \\
&= \sum_{\alpha<\beta} (a^\alpha b^\beta - a^\beta b^\alpha)((\xi_\alpha\lambda_0)\cdot(\xi_\beta\lambda_1) - (\xi_\alpha\lambda_1)\cdot(\xi_\beta\lambda_0)) \\
&= \sum_{\alpha<\beta} \sum_{\gamma<\delta} (a^\alpha b^\beta - a^\beta b^\alpha)(c^\gamma d^\delta - c^\delta d^\gamma)((\xi_\alpha\eta_\gamma)\cdot(\xi_\beta\eta_\delta) - (\xi_\alpha\eta_\delta)\cdot(\xi_\beta\eta_\gamma)) \ .
\end{aligned}$$

Q.E.D.

Lemma 2.5.13. For any integer n with $n < g-1$ and any complete fixed point free linear pencil g^1_n, there exists a proper closed complex subspace Y in $S^{g-1-n}V$ such that

$$g_n^1 + \hat{D} \text{ is complete for all } \hat{D} \in S^{g-1-n}V - Y.$$

Proof. We first show that there exists $\hat{D} \in S^{g-1-n}V$ such that $g_n^1 + \hat{D}$ is complete. Take $D \in g_n^1$. $(|D| = g_n^1.)$ For any $\hat{D} \in S^{g-1-n}V$,

$$\dim |D + \hat{D}| = \dim |K - [D + \hat{D}]|$$

by Riemann-Roch theorem. Thus, it is enough to find $\hat{D} \in S^{g-1-n}V$ with $\dim |K - [D+\hat{D}]| = 1$. Take a point $P_1 \in V$ such that P_1 is not a fixed point of $|K - [D]|$. Then

$$\dim |K - [D + P_1]| = \dim |K - [D]| - 1 = g-n-1 .$$

Take a point $P_2 \in V$ such that P_2 is not a fixed point of $|K - [D+P_1]|$. Then

$$\dim |K - [D + P_1 + P_2]| = \dim |K - [D + P_1]| - 1 = g-n-2 .$$

Repeating this process, we find

$$\hat{D} = P_1 + \cdots + P_{g-1-n} \in S^{g-1-n}V$$

such that $\dim |K - [D+\hat{D}]| = 1$.

Now, consider the holomorphic map

$$\sigma : \hat{D} \in S^{g-1-n}V \longmapsto D + \hat{D} \in G_{g-1}^1 .$$

Then, $Y = \sigma^{-1}(G_{g-1}^2)$ is a proper closed complex subspace of $S^{g-1-n}V$ and satisfies the requirement. Q.E.D.

A similar proof shows the following lemma.

Lemma 2.5.14. For any integer n with $n < g-1$ and any effective divisor D with $\deg D = n$ and $\dim |D| = 1$, there exists a proper closed complex subspace $Z \subset S^{g+1-n}V$ such that

$$|K - [D + \hat{D}]| \text{ is empty for all } \hat{D} \in S^{g+1-n_V} - Z .$$

For the proof of Proposition 5.2.10, we need one more lemma.

Lemma 2.5.15. Under the same notations as in Proposition 2.5.3, every $Q \in \mathbb{Q}_f$ is written as a linear combination of $Q_{\alpha\beta}$, $1 \leqq \alpha < \beta \leqq r$.

Proof. Let $Q \in \mathbb{Q}_f$ corresponds to a line $l \in G^1(|F - D_\infty(f)|)$. Take distinct points (ξ) and (ξ') on l. Then

$$\xi = \Sigma\, a^\alpha \xi_\alpha \quad \text{and} \quad \xi' = \Sigma\, b^\alpha \xi_\alpha .$$

Put $\eta = f\xi$ and $\eta' = f\xi'$. Then Q is given by the equation:

$$\xi\eta' - \xi'\eta = 0 .$$

But

$$\xi \cdot \eta' - \xi' \cdot \eta = \sum_{\alpha<\beta} (a^\alpha b^\beta - a^\beta b^\alpha)(\xi_\alpha \cdot \eta_\beta - \xi_\beta \cdot \eta_\alpha).$$

Q.E.D.

Now, we are ready to prove Proposition 2.5.10.

Proof of Proposition 2.5.10. Note that

$$\mathbb{Q}_K \supset \mathbb{Q}''_K \supset \mathbb{Q}'_K .$$

By Lemma 2.5.12, it is enough to show that every $Q \in \mathbb{Q}''_K$ is written as a linear combination of elements of $\mathbb{Q}'_K$. By Proposition 2.5.9, one of meromorphic functions, say f, determined by Q, has the order $n \leqq g-1$. By the assumption,

$$\dim |D_\infty(f)| = 1 .$$

If $n = g-1$, then, by Riemann-Roch theorem, $Q \in \mathbb{Q}'_K$.

Assume that $n < g-1$. Since $Q \in \mathbb{Q}_f$, it is enough, by Lemma 2.5.15, to find a basis

$$\{\omega_1, \cdots, \omega_{g+1-n}\}$$

of $H^0(V, \mathcal{O}(K-D_\infty(f)))$ such that the quadrics $Q_{\alpha\beta} \in \mathbb{Q}_f$, $1 \leqq \alpha < \beta \leqq g+1-n$, corresponding to the lines $\overline{(\omega_\alpha)(\omega_\beta)}$, belong to $\mathbb{Q}_K'$.

Let V^k be the k-times Cartesian product of V and let

$$\pi : V^k \longrightarrow S^k V$$

be the natural projection. For $1 \leqq \alpha < \beta \leqq g+1-n$, let

$$\lambda_{\alpha\beta} : V^{g+1-n} \longrightarrow V^{g-1-n}$$

be the projection defined by

$$\lambda_{\alpha\beta}(P_1, \cdots, P_{g+1-n}) = (P_1, \cdots, \check{P}_\alpha, \cdots, \check{P}_\beta, \cdots, P_{g+1-n}) .$$

Let $Y \subset S^{g-1-n}V$ be the proper closed complex subspace in Lemma 2.5.13 with respect to $g_n^1 = |D_\infty(f)|$. Let $Z \subset S^{g+1-n}V$ be the proper closed complex subspace in Lemma 2.5.14 with respect to $D = D_\infty(f)$. Put

$$X = \bigcup_{\alpha<\beta} (\pi\lambda_{\alpha\beta})^{-1}(Y) \cup \pi^{-1}(Z) ,$$

$$\Delta = \{(P_1, \cdots, P_{g+1-n}) \in V^{g+1-n} \mid P_\alpha = P_\beta \text{ for some } 1 \leqq \alpha < \beta \leqq g+1-n \} .$$

Then they are proper closed complex subspaces of V^{g+1-n}.

Take any point $(P_1, \cdots, P_{g+1-n}) \in V^{g+1-n} - X - \Delta$. Then it satisfies:

(1) $P_\alpha \neq P_\beta$ for $1 \leqq \alpha < \beta \leqq g+1-n$.

(2) $|D_\infty(f)| + P_1 + \cdots + \check{P}_\alpha + \cdots + \check{P}_\beta + \cdots + P_{g+1-n}$ is a <u>complete</u> linear pencil for all $1 \leqq \alpha < \beta \leqq g+1-n$.

(3) $h^1(D_\infty(f) + P_1 + \cdots + P_{g+1-n}) = 0$.

Hence, we get, for $1 \leqq \alpha \leqq g+1-n$,

(4) $h^0(K-[D_\infty(f) + P_1 + \cdots + \check{P}_\alpha + \cdots + P_{g+1-n}]) = 1$.

In fact, by (2),

$$h^0(K - [D_\infty(f)+P_1+\cdots+\check{P}_\alpha+\cdots+\check{P}_\beta+\cdots+P_{g+1-n}]) = 2 .$$

Thus $h^0(K-[D_\infty(f)+P_1+\cdots+\check{P}_\alpha+\cdots+P_{g+1-n}])$ must be 2 or 1. If it is 2, then $h^0(K-[D_\infty(f)+P_1+\cdots P_{g+1-n}])$ is 2 or 1, which contradicts to (3). Thus we get (4).

By (4), there exists a unique (up to constant) $\omega_\alpha \in H^0(V, \mathcal{O}(K))$ such that

$$(\omega_\alpha) \geqq D_\infty(f)+P_1+\cdots+\check{P}_\alpha+\cdots+P_{g+1-n} .$$

It is clear that

$$\{\omega_1, \cdots, \omega_{g+1-n}\}$$

is a basis of $H^0(V, \mathcal{O}(K-D_\infty(f)))$. Put

$$\omega'_\alpha = f\omega_\alpha \in H^0(V, \mathcal{O}(K)) .$$

Then, the quadric $Q_{\alpha\beta}$, corresponding to the line $\overline{(\omega_\alpha)(\omega_\beta)}$, is given by

$$\omega_\alpha\omega'_\beta - \omega_\beta\omega'_\alpha = 0 .$$

But, it is clear that, if we put

$$l_{\alpha\beta} = |D_\infty(f)|+P_1+\cdots+\check{P}_\alpha+\cdots+\check{P}_\beta+\cdots+P_{g+1-n} ,$$

then

$$Q_{\alpha\beta} = Q_{l_{\alpha\beta}, |K-[E]|} \quad \text{for} \quad E \in l_{\alpha\beta} .$$

Thus $Q_{\alpha\beta} \in \mathbb{Q}'_K$. <u>Q.E.D.</u>

<u>Theorem 2.5.16. (Andreotti-Mayer [3]).</u> If V is trigonal and $g \geqq 4$, then the elements of $\mathbb{Q}_K$ (and hence $\mathbb{Q}'_K$) span $\Sigma = \mathbb{P}^m$. In fact, the elements of $\mathbb{Q}_f$, for any $f \in R_3(V)$, span $\Sigma = \mathbb{P}^m$.

<u>Proof</u>. Let $f \in R_3(V)$. Since V is non-hyperelliptic,

$$\dim |D_\infty(f)| = 1 \quad \text{(see Theorem 2.1.9).}$$

By Riemann-Roch theorem, $h^0(K-D_\infty(f)) = g-2$. Hence, by Proposition 2.5.3, there are $\frac{(g-2)(g-3)}{2}$ $(= m+1)$ linearly independent quadrics in $\mathbb{Q}_f$. Hence the elements of $\mathbb{Q}_f$ span $\Sigma = \mathbb{P}^m$. <u>Q.E.D</u>

<u>Notes 2.5.17</u>. Studies on defining equations of projective varieties were done by Mumford [58], Saint-Donat [75] and Fujita [20]. They treat generalizations of Theorem 2.5.6.

2.6. The case of genus 5.

We apply the above consideration to the case of genus 5.

<u>Theorem 2.6.1</u>. Let V be a non-hyperelliptic compact Riemann surface of genus 5. Then

(1) If V is not trigonal, then there is a (ramified) double covering map

$$v : R_4(V)/\mathrm{Aut}(\mathbb{P}^1) \cong W_4^1 \longrightarrow C,$$

where C is a (possibly reducible) plane quintic curve. Moreover, v ramifies at $x \in W_4^1$ if and only if $2x = \kappa$ (the canonical point).

(2) If V is trigonal, then (2-i) $R_3(V)/\mathrm{Aut}(\mathbb{P}^1) \cong W_3^1 = \{y_0\}$ is one point and (2-ii) there is a plane quintic curve C with a unique singular point x_0, a double point, such that (a) C is birational to V and (b) $R_4(V)/\mathrm{Aut}(\mathbb{P}^1) \cong C - \{x_0\}$.

<u>Proof</u>. We use the same notations as in §2.5. Let $(X_1:\cdots:X_5)$ be a homogeneous coordinate system in $\mathbb{P}^4$. For $\lambda = (\lambda_{34}:\lambda_{45}:\lambda_{35}) \in \Sigma = \mathbb{P}^2$, put

$$Q_\lambda : \lambda_{34}f_{34} + \lambda_{45}f_{45} + \lambda_{35}f_{35} = 0 ,$$

where

$$f_{34} = X_3X_4 - \sum_{k=3}^{5}\alpha^1_{34k}X_1X_k - \sum_{k=3}^{5}\alpha^2_{34k}X_2X_k - \nu_{34}X_1X_2 ,$$

$$f_{45} = X_4X_5 - \sum_{k=3}^{5}\alpha^1_{45k}X_1X_k - \sum_{k=3}^{5}\alpha^2_{45k}X_2X_k - \nu_{45}X_1X_2 ,$$

$$f_{35} = X_3X_5 - \sum_{k=3}^{5}\alpha^1_{35k}X_1X_k - \sum_{k=3}^{5}\alpha^2_{35k}X_2X_k - \nu_{35}X_1X_2 .$$

Put

$$A_\lambda = \begin{pmatrix} 0 & -\Sigma\lambda_{ij}\nu_{ij} & -\Sigma\lambda_{ij}\alpha^1_{ij3} & -\Sigma\lambda_{ij}\alpha^1_{ij4} & -\Sigma\lambda_{ij}\alpha^1_{ij5} \\ -\Sigma\lambda_{ij}\nu_{ij} & 0 & -\Sigma\lambda_{ij}\alpha^2_{ij3} & -\Sigma\lambda_{ij}\alpha^2_{ij4} & -\Sigma\lambda_{ij}\alpha^2_{ij5} \\ -\Sigma\lambda_{ij}\alpha^1_{ij3} & -\Sigma\lambda_{ij}\alpha^2_{ij3} & 0 & \lambda_{34} & \lambda_{35} \\ -\Sigma\lambda_{ij}\alpha^1_{ij4} & -\Sigma\lambda_{ij}\alpha^2_{ij4} & \lambda_{34} & 0 & \lambda_{45} \\ -\Sigma\lambda_{ij}\alpha^1_{ij5} & -\Sigma\lambda_{ij}\alpha^2_{ij5} & \lambda_{35} & \lambda_{45} & 0 \end{pmatrix} ,$$

where Σ is extended over $3 \leqq i < j \leqq 5$. Then $Q_\lambda \in \mathbb{Q}_K$ if and only if $\det A_\lambda = 0$, i.e.,

$$\mathbb{Q}_K \cong \{\lambda \in \mathbb{P}^2 \mid \det A_\lambda = 0\}.$$

Thus $\mathbb{Q}_K$ is biholomorphic to either $\mathbb{P}^2$ or a (possibly reducible) plane quintic curve.

(1) Assume that V is not trigonal.

In this case, f_{34}, f_{45} and f_{35} generate the ideal of the canonical curve C_V. Hence C_V is the complete intersection of quadrics

$$Q_{(1:0:0)} = \{f_{34} = 0\} ,$$

$$Q_{(0:1:0)} = \{f_{45} = 0\} ,$$

$$Q_{(0:0:1)} = \{f_{35} = 0\} .$$

We know that $R_4(V)$ is non-empty, for it is a classically known fact that there is a meromorphic function on V of order at most $[\frac{g+3}{2}] = 4$ (see Chapter 5). By Theorem 2.1.9,

$$R_4(V)/\mathrm{Aut}(\mathbb{P}^1) \cong W_4^1 . \quad (W_4^2 \text{ is empty.})$$

W_4^1 is a translation in $J(V)$ of the singular locus of the theta-divisor and is pure $g-4 = 1$ dimensional (see Martens [52,I]). Hence $R_4(V)/\mathrm{Aut}(\mathbb{P}^1)$ is pure 1-dimensional. For every $f \in R_4(V)$, $h^0(D_\infty(f)) = 2$. By Riemann-Roch theorem, $h^1(D_\infty(f)) = 2$. Hence, by Proposition 2.5.2, there is a unique $Q_\lambda \in \mathbb{Q}_f$.

We consider the map

$$v : \tilde{\omega}(f) = f \ (\mathrm{mod}\ \mathrm{Aut}(\mathbb{P}^1)) \in R_4(V)/\mathrm{Aut}(\mathbb{P}^1) \longmapsto Q_\lambda \in \mathbb{Q}_K ,$$

($\tilde{\omega} : R_4(V) \longrightarrow R_4(V)/\mathrm{Aut}(\mathbb{P}^1)$ is the projection). We show that v is holomorphic. Let f depend holomorphically on $\tilde{\omega}(f)$ by choosing a holomorphic local cross section of $\tilde{\omega}$. It is clear that $\{K_V \otimes [D_\infty(f)]^{-1}\}_{\tilde{\omega}(f) \in U}$ forms a family of line bundles (see Chapter 4), where $\tilde{\omega}(f)$ moves in a small open subset U of $R_4(V)/\mathrm{Aut}(\mathbb{P}^1)$. Hence the disjoint union $\bigcup_{\tilde{\omega}(f) \in U} |K_V - D_\infty(f)|$ is a $\mathbb{P}^1$-bundle (see Chapter 4). Thus we may choose distinct (ω_0), $(\eta_0) \in |K_V - D_\infty(f)|$ so that they depend holomorphically on $\tilde{\omega}(f) \in U$. We write

$$(\omega_0) = D_\infty(f) + D \quad \text{and} \quad (\eta_0) = D_\infty(f) + D' .$$

Then effective divisors $D, D' \in S^4V$ also depend holomorphically on $\tilde{\omega}(f) \in U$. We put

$$(\omega_1) = D_0(f) + D \quad \text{and} \quad (\eta_1) = D_0(f) + D' .$$

Then $Q_\lambda = v(\tilde{\omega}(f))$ is given by

$$Q_\lambda : \omega_0\eta_1 - \omega_1\eta_0 = 0 .$$

Hence Q_λ depends holomorphically on $\tilde{\omega}(f) \in U$.

Now, v is a (ramified) double covering. In fact, if $Q_\lambda \in \mathbb{Q}_K$ has the rank 4, then two different 1-parameter families of planes on Q_λ define different meromorphic functions f (mod $\mathrm{Aut}(\mathbb{P}^1)$) and h (mod $\mathrm{Aut}(\mathbb{P}^1)$) such that

$$\mathrm{ord}\, f + \mathrm{ord}\, h \leqq 2g-2 = 8$$

(see Proposition 2.5.9). But we assumed that both $R_2(V)$ and $R_3(V)$ are empty. Hence

$$\mathrm{ord}\, f = \mathrm{ord}\, h = 4 .$$

In a similar way, if $Q_\lambda \in \mathbb{Q}_K$ has the rank 3, then f (mod $\mathrm{Aut}(\mathbb{P}^1)$) determined by Q_λ also has the order 4. This means that v is a (ramified) double covering. In particular, it is surjective. Hence $\mathbb{Q}_K$ must be a plane quintic curve.

v branches at $Q_\lambda \in \mathbb{Q}_K$ if and only if Q_λ has the rank 3. If Q_λ has the rank 3, then, by Proposition 2.5.4, f (mod $\mathrm{Aut}(\mathbb{P}^1)$) determined by Q_λ satisfies $h^1(2D_\infty(f)) \geqq 1$. But $\deg 2D_\infty(f) = 8 = 2g-2$. Hence

$$h^1(2D_\infty(f)) \geqq 1$$

holds if and only if

$$[2D_\infty(f)] = K_V \quad (\text{and} \quad h^1(2D_\infty(f)) = 1) .$$

This proves (1) of the theorem.

(2) Assume that V is trigonal.

By Corollary 2.4.5, $R_3(V)/\mathrm{Aut}(\mathbb{P}^1) \cong W_3^1$ is one point $\{y_o\}$. Now,

$$M^{(1)} = Q_{(1:0:0)} \cap Q_{(0:1:0)} \cap Q_{(0:0:1)}$$

is a rational ruled surface having a unique exceptional curve E of the first kind (see §2.5).

Let

$$f_o : M^{(1)} \longrightarrow \mathbb{P}^1$$

be the ruling and put

$$l_\lambda = f_o^{-1}(\lambda) \quad \text{for} \quad \lambda \in \mathbb{P}^1.$$

Then,

$$l_\infty E = 1, \; l_\infty^2 = 0 \quad \text{and} \quad E^2 = -1 .$$

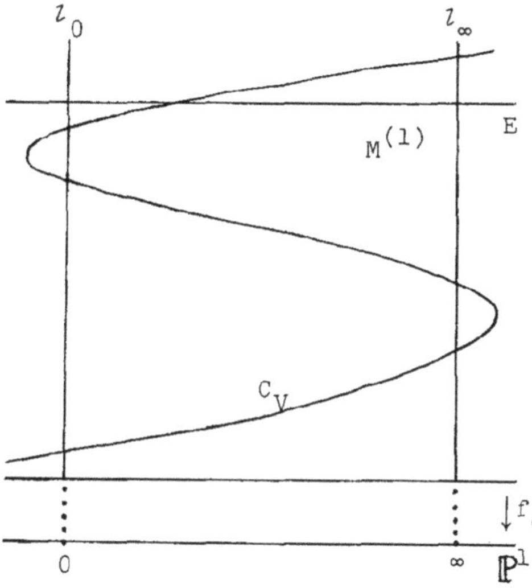

Moreover, the imbedding $M^{(1)} \subset \mathbb{P}^4$ is obtained by $\Phi_{|E+2l_\infty|}$. C_V is contained in $M^{(1)}$ and cuts every fiber l_λ of f_o by just 3 points. Hence

$$f_o = f_o \Big|_{C_V} : C_V \cong V \longrightarrow \mathbb{P}^1$$

is an element of $R_3(V)$. Since $h^0(D_\infty(f_o)) = 2$, we have $h^1(D_\infty(f_o)) = 3$. Hence $|K_V - D_\infty(f_o)|$ is biholomorphic to $\mathbb{P}^2$, so that the set of all lines $G^1(|K_V - D_\infty(f_o)|)$ in $|K_V - D_\infty(f_o)|$ is also biholomorphic to $\mathbb{P}^2$. By Proposition 2.5.5, dim $\mathbb{Q}_{f_o} = 2$. But we know that $\mathbb{Q}_K$ itself is a closed complex subspace of $\mathbb{P}^2$. Hence

$$\mathbb{Q}_{f_o} = \mathbb{Q}_K = \mathbb{P}^2 .$$

Next, we have $h^1(2D_\infty(f_o)) = 1$. In fact, l_∞ and l_0 do not intersect in $\mathbb{P}^4$. Hence they span a hyperplane in $\mathbb{P}^4$. This means that

$$h^1(2D_\infty(f_o)) = h^0(K_V - D_0(f_o) - D_\infty(f_o)) = 1 .$$

Hence, by Proposition 2.5.4, there is a unique $Q_o \in \mathbb{Q}_{f_o} = \mathbb{Q}_K$ with the rank 3. By Proposition 2.5.9, the vertex $V(Q_o)$, a line, intersects with C_V in 2 points.

Now, we consider the meromorphic map

$$\Phi = \Phi_{|K_V-[D_\infty(f_o)]|} : V \longrightarrow \mathbb{P}^2 .$$

We denote by C its image. Note that Φ is a holomorphic map. In fact, if $P \in V$ is a fixed point of $|K_V-D_\infty(f_o)|$, then

$$h^0(K_V - [D_\infty(f_o)+P]) = h^0(K_V - [D_\infty(f_o)]) = 3 .$$

This means that $D_\infty(f_o)+P$ lies on a line in $\mathbb{P}^4$. But this line must be l_∞ and $C_V \cap l_\infty = D_\infty(f_o)$, a contradiction. Hence $|K_V-[D_\infty(f_o)]|$ has no fixed point, so that Φ is a holomorphic map. Let m be the mapping order of Φ and let d be the degree of the curve C. Then

$$md = (2g-2) - \deg D_\infty(f_o) = 5 .$$

C can not be a straight line. Hence $d = 5$ and $m = 1$. Thus Φ is a birational holomorphic map of V onto a plane quintic curve C.

If we choose a basis $\{\phi_1, \cdots, \phi_5\}$ of $H^0(V, \mathcal{O}(K_V))$ so that $\{\phi_1, \phi_2, \phi_3\}$ forms a basis of

$$H^0(V, \mathcal{O}(K_V-D_\infty(f_o))) = \{\omega \in H^0(V, \mathcal{O}(K_V)) \mid (\omega) \geqq D_\infty(f_o) \},$$

then the map Φ is factored as follows:

$$\begin{array}{ccc} V & \xrightarrow{\ \Phi_K\ } & C_V \subset \mathbb{P}^4 \\ & \searrow^{\Phi} & \downarrow p \\ & & C \subset \mathbb{P}^2 , \end{array}$$

where p is the projection with the center l_∞.

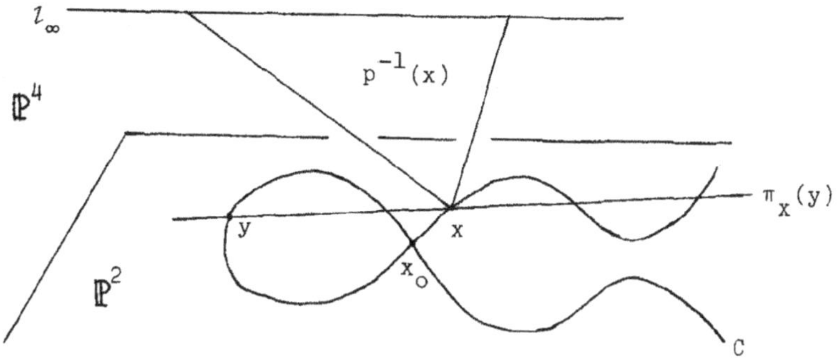

To each point $x \in \mathbb{P}^2$, we associate the plane $p^{-1}(x)$ of $\mathbb{P}^4$ containing l_∞. Let $\pi_x : C \longrightarrow \mathbb{P}^1$ be the projection with the center x. Then the projection $\pi_{p^{-1}(x)} : C_V \longrightarrow \mathbb{P}^1$ with the center $p^{-1}(x)$ is the composition of Φ and π_x.

If $x \notin C$, then $\pi_{p^{-1}(x)}$ has the order 5. If $x \in C$, then $\pi_{p^{-1}(x)}$ has the order $\leqq 4$ and has the order < 4 if and only if x is a singular point of C. In this last case, x must be a double point, for the order of $\pi_{p^{-1}(x)}$ must be 3.

Let Q_o be as above the unique quadric of rank 3 in $\mathbb{Q}_{f_o} = \mathbb{Q}_K$. Q_o has a 1-parameter family $\{S_\lambda\}_{\lambda \in \mathbb{P}^1}$ of planes such that

$$C_V \cap S_\lambda = D_\lambda(f_o) \quad \text{for all} \quad \lambda \in \mathbb{P}^1.$$

Then, $S_\infty \supset l_\infty$. S_∞ and $\mathbb{P}^2$ meet in just one point x_o, for, otherwise they are contained in a hyperplane in $\mathbb{P}^4$. Then l_∞ and $\mathbb{P}^2$ must join, a contradiction. Since $S_\infty = p^{-1}(x_o)$, x_o is a double point of C.

There is no other double point on C. In fact, if x is another double point of C, then $\pi_{p^{-1}(x)}$ has the order 3. Hence

$$\pi_{p^{-1}(x)} = f_o \pmod{\mathrm{Aut}(\mathbb{P}^1)}.$$

Thus $p^{-1}(x)$ is contained in Q_o (see §2.5). Hence $p^{-1}(x) = S_\lambda$ for some $\lambda \in \mathbb{P}^1$. This plane must contain both l_∞ and l_λ. This is impossible, unless $\lambda = \infty$, i.e., $x = x_o$.

Finally, we show that every $f \in R_4(V)$ is obtained by the projection π_x for some $x \in C - \{x_o\}$. In fact, by Proposition 2.5.8, $\mathbb{Q}_{f,f_o}$ is non-empty, i.e., there is $Q \in \mathbb{Q}_K$ such that Q contains planes S and T of $\mathbb{P}^4$ such that

$$\pi_S = f \ (\text{mod Aut}(\mathbb{P}^1)) \quad \text{and} \quad \pi_T = f_o \ (\text{mod Aut}(\mathbb{P}^1)) .$$

We may choose S so that S contains l_∞ (see the proof of Proposition 2.5.4). Note that S and $\mathbb{P}^2$ meet in just one point x. Then $S = p^{-1}(x)$. Hence $x \in C - \{x_o\}$. Hence

$$\pi_x \circ \Phi = \pi_S = f \ (\text{mod Aut}(\mathbb{P}^1)) .$$

By Proposition 2.3.3,

$$x \in C - \{x_o\} \longmapsto \pi_x \circ \Phi \in R_4(V)/\text{Aut}(\mathbb{P}^1)$$

is a holomorphic imbedding. We have seen that it is surjective. Hence it is biholomorphic.

Q.E.D.

Similar methods can be applied to V of genus 6. We summarize our results without proof:

Theorem 2.6.2. Let V be a non-hyperelliptic compact Riemann surface of genus 6. Then

(1) If V is not trigonal and W_5^2 is empty, then either (1-i) V is birational to a plane sextic curve having at most 4 double points and $R_4(V)/\text{Aut}(\mathbb{P}^1) \cong W_4^1$ is a finite set, whose number m of points satisfies $1 \leqq m \leqq 5$, or (1-ii) there is a (ramified) double covering $v : V \longrightarrow C$, where C is a non-singular plane cubic and is biholomorphic to $R_4(V)/\text{Aut}(\mathbb{P}^1) \cong W_4^1$.

(2) If W_5^2 is non-empty, then V is not trigonal but is biholomorphic to a non-singular plane quintic curve C. Moreover, $R_4(V)/\text{Aut}(\mathbb{P}^1) \cong W_4^1$ is biholomorphic to V and $R_5(V)/\text{Aut}(\mathbb{P}^1)$ is biholomorphic to

$\mathbb{P}^2 - C$. W_5^2 consists of just one point. Every element of $R_5(V)$ is obstructed, by Corollary 2.1.7.

(3) If V is trigonal, then $R_3(V)/\mathrm{Aut}(\mathbb{P}^1) \cong W_3^1$ is just one point. Moreover, $R_4(V)/\mathrm{Aut}(\mathbb{P}^1)$ is either empty or one point.

Remark 2.6.3.

(1) In (1) of the theorem, for a generic V, W_4^1 consists of 5 points (see Griffiths-Harris [25, p.299]).

(2) In (3) of the theorem, W_5^1 has just 2 components. However, Andreotti-Mayer [3] showed that this holds for any trigonal compact Riemann surface of genus $\geqq 5$, i.e., if V is trigonal and of genus $g \geqq 5$, then W_{g-1}^1 has just 2 components. They are, in fact, $x+W_{g-4}$ and $\kappa - (x+W_{g-4})$, where $x = W_3^1$ (one point).

As an application of Theorem 2.6.1,

Theorem 2.6.4. (Murasawa-Namba). Let V be a non-hyperelliptic compact Riemann surface of genus 5. Then the elements of $\mathbb{Q}_K$ (hence $\mathbb{Q}_K'$) span $\Sigma = \mathbb{P}^2$ (see the notations in §2.5 $(m = 2)$).

Proof. By Theorem 2.5.16, we may assume that V is not trigonal. Then, $\mathbb{Q}_K = \mathbb{Q}_K'$ and is identified with a (possibly reducible) plane quintic curve:

$$\mathbb{Q}_K = \{\lambda \in \mathbb{P}^2 \mid \det A_\lambda = 0\}$$

(see the proof of Theorem 2.6.1). By (1) of Theorem 2.6.1, there is a (ramified) double covering map $v : W_4^1 \longrightarrow \mathbb{Q}_K$. A quadric $Q \in \mathbb{Q}_K$ is a branch point of v if and only if it has the rank 3. The elements of $\mathbb{Q}_K$ do not span $\mathbb{P}^2$ if and only if $\mathbb{Q}_K$ is a (5-ple) straight line. We show that this does not happen.

Assume that $\mathbb{Q}_K$ is a (5-ple) straight line. Then, there are λ

and $\mu \in \mathbb{P}^2$ such that

$$\mathbb{Q}_K = \{s\lambda + t\mu \mid (s:t) \in \mathbb{P}^1\} .$$

Now, the quadric $Q_{(s:t)}$, corresponding to a point $s\lambda + t\mu$, has the rank 3 if and only if all (4×4)-minors $B(s,t)$ of the matrix $A_{s\lambda+t\mu} = sA_\lambda + tA_\mu$ are zero. $B(s,t) = 0$ has at most 4 solutions $(s:t) \in \mathbb{P}^1$, unless $B(s,t) \equiv 0$. If all (4×4)-minors of $A_{s\lambda+t\mu}$ are identically zero as functions of (s,t), then every elements of $\mathbb{Q}_K$ has the rank 3. This is impossible. (If it is so, then every point x of W_4^1 must satisfy $2x = \kappa$, by (1) of Theorem 2.6.1. But there are only a finite number of such points, a contradiction.) Thus

(1) $\{(s:t) \in \mathbb{P}^1 \mid Q_{(s:t)}$ has the rank $3\}$ consists of at most 4 points.

Next, we show that W_4^1 is irreducible. In fact, since $v : W_4^1 \longrightarrow \mathbb{Q}_K$ is a double covering and $\mathbb{Q}_K$ is a straight line, every irreducible component of W_4^1 must be mapped by v onto $\mathbb{Q}_K$. Hence the number of irreducible components of W_4^1 is at most 2. If W_4^1 has 2 components U_1 and U_2, then $v : U_1 \longrightarrow \mathbb{Q}_K$ is a birational holomorphic map. Hence U_1 is a rational curve in $J(V)$, a contradiction. (A complex torus can not contain a rational curve.) Thus W_4^1 is irreducible. Let

$$w : W \longrightarrow W_4^1$$

be a non-singular model of the curve W_4^1. Note that the genus g' of W is positive: $g' \geqq 1$.

The map $v \circ w$ is regarded as a meromorphic function of order 2 on W. By Riemann-Hurwitz formula (see §2.1),

(2) $$\sum_{P \in W}(e_p - 1) = 2g' + 2 \geqq 4 .$$

By (1), there are at most 4 branch points of v. If $\lambda \in \mathbb{Q}_K$ is not a branch point of v, then $v^{-1}(\lambda)$ consists of 2 distinct points x_1 and x_2 with $x_1+x_2 = \kappa$, $2x_1 \neq \kappa$ and $2x_2 \neq \kappa$. By Martens [52, II, p.97], x_1 and x_2 are non-singular points of W_4^1. Hence λ is not a branch point of $v \cdot w$. Thus there are at most 4 branch points of $v \cdot w$. By (2),

(3) There are just 4 branch points of $v \cdot w$ on $\mathbb{Q}_K$ and

$$g' = 1 .$$

Thus W is a complex torus. But a holomorphic map between complex tori must be a homomorphism (as Lie groups) up to translations. Hence

$$w : W \longrightarrow W_4^1 \subset J(V)$$

must be a holomorphic imbedding of W into $J(V)$, so that W_4^1 itself is a non-singular elliptic curve in $J(V)$.

Now, by (3), there are just 4 points x_i, $(i = 1,2,3,4)$, on W_4^1 such that $2x_i = \kappa$. By Martens [52, II, p.97], again, such points x_i must be singular points of W_4^1, a contradiction. Q.E.D.

Remark 2.6.5. This theorem gives a proof of Torelli's theorem for a non-hyperelliptic, non-trigonal compact Riemann surface of genus 5 (see Mumford [59, p.89]).

Chapter 3. Families of holomorphic maps of compact complex manifolds.

3.0. Banach spaces $C^p(F, |\ |)$.

The main purpose of this chapter is to prove a special form of relative Douady's theorem (see Pourcin [70]) in the case of holomorphic maps of compact complex manifolds. The method of the proof is taken from Namba [61]. §3.0-§3.2 may be the most tedious part of the present lecture notes. But a patient reader will find that the method of the proof is quite simple and natural.

In this section, we recall some results in Namba [61, §2].

Let V be a compact complex manifold of dimension d. Let F be a holomorphic vector bundle on V of rank m. We cover V by a finite number of Stein open subsets $\{U_i\}_{i\in I}$ such that

(1) there is a coordinate system $(z_i) = (z_i^1, \cdots, z_i^d)$ in $\tilde{U}_i$, an open subset of V containing U_i, such that

$$U_i = \{z_i \in \tilde{U}_i \mid |z_i| \ (= \max_\alpha |z_i^\alpha|) < 1\},$$

(2) F is trivial on U_i.

We take a small positive number e, $0 < e < 1$, such that $\{U_i^e\}$ again covers V, where

$$U_i^e = \{z_i \in U_i \mid |z_i| < 1 - e\}.$$

We define additive groups $C^p(F)$, $p = 0, 1, \cdots$, which are a little different from the usual cochain groups $C^p(\mathcal{U}, \mathcal{O}(F))$, $(\mathcal{U} = \{U_i\}_{i\in I})$. An element $\xi = \{\xi_{i_0\cdots i_p}\} \in C^p(F)$ is a function which associate to each (p+1)-ple $(i_0, \cdots, i_p)$ of indices in I, a holomorphic section $\xi_{i_0\cdots i_p}$ of F on $U_{i_0}^e \cap \cdots \cap U_{i_{p-1}}^e \cap U_{i_p}$. (If this open subset of V is empty, then we put $\xi_{i_0\cdots i_p} = 0$.) In particular, an element $\xi = \{\xi_i\} \in C^0(F)$ is a function which associates to each index $i \in I$ a holomorphic section ξ_i of F on U_i.

We define a coboundary map $\delta : C^p(F) \longrightarrow C^{p+1}(F)$ by

$$(\delta\xi)_{i_0 \cdots i_{p+1}}(z) = \sum_{\nu} (-1)^{\nu} \xi_{i_0 \cdots i_{\nu-1} i_{\nu+1} \cdots i_{p+1}}(z)$$

for $z \in U^e_{i_0} \cap \cdots \cap U^e_{i_p} \cap U_{i_{p+1}}$. It is easy to see that $\delta^2 = 0$.

We introduce a norm $| \ |$ in $C^p(F)$. For $\xi = \{\xi_{i_0 \cdots i_p}\} \in C^p(F)$, we define $|\xi|$ by

$$|\xi| = \sup \{ |\xi^{\lambda}_{i_0 \cdots i_p}(z)| \;\Big|\; \lambda = 1, \cdots, m,$$

$$z \in U^e_{i_0} \cap \cdots \cap U^e_{i_{p-1}} \cap U_{i_p}, \quad (i_0, \cdots, i_p) \in I^{p+1}\},$$

where $\xi^{\lambda}_{i_0 \cdots i_p}$ is the representation of the component $\xi_{i_0 \cdots i_p}$ of ξ with respect to the local trivialization of F on U_{i_0}. In particular, we define $|\xi|$ for $\xi \in C^0(F)$ by

$$|\xi| = \sup \{ |\xi^{\lambda}_i(z)| \;\Big|\; \lambda = 1, \cdots, m, \quad i \in I, \quad z \in U_i \} .$$

We put

$$C^p(F, | \ |) = \{ \xi \in C^p(F) \;\Big|\; |\xi| < +\infty \} .$$

It is easy to see that $C^p(F, | \ |)$ is a Banach space and the coboundary map δ maps $C^p(F, | \ |)$ continuously into $C^{p+1}(F, | \ |)$. Put

$$Z^p(F, | \ |) = \{ \xi \in C^p(F, | \ |) \;\Big|\; \delta\xi = 0 \} ,$$

$$B^p(F, | \ |) = (\delta C^{p-1}(F)) \cap C^p(F, | \ |) ,$$

$$H^p(F, | \ |) = Z^p(F, | \ |) / B^p(F, | \ |) ,$$

for $p = 0, 1, \cdots$. It is clear that $H^0(F, | \ |)$ is canonically identified with the 0-th cohomology group $H^0(V, \mathcal{O}(F))$.

Proposition 3.0.1. (Kuranishi [48, Lemma 2.5'], Namba [61]). There are continuous linear maps

$$E_0 : B^1(F,|\ |) \longrightarrow C^0(F,|\ |),$$

$$E_1 : B^2(F,|\ |) \longrightarrow C^1(F,|\ |)$$

such that δE_0 and δE_1 are the identity maps of $B^1(F,|\ |)$ and $B^2(F,|\ |)$, respectively.

Proof. We construct only E_0. E_1 is constructed in a similar way. We define an additive group $C^0_\infty(F)$ as follows: An element $\xi \in C^0_\infty(F)$ is a function which associates to each $i \in I$ a C^∞-section ξ_i on U_i of F. Then we can define a norm $|\ |$ on $C^0_\infty(F)$ in a similar way to the above.

Let $\{q_i\}_{i\in I}$ be a partition of unity subordinate to the covering $\{U^e_i\}$. For $\xi \in B^1(F,|\ |)$, define $\eta = \{\eta_j\} \in C^0_\infty(F)$ by

$$\eta_j = \sum_{i\in I} q_i \xi_{ij} .$$

Then, $\eta_k - \eta_j = \xi_{jk}$ and

(1) $$|\eta| \leqq c_1|\xi|, \quad (c_1 > 0, \text{ a constant}) .$$

Note that $\bar\partial\eta_j = \bar\partial\eta_k = \omega$ is a global F-valued $(0,1)$-form. Let $|\omega|_{k+\alpha}$ $(k \geqq 1,\ 0 < \alpha < 1)$ be the <u>Hölder norm</u> (see Morrow-Kodaira [55, p.159]). By estimating it on U^e_j and using (1), we get

(2) $$|\omega|_{k+\alpha} \leqq c_2|\xi|, \quad (c_2 > 0, \text{ a constant}) .$$

We introduce Hermitian metrices on V and F and denote by ϑ and G the adjoint operator and the Green operator, respectively. Since ω corresponds to $\xi \in B^1(F,|\ |)$ by Dolbeault's isomorphism, we get $\omega = \bar\partial\vartheta G\omega$. Put $\pi = \vartheta G\omega$. Then π is a global C^∞-section of F and there is a constant $c_3 > 0$ such that

(3) $$|\pi|_{k+1+\alpha} \leqq c_3|\omega|_{k+\alpha} .$$

Hence, by (2),

(4) $$|\pi| \leqq c_4|\xi|, \quad (c_4 > 0, \text{ a constant}).$$

Put

$$\pi_i = \pi \mid U_i, \quad \beta_i = \eta_i - \pi_i \quad \text{and} \quad \beta = \{\beta_i\}.$$

Then

$$\bar{\partial}\beta_i = \bar{\partial}\eta_i - \bar{\partial}\pi_i = \omega - \omega = 0$$

and

(5) $$|\beta| \leqq c_5|\xi|, \quad (c_5 > 0, \text{ a constant}).$$

Thus $\beta \in C^0(F,|\ |)$. Note that

$$(\delta\beta)_{ij} = \beta_j - \beta_i = \eta_j - \eta_i + \pi_j - \pi_i = \eta_j - \eta_i = \xi_{ij}.$$

Hence $\delta\beta = \xi$.

Now, the map E_0 is defined by: $\xi \longmapsto \beta$. <u>Q.E.D.</u>

For the proof of the following proposition, see Namba [61, p.590].

<u>Proposition 3.0.2.</u>

(1) $B^1(F,|\ |)$ is closed in $Z^1(F,|\ |)$.

(2) $B^1(F,|\ |) = \delta C^0(F,|\ |)$.

(3) $H^1(F,|\ |)$ is canonically isomorphic to $H^1(V, \mathcal{O}(F))$.

(4) There is a splitting

$$Z^1(F,|\ |) = B^1(F,|\ |) \oplus H^1(F,|\ |),$$

where $H^1(F,|\ |)$ (abuse of notation!) is a finite dimensional linear subspace of $Z^1(F,|\ |)$ isomorphic to $H^1(V, \mathcal{O}(F))$.

Let

$$B : Z^1(F,|\ |) \longrightarrow B^1(F,|\ |)$$

$$H : Z^1(F,|\ |) \longrightarrow H^1(F,|\ |)$$

be the projections corresponding to the splitting.

We also define a map

$$\Lambda : C^1(F,|\ |) \longrightarrow Z^1(F,|\ |)$$

by $\Lambda = 1 - E_1\delta$. It is a projection map.

These continuous linear maps B, H, and Λ are used in §3.2.

3.1. Some lemmas.

Let $(X,\pi,S) = \{V_s\}_{s\in S}$ be a family of compact complex manifolds. Fix a point $o \in S$. Put $V = V_o$. Let S_1 and $\tilde{S}$ be small open neighborhoods of o in S such that $S_1 \Subset \tilde{S}$. ($A \Subset B$ means that the closure $\bar{A}$ is compact and is contained in B.)

As in §0.1, let $\tilde{\Omega}$ be an ambient space of $\tilde{S}$ such that $\dim \tilde{\Omega} = \dim T_oS$, where T_oS is the Zariski tangent space to S at o. Let $(s^1,\cdots,s^q)$ be a coordinate system in $\tilde{\Omega}$ such that $s_\lambda(o) = 0$, $\lambda = 1,\cdots,q$. For $s = (s^1,\cdots,s^q) \in \tilde{\Omega}$, put $|s| = \max_\lambda |s^\lambda|$. For a positive number ε, we put

$$\Omega_\varepsilon = \{s = (s^1,\cdots,s^q) \in \tilde{\Omega} \mid |s| < \varepsilon\} .$$

Then, we may assume that $S_1 = \tilde{S} \cap \Omega_1$. In general, we put

$$S_\varepsilon = \tilde{S} \cap \Omega_\varepsilon .$$

Now, we can find finite collections of open subsets $\{X_i\}_{i\in I}$ and $\{\tilde{X}_i\}_{i\in I}$ of X such that

(1) $X_i \Subset \tilde{X}_i$,

(2) $\{X_i\}$ covers V,

(3) there are open subsets U_i and $\tilde{U}_i$ in $\mathbb{C}^d$ with $U_i \Subset \tilde{U}_i$ and a

holomorphic isomorphism $\eta_i : \tilde{X}_i \longrightarrow \tilde{U}_i \times \tilde{S}$ which maps X_i onto $U_i \times S_1$ and makes the diagram

$$\begin{array}{ccc} \tilde{X}_i & \xrightarrow{\eta_i} & \tilde{U}_i \times \tilde{S} \\ & {\scriptstyle\pi}\searrow \quad \swarrow{\scriptstyle \text{proj}} & \\ & \tilde{S} & \end{array}$$

commutative and

(4) there is a coordinate system $z_i = (z_i^1, \cdots, z_i^d)$ in $\tilde{U}_i$ such that

$$U_i = \{z_i \in \tilde{U}_i \mid |z_i| \quad (= \max_\alpha |z_i^\alpha|) < 1 \}.$$

We often identify U_i and $\tilde{U}_i$ with $X_i \cap V$ and $\tilde{X}_i \cap V$, respectively.

When $\tilde{X}_i \cap \tilde{X}_k$ is non-empty, put

$$\eta_{ik} = \eta_i \eta_k^{-1} : \eta_k(\tilde{X}_i \cap \tilde{X}_k) \longrightarrow \eta_i(\tilde{X}_i \cap \tilde{X}_k) .$$

Then, it is written as

$$\eta_{ik}(z_k, s) = (g_{ik}(z_k, s), s).$$

We are going to extend η_{ik} to an ambient space of $\eta_k(X_i \cap X_k)$.

For a point $P \in \overline{U}_i \cap \overline{U}_k$, there are a positive number $\varepsilon(P)$ (depending on P) and an open neighborhood $U_P \times S_{\varepsilon(P)}$ of $\eta_k(P)$ in $\eta_k(\tilde{X}_i \cap \tilde{X}_k)$ such that $P \in U_P \subset \tilde{U}_i \cap \tilde{U}_k$. We cover $\eta_k(\overline{U}_i \cap \overline{U}_k)$ by such open subsets $\{U_P \times S_{\varepsilon(P)}\}$ of $\eta_k(\tilde{X}_i \cap \tilde{X}_k)$ and choose a finite subcovering $\{U_{P_\lambda} \times S_{\varepsilon(P_\lambda)}\}$. Put $U_{P_\lambda} = U_\lambda$ and $\varepsilon_o = \min_\lambda \varepsilon(P_\lambda)$.

For the proof of the following two lemmas, see Namba [62].

<u>Lemma 3.1.1</u>. There is a Stein open subset $U_o = U_{o(ik)}$ of $\tilde{U}_k$ such that

$$\overline{U}_i \cap \overline{U}_k \subset U_o \subset \bigcup_\lambda U_\lambda \subset \tilde{U}_i \cap \tilde{U}_k .$$

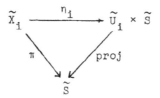

Lemma 3.1.2. Let U_o be as in Lemma 3.1.1. If ε, $0 < \varepsilon < \varepsilon_o$, is sufficiently small, then

$$\eta_k(X_i \cap X_k) \cap (\widetilde{U}_k \times S_\varepsilon) \subset U_o \times S_\varepsilon \quad (\subset \eta_k(\widetilde{X}_i \cap \widetilde{X}_k)) .$$

Now, $U_o \times S_{\varepsilon_o}$ is a closed complex subspace of $U_o \times \Omega_{\varepsilon_o}$, which is Stein. Hence the holomorphic map $\eta_{ik} : U_o \times S_{\varepsilon_o} \longrightarrow \widetilde{U}_i \times S_{\varepsilon_o}$ is extended to a holomorphic map

$$\eta_{ik} : U_o \times \Omega_{\varepsilon_o} \longrightarrow \widetilde{U}_i \times \Omega_{\varepsilon_o} .$$

The extended map η_{ik} is written as follows:

$$\eta_{ik}(z_k, s) = (g_{ik}(z_k, s), s) ,$$

where $g_{ik} : U_o \times \Omega_{\varepsilon_o} \longrightarrow \widetilde{U}_i$ is a holomorphic extension of the above map g_{ik}.

For a positive number e, $0 < e < 1$, put as in §3.0

$$U_i^e = \{z_i \in U_i \mid |z_i| < 1-e\} .$$

Take e and e' $(0 < e < e')$ sufficiently small so that $\{U_i^{e'}\}_{i \in I}$ again covers V.

The following Lemmas 3.1.3, 3.1.4 and 3.1.5, are easily proved within the general topological argument. (See Namba [62].)

Lemma 3.1.3. There is a small positive number ε with $0 < \varepsilon < \varepsilon_o$ such that, if $s \in \Omega_\varepsilon$, then $g_{ik}(z_k, s)$ is defined and is a point of U_i for all $z_k \in U_i^e \cap U_k$.

Lemma 3.1.4. There is a small positive number ε with $0 < \varepsilon < \varepsilon_o$ such that, if $s \in S_\varepsilon$, then $\eta_k^{-1}(z_k, s) \in X_i \cap X_k$ for all $z_k \in U_i^e \cap U_k$.

Lemma 3.1.5. There is a small positive number ε with $0 < \varepsilon < \varepsilon_o$ such that, if $s \in S_\varepsilon$ and if $\eta_k^{-1}(z_k, s) \in X_i^{e'} \cap X_k$, then $z_k \in U_i^e \cap U_k$. $(X_i^{e'} = \eta_i^{-1}(U_i^{e'} \times S_1).)$

These lemmas will be used in the next section.

Next, let $(Y, \mu, S) = \{W_s\}_{s \in S}$ be another family of compact complex manifolds with the same parameter space S. Put $W = W_o$. Let f be a holomorphic map of V into W. We may assume that there are collections of open subsets $\{Y_i\}_{i \in I}$ and $\{\tilde{Y}_i\}_{i \in I}$ (the same set I of indices) of Y such that

(1)' $Y_i \Subset \tilde{Y}_i$,

(2)' $\{Y_i\}_{i \in I}$ covers $f(V)$,

(3)' there are open subsets W_i and $\tilde{W}_i$ in $\mathbb{C}^m$ with $W_i \Subset \tilde{W}_i$ and a holomorphic isomorphism $\zeta_i : \tilde{Y}_i \longrightarrow \tilde{W}_i \times \tilde{S}$ which maps Y_i onto $W_i \times S_1$ and makes the diagram

$$\begin{array}{ccc} \tilde{Y}_i & \xrightarrow{\ \zeta_i\ } & \tilde{W}_i \times \tilde{S} \\ & {\scriptstyle \mu} \searrow \quad \swarrow {\scriptstyle \text{proj}} & \\ & \tilde{S} & \end{array}$$

commutative,

(4)' there is a coordinate system $w_i = (w_i^1, \cdots, w_i^m)$ in $\tilde{W}_i$ such that $W_i = \{w_i \in \tilde{W}_i \mid |w_i| < 1\}$ and

(5) $f(U_i) \Subset W_i$ and $\tilde{f}(U_i) \Subset \tilde{W}_i$, where W_i and $\tilde{W}_i$ are identified with $Y_i \cap W$ and $\tilde{Y}_i \cap W$, respectively.

When $\tilde{Y}_i \cap \tilde{Y}_k$ is non-empty, put

$$\zeta_{ik} = \zeta_i \zeta_k^{-1} : \zeta_k(\tilde{Y}_i \cap \tilde{Y}_k) \longrightarrow \zeta_i(\tilde{Y}_i \cap \tilde{Y}_k) .$$

Then it is written as

$$\zeta_{ik}(w_k, s) = (h_{ik}(z_k, s), s) .$$

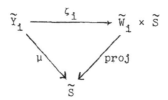

Lemma 3.1.1 and 3.1.2 are applied to this case so that ζ_{ik} (hence h_{ik}) is extended holomorphically to an ambient space $W_o \times \Omega_{\varepsilon_o}$.

Let e, $0 < e < 1$, be sufficiently small such that

(6) $\{W_i^e\}_{i \in I}$ covers $f(V)$ and

(7) $f(\overline{U}_i) \subset W_i^e$, ($\overline{U}_i$ is the closure of U_i in V.)

where

$$W_i^e = \{w_i \in W_i \mid |w_i| < 1-e\} .$$

Take ε sufficiently small so that Lemmas 3.1.3, 3.1.4 and 3.1.5 also hold for these data of (Y,μ,S).

Let A be a compact subset of W_k. We regard W_k as a polydisc $W_k = \{w_k \in \mathbb{C}^m \mid |w_k| < 1\}$ in $\mathbb{C}^m$. We consider the subset

$$A_\varepsilon = \{w_k + x_k \mid w_k \in A, \ |x_k| \leqq \varepsilon\}$$

of $\mathbb{C}^m$ for a small positive number ε. A_ε is compact, since the summation is a continuous operation. The following lemma is easy to prove.

Lemma 3.1.6. Let A be a compact subset of W_k. Then, there is a positive number ε such that $A_\varepsilon \subset W_k$.

Since $f(\overline{U}_i) = \overline{f(U_i)}$ is a compact subset of W_i^e, $f(\overline{U}_i) \cap f(\overline{U}_k)$ is a compact subset of $W_i^e \cap W_k$, which is open in W_k. By Lemma 3.1.6, there is ε such that $(f(\overline{U}_i) \cap f(\overline{U}_k))_\varepsilon \subset W_k$. The following lemma is easy to show.

Lemma 3.1.7. There is a positive number ε such that

$$(f(\overline{U}_i) \cap f(\overline{U}_k))_\varepsilon \subset W_i^e \cap W_k \subset W_k .$$

Now, f maps $\widetilde{U}_i$ into $\widetilde{W}_i$ by (5). Using the local coordinate systems, it is expressed by the equations:

$$w_i = f_i(z_i), \quad i \in I,$$

where f_i is a vector valued holomorphic functions on U_i.

Let TW be the holomorphic tangent bundle of W. Put $F = f^*TW$, the induced bundle on V of TW over f. Then F is given by the transition matrices $\{(\frac{\partial h_{ik}}{\partial w_k})(f_k(z_k), o)\}$. As in §0.1, we put

$$\Theta_V = \mathcal{O}(TV) \quad \text{and} \quad \Theta_W = \mathcal{O}(TW) .$$

Note that

$$\mathcal{O}(F) = \mathcal{O}(f^*TW) = f^*\Theta_W .$$

f induces natural homomorphisms between cochain groups:

$$f_* : C^p(\mathcal{U}, \Theta_V) \longrightarrow C^p(\mathcal{U}, f^*\Theta_W) ,$$

$$f^* : C^p(\mathcal{W}, \Theta_W) \longrightarrow C^p(\mathcal{U}, f^*\Theta_W) ,$$

($f_*\theta = (df)\theta$ and $f^*\eta = \eta \circ f$.) which commute with the coboundary maps δ. Here, $\mathcal{U} = \{U_i\}_{i\in I}$ and $\mathcal{W} = \{W_i\}_{i\in I} \cup \{W_\alpha\}_{\alpha\in A}$, are finite open Stein coverings of V and W, respectively. (We assume that the set of indices I and A are disjoint and $\overline{W}_\alpha \cap f(V) =$ empty for all $\alpha \in A$. Of course, A is empty if f is surjective.)
Hence, we get linear maps

$$f_* : H^p(V, \Theta_V) \longrightarrow H^p(V, \mathcal{O}(F)) ,$$

$$f^* : H^p(W, \Theta_W) \longrightarrow H^p(V, \mathcal{O}(F)) .$$

Let $\rho_o^1 : T_oS \longrightarrow H^1(V, \Theta_V)$ and $\rho_o^2 : T_oS \longrightarrow H^1(W, \Theta_W)$ be the Kodaira-Spencer map at o for the families $\{V_s\}_{s\in S}$ and $\{W_s\}_{s\in S}$, respectively (see §0.1). We define a linear map

$$\tau : T_oS \longrightarrow H^1(V, \mathcal{O}(F))$$

by $\tau(\frac{\partial}{\partial s}) = f^*\rho_o^2(\frac{\partial}{\partial s}) - f_*\rho_o^1(\frac{\partial}{\partial s})$. Then, $\tau(\frac{\partial}{\partial s})$ has the following element in $Z^1(\mathcal{U}, \Theta(F))$ as one of its representatives:

$$\{(\frac{\partial h_{ik}}{\partial s})(f_k(z_k),o) - (\frac{\partial f_i}{\partial z_i})_{z_i}(\frac{\partial g_{ik}}{\partial s})(z_k,o)\} .$$

Lemma 3.1.8. The 1-cocycle

$$\tilde{\tau}(\frac{\partial}{\partial s}) = \{(\frac{\partial h_{ik}}{\partial s})(f_k(z_k),o) - (\frac{\partial f_i}{\partial z_i})_{z_i}(\frac{\partial g_{ik}}{\partial s})(z_k,o)\}$$

is an element of $Z^1(F,|\ |)$ (see §3.0) and

$$\tilde{\tau} : T_oS \longrightarrow Z^1(f,|\ |)$$

is a continuous linear map which induces τ.

Proof. This follows from the fact: $\left|(\frac{\partial f_i}{\partial z_i})_{z_i}\right|$, $z_i \in U_i^e$, is estimated by $\sup\{|f_i(z_i)| \mid z_i \in U_i\}$, (< 1). Q.E.D.

3.2. Relative Douady space of holomorphic maps.

Let $(X,\pi,S) = \{V_s\}_{s\in S}$ be a family of compact complex manifolds. Let T be a complex space and b be a holomorphic map of T into S. Let (b^*X,π^*,T) be the induced family over b (see §0.1).

Definition 3.2.1. Let $(X,\pi,S) = \{V_s\}_{s\in S}$ and $(Y,\mu,S) = \{W_s\}_{s\in S}$ be families of compact complex manifolds with the same parameter space S. Let T be a complex space. A triple $(\mathfrak{F},T,b)$ is called a family of holomorphic maps of (X,π,S) into (Y,μ,S) if

(1) b is a holomorphic map of T into S and

(2) $\mathfrak{F}$ is a holomorphic map of b^*X into b^*Y such that the diagram

$$\begin{array}{ccc} b^*X & \xrightarrow{\mathcal{F}} & b^*Y \\ & {\scriptstyle \pi^*}\searrow \quad \swarrow{\scriptstyle \mu^*} & \\ & T & \end{array}$$

commutes. T is called the <u>parameter space of</u> $(\mathcal{F},T,b)$.

For every $t \in T$, the fibers $V_t = \pi^{*-1}(t)$ and $W_t = \mu^{*-1}(t)$ are identified with $V_{b(t)}$ and $W_{b(t)}$, respectively. Hence we may regard $(\mathcal{F},T,b)$ as a collection $\{f_t\}_{t\in T}$ of holomorphic maps, where

$$f_t : V_{b(t)} \longrightarrow W_{b(t)} .$$

We sometimes write $\{f_t\}_{t\in T}$ instead of $(\mathcal{F},T,b)$.

For the purpose of this section, we <u>do not</u> in fact need the compactness of each W_t. But we assume it for simplicity.

<u>Definition 3.2.2</u>. $(\mathcal{F},T,b) = \{f_t\}_{t\in T}$ is said to be <u>maximal at</u> $t_o \in T$ if, for any family $(\mathcal{G},R,c) = \{g_r\}_{r\in R}$ with a point $r_o \in R$ such that $b(t_o) = c(r_o)$ and $f_{t_o} = g_{r_o} : V_{b(t_o)} \longrightarrow W_{b(t_o)}$, there are an open neighborhood U of r_o in R and a holomorphic map $a : U \longrightarrow T$ such that

(1) $a(t_o) = r_o$,

(2) $ba = c$ and

(3) $g_r = f_{a(r)} : V_{c(r)} \longrightarrow W_{c(r)}$ for all $r \in U$.

A <u>maximal family</u> is a family which is maximal at every point of its parameter space.

<u>Theorem 3.2.3</u>. Let $(X,\pi,S) = \{V_s\}_{s\in S}$ and $(Y,\mu,S) = \{W_s\}_{s\in S}$ be families of compact complex manifolds with the same parameter space S. For a point $o \in S$, let f be a holomorphic map of V_o into W_o. Then, there exists a maximal family $(\mathcal{F},T,b) = \{f_t\}_{t\in T}$ of holomorphic maps

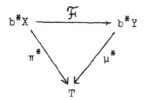

of (X,π,S) into (Y,μ,S) with a point $t_o \in T$ such that (1) $b(t_o) = o$ and (2) $f_{t_o} = f$.

Moreover, the parameter space T is given as follows: There are an open neighborhood U of 0 in $H^0(V, \Theta(F))$ $(F = f^*TW_o)$, an open neighborhood S' of o in S and a holomorphic map $\alpha : U \times S' \longrightarrow H^1(V, \Theta(F))$ such that (3) $T = \{(\xi,s) \in U \times S' \mid \alpha(\xi,s) = 0\}$, (4) $t_o = (0,o)$ and (5) $(d\alpha)_{(0,o)} = (0,\tau)$, where τ is the map defined in §3.1.

In the sequel, we give a proof of the theorem.

We use the same notations as in §3.0 and §3.1. Using the local coordinate systems, f is expressed by the equations:

$$w_i = f_i(z_i), \quad i \in I .$$

Then, the vector valued holomorphic functions f_i, $i \in I$, must satisfy the following compatibility conditions:

$$h_{ik}(f_k(z_k),o) = f_i(g_{ik}(z_k,o)) \quad \text{for all} \quad z_k \in U_i \cap U_k .$$

Put $F = f^*TW$. $(V = V_o$ and $W = W_o$.) Consider the product $C^0(F,|\ |) \times T_oS$. We introduce a norm $|\ |$ in $C^0(F,|\ |) \times T_oS$ as follows:

$$|(\phi,s)| = \max \{|\phi|,|s|\} \quad \text{for} \quad (\phi,s) \in C^0(F,|\ |) \times T_oS ,$$

where $|s| = \max_\lambda |a^\lambda|$, $s = \sum_\lambda a^\lambda(\frac{\partial}{\partial s^\lambda})$. Then $C^0(F,|\ |) \times T_oS$ is a Banach space. We identify $\tilde{\Omega}$ with an open subset of T_oS by

$$(a^1,\cdots,a^q) \in \tilde{\Omega} \longmapsto \sum_\lambda a^\lambda(\frac{\partial}{\partial s^\lambda}) \in T_oS .$$

Let f' be a holomorphic map of V_s into W_s, $s \in S_1$, such that $f'(V_s \cap X_i) \Subset W_s \cap Y_i$ for all $i \in I$. We express f' by the equations:

$$w_i = f_i'(z_i), \quad i \in I ,$$

using the holomorphic isomorphisms η_i and ζ_i. Then the vector valued holomorphic functions f_i' satisfy $f_i'(U_i) \Subset W_i$. We write

$$f'_i = f_i + \phi_i$$

and regard $\phi = \{\phi_i\}_{i \in I}$ as an element of $C^0(F,|\ |)$. We associate $(\phi,s) \in C^0(F,|\ |) \times T_oS$ to f'. ($s \in S_1 \subset \Omega_1 \subset T_oS$.) Then it is clear that (ϕ,s) must satisfy the following compatibility conditions:

$$(1) \qquad s \in S_1 ,$$

$$(2) \qquad h_{ik}(f_k(z_k) + \phi_k(z_k),s) = f_i(g_{ik}(z_k,s)) + \phi_i(g_{ik}(z_k,s)) ,$$

for $(z_k,s) \in \eta_k(X_i \cap X_k \cap V_s)$ and $(f_k(z_k)+\phi_k(z_k),s) \in \zeta_k(Y_i \cap Y_k \cap W_s)$.

Conversely, if ε satisfies Lemma 3.1.6 for $A = f_k(\overline{U}_k)$ and an element $(\phi,s) \in C^0(F,|\ |) \times T_oS$ satisfies $|(\phi,s)| < \varepsilon$ and the conditions (1) and (2) above, then the equations:

$$w_i = f'_i(z_i) = f_i(z_i) + \phi_i(z_i), \quad z_i \in U_i, \quad i \in I,$$

define a holomorphic map f' of V_s into W_s. By Lemma 3.1.6, f' satisfies

$$f'(V_s \cap X_i) \Subset W_s \cap Y_i, \quad i \in I.$$

Now, let ε be a sufficiently small positive number. Let

$$B_\varepsilon = \{\phi \in C^0(F,|\ |) \;\Big|\; |\phi| < \varepsilon\}$$

be the ε-ball. We define a map

$$K : B_\varepsilon \times \Omega_\varepsilon \longrightarrow C^1(F,|\ |)$$

by

$$K(\phi,s)_{ik}(z_i) = h_{ik}(f_k(z_k) + \phi_k(z_k),s) - f_i(g_{ik}(z_k,s)) - \phi_i(g_{ik}(z_k,s))$$

for $z_i \in U_i^e \cap U_k$, where $z_k = g_{ki}(z_i,o)$. (We write $K(\phi,s)_{ik}(z_i)$ instead of $K(\phi,s)_{ik}(z_k)$ in order to fit the definition of the norm in $C^1(F,|\ |)$.) We show that K is well defined.

If $z_k \in U_i^e \cap U_k$ and $s \in \Omega_\varepsilon$, then $g_{ik}(z_k,s)$ is defined and is a point of U_i by Lemma 3.1.3. Hence, $f_i(g_{ik}(z_k,s))$ and $\phi_i(g_{ik}(z_k,s))$ are defined. On the other hand, $f_k(z_k)+\phi_k(z_k) \in W_i^e \cap W_k$ for $z_k \in U_i^e \cap U_k$ by Lemma 3.1.7. Hence $h_{ik}(f_k(z_k)+\phi_k(z_k),s)$ is defined and is a point of W_i by Lemma 3.1.3 (with respect to the data of (Y,μ,S)). Moreover, if $|(\phi,s)| < \varepsilon$, then $|K(\phi,s)| < 2+\varepsilon$. Thus K maps $B_\varepsilon \times \Omega_\varepsilon$ into $C^1(F,|\ |)$. Note that $K(0,o) = 0$.

Let $v_1,\cdots,v_p$ be defining holomorphic functions of S_ε in Ω_ε, i.e.,

$$S_\varepsilon = \{s \in \Omega_\varepsilon \mid v_1(s) = \cdots = v_p(s) = 0\} .$$

Let $v : \Omega_\varepsilon \longrightarrow \mathbb{C}^p$ be the holomorphic map defined by

$$s \in \Omega_\varepsilon \longmapsto v(s) = (v_1(s),\cdots,v_p(s)) \in \mathbb{C}^p .$$

Put

$$\begin{aligned} M &= \{(\phi,s) \in B_\varepsilon \times S_\varepsilon \mid K(\phi,s) = 0\} \\ &= \{(\phi,s) \in B_\varepsilon \times \Omega_\varepsilon \mid K(\phi,s) = 0,\ \ v(s) = 0\} . \end{aligned}$$

Let $(\phi,s) \in B_\varepsilon \times \Omega_\varepsilon$ satisfy the compatibility conditions (1) and (2). Fix a point $z_i \in U_i^e \cap U_k$. Then, $(z_k,s) \in \eta_k(X_i \cap X_k)$ by Lemma 3.1.4. $(z_k = g_{ki}(z_i,o).)$ By Lemma 3.1.7 and 3.1.4, $(f_k(z_k)+\phi_k(z_k),s) \in \zeta_k(Y_i \cap Y_k)$. Hence, by (2), $K(\phi,s)_{ik}(z_i) = 0$. Since $z_i \in U_i^e \cap U_k$ is arbitrary, $K(\phi,s) = 0$, so that $(\phi,s) \in M$.

Conversely, let $(\phi,s) \in M$. Let $z_k \in U_k$ satisfy $(z_k,s) \in \eta_k(X_i^{e'} \cap X_k^{e'})$ and $(f_k(z_k)+\phi_k(z_k),s) \in \zeta_k(Y_i^{e'} \cap Y_k^{e'})$. Then, by Lemma 3.1.5, $z_k \in U_i^e \cap U_k$. Since $K(\phi,s) = 0$,

$$h_{ik}(f_k(z_k) + \phi_k(z_k),s) = f_i(g_{ik}(z_k,s)) + \phi_i(g_{ik}(z_k,s)) .$$

Hence the equations:

$$w_i = f_i(z_i) + \phi_i(z_i) \quad \text{for} \quad z_i \in U_i^{e'}, \quad i \in I ,$$

define a holomorphic map f' of V_s into W_s. Thus, by the principle of analytic continuation, the equations

$$w_i = f_i(z_i) + \phi_i(z_i) \quad \text{for} \quad z_i \in U_i, \quad i \in I,$$

define f'. (Hence (ϕ,s) satisfies (2) of the compatibility conditions.)

Thus the problem is reduced to analyze the set M.

Lemma 3.2.4. Let ε be sufficiently small. Then

$$K : B_\varepsilon \times \Omega_\varepsilon \longrightarrow C^1(F,|\ |)$$

is an analytic map and

$$K'(0,o) = (\delta,\tilde{\tau}) : C^0(F,|\ |) \times T_oS \longrightarrow C^1(F,|\ |),$$

where $(\delta,\tilde{\tau})(\phi,s) = \delta\phi + \tilde{\tau}s$, for $(\phi,s) \in C^0(F,|\ |) \times T_oS$.

Proof. The proof of the first part is similar to that of Lemma 3.4 of Namba [61], so we omit it. We prove the second part.

We denote by $o(\phi,s)$ some function of (ϕ,s) (and of z_i) such that

$$|o(\phi,s)|/|(\phi,s)| \longrightarrow 0 \quad \text{as} \quad |(\phi,s)| \longrightarrow 0 .$$

Let $z_i \in U_i^e \cap U_k$. Put $z_k = g_{ki}(z_i,o)$. Then

$$\begin{aligned}
K(\phi,s)_{ik}(z_i) &= K(\phi,s)_{ik}(z_i) - K(0,o)_{ik}(z_i) \\
&= (\partial h_{ik}/\partial w_k)_{(f_k(z_k),o)}\phi_k(z_k) + (\partial h_{ik}/\partial s)_{(f_k(z_k),o)}s \\
&\quad - (f_i(g_{ik}(z_k,s)) - f_i(g_{ik}(z_k,o))) \\
&\quad - (\phi_i(g_{ik}(z_k,s)) - \phi_i(g_{ik}(z_k,o))) - \phi_i(z_i) + o(\phi,s) \\
&= (\partial h_{ik}/\partial w_k)_{(f_k(z_k),o)}\phi_k(z_k) + (\partial h_{ik}/\partial s)_{(f_k(z_k),o)}s \\
&\quad - (\partial f_i/\partial z_i)_{z_i}(\partial g_{ik}/\partial s)_{(z_k,o)}s
\end{aligned}$$

$$- (\partial\phi_i/\partial z_i)_{z_i}(\partial g_{ik}/\partial s)_{(z_k,o)}s - \phi_i(z_i) + o(\phi,s) .$$

Since $|(\partial\phi_i/\partial z_i)_{z_i}|$, $z_i \in U_i^e$, is estimated by $|\phi|$, we may put

$$- (\partial\phi_i/\partial z_i)_{z_i}(\partial g_{ik}/\partial s)_{(z_k,o)}s = o(\phi,s) .$$

Hence

$$K(\phi,s)_{ik}(z_i) = (\delta\phi)_{ik}(z_i) + (\partial h_{ik}/\partial s)_{(f_k(z_k),o)}s$$
$$- (\partial f_i/\partial z_i)_{z_i}(\partial g_{ik}/\partial s)_{(z_k,o)}s + o(\phi,s)$$
$$= (\delta\phi)_{ik}(z_i) + (\tilde{\tau}s)_{ik}(z_i) + o(\phi,s) .$$

Hence

$$K(\phi,s) = \delta\phi + \tilde{\tau}s + o(\phi,s) .$$ Q.E.D.

Now, we define a map

$$L : B_\varepsilon \times \Omega_\varepsilon \longrightarrow C^0(F,|\ |) \times T_oS$$

by

$$L(\phi,s) = (\phi + E_0 B\Lambda K(\phi,s) - E_c\delta\phi,\ s),$$

where E_o, B and Λ are the continuous linear maps defined in §3.1. Then L is analytic and $L(0,o) = (0,o)$. Moreover, by Lemma 3.2.4,

$$(3) \qquad L'(0,o) = \begin{pmatrix} 1 + E_0 B\Lambda\delta - E_0\delta & E_0 B\Lambda\tilde{\tau} \\ & \\ 0 & 1 \end{pmatrix} = \begin{pmatrix} 1 & E_0 B\tilde{\tau} \\ & \\ 0 & 1 \end{pmatrix}.$$

(Note that $B\Lambda\delta = \delta$ and $\Lambda\tilde{\tau} = \tilde{\tau}$.) Thus $L'(0,o)$ is a continuous linear isomorphism. By the inverse mapping theorem, there are a small positive number ε', an open neighborhood U of $(0,o)$ in $B_\varepsilon \times \Omega_\varepsilon$ and an analytic isomorphism Φ of $B_{\varepsilon'} \times \Omega_{\varepsilon'}$ onto U such that

$L \mid U = \Phi^{-1}$. Put

$$T = \{(\xi,s) \in (H^0(F,|\ |) \cap B_{\varepsilon'}) \times S_{\varepsilon'} \mid K\Phi(\xi,s) = 0\} .$$

Lemma 3.2.5. $T = L(M \cap U)$.

Proof. If $(\phi,s) \in M \cap U$, then

$$L(\phi,s) = (\phi + E_0 B \Lambda K(\phi,s) - E_0\delta\phi, s) = (\phi - E_0\delta\phi, s) .$$

But $\delta(\phi - E_0\delta\phi) = \delta\phi - \delta\phi = 0$. Q.E.D.

Now, let $(\xi,s) \in (H^0(F,|\ |) \cap B_{\varepsilon'}) \times S_{\varepsilon'}$. Put $(\phi,s) = \Phi(\xi,s)$.

Then

$$0 = \delta\xi = \delta(\phi + E_0 B\Lambda K(\phi,s) - E_0\delta\phi) = B\Lambda K(\phi,s) = B\Lambda K\Phi(\xi,s) .$$

Hence

$$\begin{aligned} K\Phi(\xi,s) &= H\Lambda K\Phi(\xi,s) + B\Lambda K\Phi(\xi,s) + E_1\delta K\Phi(\xi,s) \\ &= H\Lambda K\Phi(\xi,s) + E_1\delta K\Phi(\xi,s) , \end{aligned}$$

where H and E_1 are continuous linear maps defined in §3.0.

We give only a short sketch for the proof of the following lemma. See Lemma 3.6 of Namba [61] for detail. (Its idea is originally due to Kuranishi [49].)

Lemma 3.2.6. If ε' is sufficiently small, then

$$T = \{(\xi,s) \in (H^0(F,|\ |) \cap B_{\varepsilon'}) \times S_{\varepsilon'} \mid H\Lambda K\Phi(\xi,s) = 0\} .$$

Proof. (Sketch.) By some calculations, we have the following estimate: If $|(\phi,s)|$ is sufficiently small, then

$$|\delta K(\phi,s)| \leqq c_1 |K(\phi,s)| \cdot |(\phi,s)|, \quad (c_1 > 0, \text{ a constant}).$$

Hence

$$|E_1 \delta K(\phi,s)| \leqq c_2 |K(\phi,s)| \cdot |(\phi,s)|, \quad (c_2 > 0, \text{ a constant}).$$

Let $(\xi,s) \in (H^0(F,|\ |) \cap B_{\varepsilon'}) \times S_{\varepsilon'}$. Taking ε' sufficiently small, we may assume that

$$|\phi(\xi,s)| < 1/c_2 .$$

Assume that $H\Lambda K\Phi(\xi,s) = 0$. Then

$$|K\Phi(\xi,s)| = |H\Lambda K\Phi(\xi,s) + E_1 \delta K\Phi(\xi,s)| = |E_1 \delta K\Phi(\xi,s)|$$

$$\leqq c_2 |K\Phi(\xi,s)| \cdot |\Phi(\xi,s)| .$$

If $K\Phi(\xi,s) \neq 0$, then $1 \leqq c_2 |\Phi(\xi,s)|$, a contradiction. Hence $K\Phi(\xi,s) = 0$. Q.E.D.

Lemma 3.2.7. $d(H\Lambda K\Phi)_{(0,o)} = (0,\tau)$.

Proof. By (3),

$$d(H\Lambda K\Phi)_{(0,o)} = H\Lambda(\delta,\tilde{\tau}) \begin{pmatrix} 1 & -E_0 B\tilde{\tau} \\ 0 & 1 \end{pmatrix} = H\Lambda(\delta, -\delta E_0 B\tilde{\tau} + \tilde{\tau})$$

$$= (0, H\tilde{\tau}) = (0,\tau) .$$ Q.E.D.

Now, for every $t = (\xi,s) \in T$, we put

$$\Phi(t) = (\phi(t), b(t)) .$$

Then $\phi : T \longrightarrow C^0(F,|\ |)$ and $b : T \longrightarrow S$ are analytic maps. The map b is in fact the projection map $t = (\xi,s) \longmapsto s$. If we write $\phi(t) = \{\phi_i(z_i,t)\}_{i \in I}$, then

$$\phi_i : U_i \times T \longrightarrow \mathbb{C}^m$$

is a holomorphic map. We define a holomorphic map

$$\mathcal{F} : b^*X \longrightarrow b^*Y$$

by the equations

$$w_i = f_i(z_i) + \phi_i(z_i,t) \quad \text{for} \quad z_i \in U_i .$$

Then $(\mathcal{F},T,b) = \{f_t\}_{t\in T}$ is a family of holomorphic maps of (X,π,S) into (Y,μ,S). Note that $f_{(0,o)} = f$.

We show that $(\mathcal{F},T,b)$ is a maximal family. Let $t = (\xi,s)$ be a point of T. Let $(\mathcal{G},R,c) = \{g_r\}_{r\in R}$ be a family of holomorphic maps of (X,π,S) into (Y,μ,S) with a point $r_o \in R$ such that $c(r_o) = s$ and $g_{r_o} = f_t : V_s \longrightarrow W_s$. Then $g_{r_o} = f_t$ is defined by the equations

$$w_i = f_i(z_i) + \phi_i(z_i,t) \quad \text{for} \quad z_i \in U_i .$$

Then, it is easy to see that there are an open neighborhood R' of r_o, an ambient space $\widetilde{R}'$ of R' and a vector valued holomorphic function ψ_i on $U_i \times \widetilde{R}'$ such that, for each fixed $r \in R'$, g_r is defined by the equations

$$w_i = f_i(z_i) + \phi_i(z_i,t) + \psi_i(z_i,r) \quad \text{for} \quad z_i \in U_i .$$

We put

$$\phi_i'(z_i,r) = \phi_i(z_i,t) + \psi_i(z_i,r) ,$$

$$\phi'(r) = \{\phi_i'(z_i,r)\}_{i\in I} ,$$

for $r \in \widetilde{R}'$. We extend the map c to $\widetilde{R}'$. Then

$$(\phi'(r),c(r)) \in C^0(F,|\ |) \times \Omega_1 .$$

Note that $(\phi'(r_o),c(r_o)) = \Phi(t)$. It is easy to see that ϕ' is an analytic map of $\widetilde{R}'$ into $C^0(F,|\ |)$. We may assume that

$$(\phi'(r),c(r)) \in U = \Phi(B_{\varepsilon'} \times \Omega_{\varepsilon'}) \quad \text{for all} \quad r \in \widetilde{R}' .$$

Let $r \in R'$. Since the equations

$$w_i = \phi_i'(z_i, r) \quad \text{for} \quad z_i \in U_i ,$$

define a holomorphic map of $V_{c(r)}$ into $W_{c(r)}$, $(\phi'(r), c(r)) \in M \cap U$. Hence $L(\phi'(r), c(r)) \in T$ for all $r \in R'$.

Put

$$a(r) = L(\phi'(r), c(r)) \quad \text{for} \quad r \in R' .$$

Then a is a holomorphic map of R' into T. Note that

$$a(r_o) = L\Phi(t) = t .$$

We have

$$\Phi(a(r)) = (\phi'(r), h(r)), \quad r \in R' .$$

Hence $c = ba$ and $\phi' = \phi a$. From these relations, we have

$$g_r = f_{a(r)} : V_{c(r)} \longrightarrow W_{c(r)} \quad \text{for all} \quad r \in R' .$$

Thus $(\mathfrak{F}, T, b) = \{f_t\}_{t \in T}$ is a maximal family.

This completes the proof of Theorem 3.2.3.

Remark 3.2.8. If the parameter space S is one point, then it is easy to see that Theorem 3.2.3 reduces to Theorem 0.2.2 for the case of holomorphic maps of V into W.

Using the maximality of the family, we can easily patch up the local data given in Theorem 3.2.3 and get the following global theorem, which is a special form of relative Douady's theorem given by Pourcin [70].

Theorem 3.2.9. Let $(X, \pi, S) = \{V_s\}_{s \in S}$ and $(Y, \mu, S) = \{W_s\}_{s \in S}$ be families of compact complex manifolds with the same parameter space

S. Then, the disjoint union

$$\mathbb{H}\mathrm{ol} = \underset{s\in S}{\cup} \mathrm{Hol}(V_s, W_s)$$

admits a complex space structure such that

(1) the canonical projection $b : \mathbb{H}\mathrm{ol} \longrightarrow S$ is holomorphic,

(2) the map $(P,f) \in X \times_S \mathbb{H}\mathrm{ol} \longmapsto f(P) \in Y$ is holomorphic.

Moreover, $\{f_t\}_{t\in \mathbb{H}\mathrm{ol}}$ (f_t = the holomorphic map corresponding to t) is a family of holomorphic maps of $\{V_s\}_{s\in S}$ into $\{W_s\}_{s\in S}$ having the following universal property:

For any family $(\mathcal{G}, R, c) = \{g_r\}_{r\in R}$ of holomorphic maps of $\{V_s\}_{s\in S}$ into $\{W_s\}_{s\in S}$, there is a unique holomorphic map $a : R \longrightarrow \mathbb{H}\mathrm{ol}$ such that

(3) $ba = c$,

(4) $g_r = f_{a(r)} : V_{c(r)} \longrightarrow W_{c(r)}$ for all $r \in R$.

Remark 3.2.10. (1) We must show that the 'complex space' $\mathbb{H}\mathrm{ol}$ thus defined is a Hausdorff space. It is done by introducing a mertic on $\mathbb{H}\mathrm{ol}$ whose topology is weaker than that of $\mathbb{H}\mathrm{ol}$. See Namba [62, p.26], for detail.

(2) Theorem 3.2.9 implies in particular that $\mathrm{Hol}(V,W)$ is a complex space, where V and W are comact complex manifolds. (S = one point.) In this case, the underlying topology is eventually the compact-open topology. Every fiber $b^{-1}(s)$ of the map b in the theorem can be identified with the complex space $\mathrm{Hol}(V_s, W_s)$.

Now, the composition of holomorphic maps $g\circ f$ should depend holomorphically on (f,g). We formulate this assertion as follows. The proof is omitted. (See Namba [62].)

Theorem 3.2.11. Let $\{V_s\}_{s\in S}$, $\{W_s\}_{s\in S}$ and $\{U_s\}_{s\in S}$ be families of compact complex manifolds. Let

$$\mathbb{H}\mathrm{ol}_1 = \cup_s \mathrm{Hol}(V_s, W_s) ,$$

$$\mathbb{H}\mathrm{ol}_2 = \cup_s \mathrm{Hol}(W_s, U_s),$$

$$\mathbb{H}\mathrm{ol}_3 = \cup_s \mathrm{Hol}(V_s, U_s),$$

be the complex spaces given in Theorem 3.2.9. Then the map

$$(f,g) \in \mathbb{H}\mathrm{ol}_1 \times_S \mathbb{H}\mathrm{ol}_2 \longmapsto g \circ f \in \mathbb{H}\mathrm{ol}_3$$

is holomorphic.

As an application of this theorem, we get the following theorem (see Namba [62, p.237]; see also Schuster [77] for its general form).

Theorem 3.2.12. Let $(X,\pi,S) = \{V_s\}_{s\in S}$ be a family of compact complex manifolds. Then the disjoint union

$$\mathbb{A}\mathrm{ut} = \cup_s \mathrm{Aut}(V_s)$$

admits a complex space structure such that the following maps are holomorphic:

(1) the canonical projection $\mathbb{A}\mathrm{ut} \longrightarrow S$,

(2) $(P,a) \in X \times_S \mathbb{A}\mathrm{ut} \longmapsto a(P) \in X$,

(3) $s \in S \longmapsto 1_s \in \mathbb{A}\mathrm{ut}$, where 1_s is the identity of $\mathrm{Aut}(V_s)$,

(4) $(a,b) \in \mathbb{A}\mathrm{ut} \times_S \mathbb{A}\mathrm{ut} \longmapsto b^{-1}a \in \mathbb{A}\mathrm{ut}$.

In particular, if S is one point, then we get

Corollary 3.2.13. (Bochner-Montgomery [7]). Let V be a compact complex manifold. Then $\mathrm{Aut}(V)$ is a complex Lie group.

Remark 3.2.14. (1) The corresponding theorem for a compact complex space was proved by Kaup [35].

(2) Note that $\mathrm{Aut}(V)$ is open and closed in $\mathrm{Hol}(V,V)$. This holds even for a compact complex space V (see Namba[63]).

3.3. Non-singularity of the space R_n.

Let $\{V_s\}_{s\in S}$ and $\{W_s\}_{s\in S}$ be families of compact complex manifolds. Let $o \in S$ and $f \in \mathrm{Hol}(V_o, W_o)$. We want to find sufficient conditions for f to be a non-singular point of $|\mathrm{Hol}| = \cup_s \mathrm{Hol}(V_s, W_s)$.

The following proposition is an easy consequence of Theorem 3.2.3.

Proposition 3.3.1. (c.f., Corollary 0.2.3). Assume that

$$H^1(V_o, \Theta(f^*TW_o)) = 0 .$$

Then

(1) If o is a non-singular point of S, then f is a non-singular point of $|\mathrm{Hol}|$ and

$$\dim_f |\mathrm{Hol}| = h^0(f^*TW_o) + \dim_o S .$$

(2) f is a non-singular point of $\mathrm{Hol}(V_o, W_o)$ and

$$\dim_f \mathrm{Hol}(V_o, W_o) = h^0(f^*TW_o) .$$

However, the requirement of the vanishing of the first cohomology group in the proposition seems sometimes too strong.

When the parameter space S is big enough, the following theorem seems to be useful.

Theorem 3.3.2. Let $\{V_s\}_{s\in S}$ and $\{W_s\}_{s\in S}$ be families of compact complex manifolds. Let o be a non-singular point of S. For $f \in \mathrm{Hol}(V_o, W_o)$, assume that the linear map

$$\tau : T_oS \longrightarrow H^1(V_o, \Theta(f^*TW_o))$$

(see §3.1) is surjective. Then f is a non-singular point of $|\mathrm{Hol}|$ and

$$\dim_f |\mathrm{Hol}| = h^0(f^*TW_o) - h^1(f^*TW_o) + \dim_o S .$$

Proof. By (5) of Theorem 3.2.3, $(d\alpha)_{(0,o)} = (0,\tau)$. If τ is surjective, then the holomorphic map α is of maximal rank at $(0,o)$. Hence the complex space T in Theorem 3.2.3 is non-singular at $(0,o)$ and $\dim_o T = h^0(f^*TW_o) - h^1(f^*TW_o) + \dim_o S$. Q.E.D.

Now, let T_g, $g \geqq 2$, be the Teichmüller space of compact Riemann surfaces of genus g (see Example 0.1.2). Let V_t be the compact Riemann surface corresponding to $t \in T_g$. By Theorem 3.2.9,

$$|\mathrm{Hol}| = \cup_t \mathrm{Hol}(V_t, \mathbb{P}^1)$$

is a complex space. (Take the trivial family $\{\mathbb{P}^1\}_{t\in T_g}$ for $\{W_t\}_{t\in T_g}$. For $n \geqq 2$, we put

$$R_n = \cup_t R_n(V_t) .$$

The following lemma is proved in a similar way to Lemma 1.1.1.

Lemma 3.3.3. R_n is open and closed in $|\mathrm{Hol}|$.

We have to check

Lemma 3.3.4. For any $n \geqq 2$, R_n is non-empty.

Proof. (following Mumford [59, p.14]). We choose

(a) distinct $r = 2n+2g-2$ points $\{Q_\lambda\}$ on $\mathbb{P}^1$,

(b) a base point $Q_o \in \mathbb{P}^1$ and a set of "cuts" joining Q_λ to Q_o,

(c) r transpositions A_λ acting on $\{1,\cdots,n\}$ such that

$$(A_1 \cdots A_r) = 1 \quad \text{(the identity permutation)}.$$

Next, by glueing together n copies of $\mathbb{P}^1$ via the transposition A_λ on the λ-th cut, we have a topological covering space V of $\mathbb{P}^1$. Then the projection $f : V \longrightarrow \mathbb{P}^1$ is a local homeomorphism outside of the points of ramification and is equivalent to

$$z \in \mathbb{C} \longmapsto z^2 \in \mathbb{C}$$

around a point of ramification. Hence we can introduce a unique complex manifold structure on V so that f becomes a holomorphic map. Thus V is a compact Riemann surface of genus g (see Riemann-Hurwitz formula in §2.1) and $f \in R_n(V)$. Q.E.D.

Now, we show

Theorem 3.3.5. For any integer $n \geqq 2$, R_n is non-singular and of dimencion $2n+2g-2$.

Proof. Take any $o \in T_g$ and $f \in R_n(V_o)$. In order to prove the theorem, we make use of Theorem 3.3.2. We show that the map

$$\tau : T_oT_g \text{ (the tangent space at } o) \longrightarrow H^1(V_o, \mathcal{O}(f^*T\mathbb{P}^1))$$

is surjective. We use the same notations as in §3.2. Let $t = (t^1, \cdots, t^{3g-3})$ be a coordinate system in T_g. Then τ is given by

$$(1) \quad \tau(\frac{\partial}{\partial t}) = -(\frac{\partial f_i}{\partial z_i})_{z_i}(\frac{\partial g_{ik}}{\partial t})_{(z_i, o)} = -(\rho(\frac{\partial}{\partial t}))(f_i) ,$$

where ρ is the Kodaira-Spencer map.

We define a sheaf homomorphism

$$f_* : \mathcal{O}(TV_o) \longrightarrow \mathcal{O}(f^*T\mathbb{P}^1) = \mathcal{O}([2D_\infty(f)])$$

by

$$f_*(a_i(z_i)\frac{\partial}{\partial z_i}) = a_i(z_i)(\frac{\partial f_i}{\partial z_i}) .$$

Then f_* is injective, for f is non-constant. Let $\mathcal{S}$ be the quotient sheaf, so that

$$0 \longrightarrow \mathcal{O}(TV_o) \xrightarrow{f_*} \mathcal{O}([2D_\infty(f)]) \longrightarrow \mathcal{S} \longrightarrow 0$$

is exact. Note that the support of $\mathcal{S}$ is the set of all points of ramification of $f : V_0 \longrightarrow \mathbb{P}^1$. It is a finite set. If $P \in V_0$ is such a point with the ramification exponent e_P (> 1), then $(\frac{\partial f_i}{\partial z_i})$ has zero of order $e_P - 1$ at P. Hence the stalk $\mathcal{S}_P$ at P of $\mathcal{S}$ is isomorphic to

$$\mathcal{O}_{\mathbb{C},0}/z^{e_P-1}\,\mathcal{O}_{\mathbb{C},0} \cong \mathbb{C}^{e_P-1}.$$

Hence, from the above exact sequence, we get the following exact sequence:

(2)
$$0 \longrightarrow H^0(V_0, \mathcal{O}([2D_\infty(f)])) \longrightarrow \mathbb{C}^{\Sigma(e_P-1)}$$
$$\longrightarrow H^1(V_0, \mathcal{O}(TV_0)) \xrightarrow{f_*} H^1(V_0, \mathcal{O}([2D_\infty(f)])) \longrightarrow 0$$

(Note that $H^0(V_0, \mathcal{O}(TV_0)) = 0$.)

Now (1) shows that the diagram

$$\begin{array}{ccc} T_0T_g & \xrightarrow{\ \rho\ } & H^1(V_0, \mathcal{O}(TV_0)) \\ & {\scriptstyle -\tau}\searrow \quad \swarrow{\scriptstyle f_*} & \\ & H^1(V_0, \mathcal{O}([2D_\infty(f)])) & \end{array}$$

is commutative. Since ρ is isomorphic and f_* is surjective, τ must be surjective. Hence, by Theorem 3.3.2, f is a non-singular point of R_n and

$$\dim_f R_n = h^0(2D_\infty(f)) - h^1(2D_\infty(f)) + (3g-3) = 2n+2g-2,$$

by Riemann-Roch theorem. Q.E.D.

Remark 3.3.6. If we take the alternating sum of the dimensions of vector spaces in (2) above, then we get the Riemann-Hurwitz formula in §2.1. (Note that $h^1(TV_0) = 3g-3$.)

3.4. Moduli of open holomorphic maps.

First of all, we give some remarks on families of surjective holomorphic maps. The following lemma is an easy consequence of the proper mapping theorem (see, e.g., Gunning-Rossi [29]).

Lemma 3.4.1. Let V and W be compact complex manifolds. Then $f \in \mathrm{Hol}(V,W)$ is surjective if and only if there is a point $P \in V$ such that $\mathrm{rank}\,(df)_P = \dim W$.

Now, let $(X,\pi,S) = \{V_s\}_{s\in S}$ and $(Y,\mu,S) = \{W_s\}_{s\in S}$ be families of compact complex manifolds. Let $\mathbb{H}ol = \cup_s \mathrm{Hol}(V_s,W_s)$ be the complex space in Theorem 3.2.9. We put

$$\mathrm{Surj}(V_s,W_s) = \{f \in \mathrm{Hol}(V_s,W_s) \mid f \text{ is surjective}\},$$

$$\mathcal{S}urj = \cup_s \mathrm{Surj}(V_s,W_s) \quad \text{(disjoint union)}.$$

Lemma 3.4.2. $\mathcal{S}urj$ is open and closed in $\mathbb{H}ol$.

Proof. Let $f_1, f_2, \cdots$ be a sequence of surjective holomorphic maps in $\mathbb{H}ol$ converging to $f \in \mathbb{H}ol$. Assume that f_ν maps V_{s_ν} onto W_{s_ν} for $\nu = 1,2,\cdots$ and $f \in \mathrm{Hol}(V_o,W_o)$. Then the sequence $s_1, s_2, \cdots$ converges to $o \in S$. Take any point $Q \in W_o$. Take a sequence $Q_1, Q_2, \cdots$ of points in Y converging to Q such that $Q_\nu \in W_{s_\nu}$ for $\nu = 1,2,\cdots$. Let $P_\nu \in V_{s_\nu}$ be such that $f_\nu(P_\nu) = Q_\nu$. Taking a subsequence, if necessary, we may assume that the sequence $P_1, P_2, \cdots$ converges to a point $P \in V_o$. Then $f(P) = Q$. Hence f is surjective. This shows that $\mathcal{S}urj$ is closed in $\mathbb{H}ol$.

Next, let $(\mathcal{F},T,b) = \{f_t\}_{t\in T}$ be a family of holomorphic maps of $\{V_s\}_{s\in S}$ into $\{W_s\}_{s\in S}$ (see Definition 3.2.1). For a point $o \in T$, assume that $f_o : V_o \longrightarrow W_o$ is surjective. By Lemma 3.4.1, there is a point $P \in V_o$ such that $\mathrm{rank}\,(df_o)_P = \dim W_o$. Let $\mathcal{U}$ (resp. $\mathcal{W}$)

be an open neighborhood of P in b^*X (resp. of $f_o(P)$ in b^*Y) such that there are biholomorphic maps

$$\eta : \mathcal{U} \longrightarrow U \times T', \quad U \text{ open} \subset V_o ,$$

$$\zeta : \mathcal{W} \longrightarrow W \times T', \quad W \text{ open} \subset W_o ,$$

such that the diagrams

$$\begin{array}{ccc} \mathcal{U} & \xrightarrow{\ \eta\ } & U \times T' \\ & {\scriptstyle \pi}\searrow \quad \swarrow{\scriptstyle \text{proj}} & \\ & T' & \end{array} \qquad\qquad \begin{array}{ccc} \mathcal{W} & \xrightarrow{\ \zeta\ } & W \times T' \\ & {\scriptstyle \mu}\searrow \quad \swarrow{\scriptstyle \text{proj}} & \\ & T' & \end{array}$$

commute, where T' is an open neighborhood of o in T. Assume that U (resp. W) has a coordinate system $(z) = (z^1,\cdots,z^d)$ (resp. $(w) = (w^1,\cdots,w^m)$). Then, via η and ζ, the map $\mathfrak{F}$ is locally expressed as

$$\mathfrak{F} : (z,t) \longmapsto (f(z,t),t) .$$

Now,

$$(df_o)_P = \text{rank}\ (\frac{\partial f}{\partial z}(P,o)) = \dim W_o = m .$$

Hence, if $\mathcal{U}$ is sufficiently small, then

$$(df_t)_Q = \text{rank}\ (\frac{\partial f}{\partial z}(z,t)) = m \quad \text{for all} \quad Q = (z,t) \in \mathcal{U} .$$

By Lemma 3.4.1, f_t is surjective for all $t \in T'$. This proves that $\mathcal{S}urj$ is open in $\mathbb{H}ol$. <u>Q.E.D.</u>

<u>Corollary 3.4.3</u>. Let V and W be compact complex manifolds. Then $Surj(V,W)$ is open and closed in $Hol(V,W)$.

<u>Proposition 3.4.4</u>. Let U, V and W be compact complex manifolds. Then, for any $h \in Surj(U,V)$, the map

$$j_h : f \in \mathrm{Hol}(V,W) \longmapsto f \circ h \in \mathrm{Hol}(U,W)$$

is a holomorphic imbedding.

Proof. It is clear that j_h is injective. It is also clear that the diagram

$$\begin{array}{ccc} T_f\mathrm{Hol}(V,W) & \xrightarrow{(dj_h)_f} & T_{fh}\mathrm{Hol}(U,W) \\ \Big\downarrow \sigma_f & & \Big\downarrow \sigma_{fh} \\ H^0(V,\Theta(f^*TW)) & \xrightarrow{h^*} & H^0(U,\Theta((fh)^*TW)) \end{array}$$

commutes for all $f \in \mathrm{Hol}(V,W)$, where σ_f and σ_{fh} are the characteristic maps defined in §0.2. Moreover, the surjectivity of h implies the injectivity of the induced linear map h^*. Hence $(dj_h)_f$ is injective.

Q.E.D.

In the rest of this section, we consider the case when $\{W_s\}_{s\in S}$ is a trivial family, i.e., $Y = W \times S$. Then $\mathbb{Hol}$ and $\mathcal{S}urj$ are written as

$$\mathbb{Hol} = \cup_s \mathrm{Hol}(V_s, W) ,$$

$$\mathcal{S}urj = \cup_s \mathrm{Surj}(V_s, W) .$$

By Theorem 3.2.11, the action of $\mathrm{Aut}(W)$:

$$(b,f) \in \mathrm{Aut}(W) \times \mathbb{Hol} \longmapsto bf \in \mathbb{Hol} ,$$

$$(b,f) \in \mathrm{Aut}(W) \times \mathcal{S}urj \longmapsto bf \in \mathcal{S}urj ,$$

are holomorphic. By Proposition 3.4.4, the action on $\mathcal{S}urj$ is free.

Problem. Is the action of $\mathrm{Aut}(W)$ on $\mathcal{S}urj$ proper? i.e., is the map

$$(b,f) \in \mathrm{Aut}(W) \times \mathcal{S}urj \longmapsto (bf,f) \in \mathcal{S}urj \times \mathcal{S}urj$$

a proper map?

If we restrict our families to families of <u>open</u> holomorphic maps, then we get an affirmative answer. In the sequel, we follow the arguments in Namba [64], generalizing to relative cases.

For the proof of the following lemma, see, e.g., Narasimhan [68, p.132].

<u>Lemma 3.4.5</u>. Let X be a pure d-dimensional complex space and let Y be a m-dimensional locally irreducible complex space. Then, a holomorphic map f of X into Y is an open map if and only if $\dim_x f^{-1}f(x) = d-m$, for all $x \in X$.

Now, we put

$$\mathrm{Open}(V_s,W) = \{f \in \mathrm{Hol}(V_s,W) \mid f \text{ is an open map}\},$$

$$\mathbb{O}\mathrm{pen} = \cup_s \mathrm{Open}(V_s,W) \quad \text{(disjoint union)}.$$

<u>Lemma 3.4.6</u>. $\mathbb{O}\mathrm{pen}$ is open in $\mathbb{H}\mathrm{ol}$. Hence it is open in $\mathcal{S}urj$.

<u>Proof</u>. Let $(\mathfrak{F},T,b) = \{f_t\}_{t\in T}$ be a family of holomorphic maps of $(X,\pi,S) = \{V_s\}_{s\in S}$ into W. Let $o \in T$. Assume that $f_o : V_o \longrightarrow W$ is open. Then, by Lemma 3.4.5,

$$\text{(1)} \qquad \dim_P f_o^{-1} f_o(P) = d-m \quad \text{for all } P \in V_o .$$

We may replace T by any small open neighborhood of o. Let $T = \bigcup_{\alpha=1}^{k} T_\alpha$ be the decomposition of T at o into irreducible components. For each α, the restriction $\mathfrak{F}_\alpha$ of $\mathfrak{F}$ to $(b^*X)|T_\alpha$:

$$\mathfrak{F}_\alpha : (b^*X)|T_\alpha \longrightarrow W \times T_\alpha$$

satisfies

$$(2)\quad \dim \mathfrak{F}_\alpha^{-1}\mathfrak{F}_\alpha(P) \geqq (d+n_\alpha) - (m+n_\alpha) = d-m \quad \text{for all} \quad P \in (b^*X)|T_\alpha ,$$

where $n_\alpha = \dim T_\alpha$. Hence, by (1), (2) and the upper semi-continuity of $\dim_P \mathfrak{F}_\alpha^{-1}\mathfrak{F}_\alpha(P)$,

$$\dim_P f_t^{-1} f_t(P) = \dim_P \mathfrak{F}_\alpha^{-1}\mathfrak{F}_\alpha(P) = d-m \quad \text{for all} \quad P \in (b^*X)|T_\alpha ,$$

where $t = \pi^*(P)$ $(\pi^* : b^*X \longrightarrow T)$, provided T_α is sufficiently small. This holds for all α. Hence, each f_t, $t \in T$, is an open map by Lemma 3.4.5. This proves the lemma. <u>Q.E.D.</u>

<u>Proposition 3.4.7</u>. Assume that $\mathfrak{Open}$ is non-empty. Then the action of $\mathrm{Aut}(W)$ on $\mathfrak{Open}$ is proper, i.e.,

$$\psi : (b,f) \in \mathrm{Aut}(W) \times \mathfrak{Open} \longmapsto (bf,f) \in \mathfrak{Open} \times \mathfrak{Open}$$

is a proper map.

In order to prove this proposition, we need the following two lemmas.

<u>Lemma 3.4.8</u>. Let $f_1, f_2, \cdots$ be a sequence of points of $\mathfrak{Open}$ converging to $f \in \mathfrak{Open}$. Let $f_\nu \in \mathrm{Open}(V_{s_\nu}, W)$ and $f \in \mathrm{Open}(V_{s_o}, W)$. Then, for any point $P \in V_{s_o}$, there is a sequence $P_1, P_2, \cdots$ of points in X converging to P such that $P_\nu \in V_{s_\nu}$ and $f_\nu(P_\nu) = f(P)$ for $\nu = 1, 2, \cdots$.

<u>Proof</u>. Let $(\tilde{\mathfrak{F}}, \tilde{T}, \tilde{b}) = \{f_t\}_{t\in\tilde{T}}$ be the universal family of holomorphic maps of $\{V_s\}_{s\in S}$ into W given in Theorem 3.2.9. ($\tilde{T} = |\mathrm{Hol}| = \cup_s \mathrm{Hol}(V_s, W)$.) Let $o \in \tilde{T}$ be the point corresponding to f, i.e., $f_o = f$. Then, by Lemma 3.4.6, there is a small open neighborhood T of o in $\tilde{T}$ such that, for every $t \in T$, $f_t \in \mathrm{Open}(V_{\tilde{b}(t)}, W)$. We may assume that

$f_\nu = f_{t_\nu}$, where $t_\nu \in T$ for $\nu = 1,2,\cdots$. Then the sequence $t_1, t_2, \cdots$ converges to o and $\tilde{b}(t_\nu) = s_\nu$ for $\nu = 1,2,\cdots$ and $\tilde{b}(o) = s_o$.

Let $T = \bigcup_{\alpha=1} T_\alpha$ be the decomposition of T at o into irreducible components. We may assume that every t belongs to a fixed component, say T_1. We consider the restriction

$$\tilde{\mathfrak{F}}_1 = \tilde{\mathfrak{F}} \mid \pi^{*-1}(T_1) : \pi^{*-1}(T_1) \longrightarrow W \times T_1 .$$

($\pi^* : \tilde{b}^* Y \longrightarrow \tilde{T}$, the projection.) Then, as was shown in the proof of Lemma 3.4.6,

$$\dim_P \tilde{\mathfrak{F}}_1^{-1} \tilde{\mathfrak{F}}_1(P) = d-m \quad \text{for all} \quad P \in \pi^{*-1}(T_1) .$$

By Lemma 3.4.5, $\tilde{\mathfrak{F}}_1$ is an open map.

Now, we fix a point $P \in V_{s_o}$ and put

$$R = \{P' \in \pi^{*-1}(T_1) \;\Big|\; \tilde{\mathfrak{F}}_1(P') = (f(P), t)\} ,$$

where $t = \pi^*(P')$. Then R is a closed complex subspace of $\pi^{*-1}(T_1)$ containing P. For an open subset $\mathcal{U}$ of $\pi^{*-1}(T_1)$, we have

$$f(P) \times (\pi^*(\mathcal{U} \cap R)) = \tilde{\mathfrak{F}}_1(\mathcal{U}) \cap (f(P) \times \pi^*(\mathcal{U})) .$$

This implies that $\pi^*(\mathcal{U} \cap R)$ $(\subset \pi^*(\mathcal{U}))$ is open in T_1.

Let $\mathcal{U}_1 \supset \mathcal{U}_2 \supset \cdots$ be a fundamental system of open neighborhoods of P in $\pi^{*-1}(T_1)$. We put

$$T^\nu = \pi^*(\mathcal{U}_\nu \cap R) \quad \text{for} \quad \nu = 1, 2, \cdots .$$

Then, each T^ν is an open neighborhood of o in T_1, which is containe in $\pi^*(\mathcal{U}_\nu)$. Hence, $T^1 = T_1 \supset T^2 \supset \cdots$ is a fundamental system of open neighborhoods of o in T_1.

Now, to each ν, we associate a positive integer $k(\nu)$ or $+\infty$ as follows: $k(\nu) = k$ when $t_\nu \in T^k - T^{k+1}$. If there is not such an integer k, then t_ν must be o itself. In this case, we put $k(\nu) = +\infty$. Then it is easy to see that $k(\nu) \longrightarrow +\infty$, as $\nu \longrightarrow +\infty$. For a positive integer ν, assume that $k(\nu) \neq +\infty$. Then, there is $P_\nu \in \mathcal{U}_{k(\nu)} \cap V_{t_\nu}$ such that $f_{t_\nu}(P_\nu) = f(P)$. If $k(\nu) = +\infty$, then we put $P_\nu = P$. Then, it is clear that the sequence $P_1, P_2, \cdots$ converges to P. Q.E.D.

Lemma 3.4.9. Let $\{f_\nu\}_{\nu=1,2,\cdots}$ and $\{g_\nu\}_{\nu=1,2,\cdots}$ be two sequences in $\mathfrak{Open}$ converging to $f, g \in \mathfrak{Open}$, respectively. Assume that, for each ν, there is $b_\nu \in \mathrm{Aut}(W)$ such that $g_\nu = b_\nu f_\nu$. Then there is $b \in \mathrm{Aut}(W)$ such that (1) $g = bf$ and (2) $\{b_\nu\}_{\nu=1,2,\cdots}$ converges to b.

Proof. Let Q be an arbitrary point of W. We take a point $P \in f^{-1}(Q)$ and put $R = g(P)$. By Lemma 3.4.8, there is a sequence $P_1, P_2, \cdots$ of points of X converging to P such that $f_\nu(P_\nu) = Q$ for $\nu = 1, 2, \cdots$. Put $R_\nu = b_\nu(Q)$. Then

$$R_\nu = b_\nu(Q) = b_\nu f_\nu(P_\nu) = g_\nu(P_\nu) .$$

Hence the sequence $R_1, R_2, \cdots$ converges to $g(P) = R$. Since each $R_\nu = b_\nu(Q)$ does not depend on the choice of $P \in f^{-1}(Q)$, R does not either. Hence we get a map

$$b : Q \in W \longmapsto R \in W .$$

By the definition of the map b, we have $g = bf$. Since the conditions are symmetric with respect to f and g, b is a bijective map.

We first show that b is a homeomorphism. Put

$$A = \{(P,Q,R) \in V_{s_o} \times W \times W \mid f(P) = Q \text{ and } g(P) = R\} ,$$

where $s_o \in S$ is the point such that $f, g \in \mathrm{Open}(V_{s_o}, W)$. Then A is a closed complex submanifold of $V_{s_o} \times W \times W$ isomorphic to V_{s_o}. Let

$$\lambda : V_{s_o} \times W \times W \longrightarrow W \times W$$

be the projection. Then $\lambda(A)$ is a closed irreducible subvariety of $W \times W$. Let λ_1 and λ_2 be the restrictions to $\lambda(A)$ of the first and the second projections $W \times W \longrightarrow W$, respectively. Then it is easy to see that λ_1 and λ_2 are bijective and $b = \lambda_2 \lambda_1^{-1}$. Since $\lambda(A)$ is irreducible and since λ_1 is a bijective holomorphic map of $\lambda(A)$ onto W, λ_1 is a homeomorphism (by Lemma 3.4.5). In a similar way, λ_2 is a homeomorphism of $\lambda(A)$ onto W. Hence b is a homeomorphism.

Next, we show that $b \in \mathrm{Aut}(W)$. We put

$$D_f = \{P \in V_{s_o} \mid (df)_P \text{ has the rank} < \dim W = m\} ,$$

$$E_f = f(D_f) ,$$

$$D_f' = f^{-1}(E_f) .$$

Then D_f and D_f' are proper closed complex subspaces of V_{s_o} and E_f is a proper closed complex subspace of W. Moreover,

$$f : V_{s_o} - D_f' \longrightarrow W - E_f$$

is a surjective proper holomorphic map of maximal rank at every point. Hence, for any point $Q \in W - E_f$, there are an open neighborhood U of Q in $W - E_f$ and a holomorphic section $h : U \longrightarrow V_{s_o} - D_f'$ of f. Then we get $b = gh$ on U. Hence b is holomorphic on $W - E_f$. Since b is a homeomorphism on W, it is holomorphic on whole W. A similar argument shows that b^{-1} is also holomorphic on W. Hence $b \in \mathrm{Aut}(W)$.

Finally, we show that the sequence $b_1, b_2, \cdots$ converges to b. Let $(\tilde{\mathcal{H}}, \tilde{T}, \tilde{b}) = \{f_t\}_{t \in \tilde{T}}$ be the universal family of holomorphic maps of $\{V_s\}_{s \in S}$ into W, given in Theorem 3.2.9. ($\tilde{T} = \mathbb{H}\mathrm{ol} = \cup_s \mathrm{Hol}(V_s, W)$.) Let $o \in \tilde{T}$ be the point corresponding to f, i.e., $f_o = f$. As in the proof of Lemma 3.4.8, we may assume that there is an open neighborhood T of o in $\tilde{T}$ such that $f_\nu = f_{t_\nu}$, where $t_\nu \in T$ for $\nu = 1, 2, \cdots$. Then $\{t_\nu\}_{\nu=1,2,\cdots}$ converges to o. Let d be a metric on W. For any point $t \in T$, we denote by $\mathrm{Cont}(V_t, W)$ the topological space of all continuous maps of V_t into W with the comact-open topology. We define a metric on $\mathrm{Cont}(V_t,W)$ as follows: For f_1, $f_2 \in \mathrm{Cont}(V_t,W)$,

$$d_t(f_1,f_2) = \sup\{d(f_1(P),f_2(P)) \mid P \in V_t\}.$$

Then it is easy to see that the topology of the metric d_t is equal to the compact-open topology. In a similar way, we define a metric $\hat{d}$ on $\mathrm{Aut}(W)$ whose topology is equal to the compact-open topology. Note that, if $h \in \mathrm{Cont}(V_t,W)$ is surjective, then

(1) $$\hat{d}(b_1,b_2) = d_t(b_1h,b_2h) \quad \text{for all } b_1, b_2 \in \mathrm{Aut}(W).$$

Now, since a family of compact complex manifolds is differentiablly trivial (see Kuranishi [50]), we may assume that there is a continuous map

$$\omega : \pi^{*-1}(T) \longrightarrow V_o$$

such that the restriction ω_t of ω to each $V_t = \pi^{*-1}(t)$, $t \in T$, is a diffeomorphism of V_t onto V_o and ω_o = the identity map on V_o. Let d' be a metric on T. We put

$$d(h_1,h_2) = d'(t_1,t_2) + d_o(h_1\omega_{t_1}^{-1},h_2\omega_{t_2}^{-1})$$

for $h_1 \in \mathrm{Hol}(V_{t_1},W)$ and $h_2 \in \mathrm{Hol}(V_{t_2},W)$. Then, it is easy to see that

(2) d is a metric on the set $\bigcup_{t\in T} \mathrm{Hol}(V_t,W)$,

(3) for $h_1, h_2 \in \mathrm{Hol}(V_t,W)$, $d(h_1,h_2) = d_t(h_1,h_2)$,

(4) if $h \longrightarrow h_o$ in $\bigcup_{t\in T} \mathrm{Hol}(V_t,W)$, then $d(h,h_o) \longrightarrow 0$.

Now, by (1) - (3), we have

$$\begin{aligned}\hat{d}(b_\nu,b) = d(b_\nu f_\nu,bf_\nu) &\leqq d(b_\nu f_\nu,bf) + d(bf_\nu,bf)\\ &= d(g_\nu,g) + d(bf_\nu,bf).\end{aligned}$$

Note that, by (4), $d(g_\nu,g) \longrightarrow 0$ as $\nu \longrightarrow +\infty$. On the other hand, since the sequence $bf_1, bf_2, \cdots$ converges to bf,

$$d(bf_\nu, bf) \longrightarrow 0 \quad \text{as} \quad \nu \longrightarrow +\infty .$$

Hence $\{b_\nu\}_{\nu=1,2,\cdots}$ converges to b. Q.E.D.

Now, we are ready to prove Proposition 3.4.7.

Proof of Proposotion 3.4.7. Let K be a compact subset of $\mathfrak{Open} \times \mathfrak{Open}$. We show that $\psi^{-1}(K)$ is a compact subset of $\mathrm{Aut}(W) \times \mathfrak{Open}$. It suffices to show that, for any sequence $(b_1, f_1), (b_2, f_2), \cdots$ of points of $\psi^{-1}(K)$, we can find a subsequence converging to a point of $\psi^{-1}(K)$. Put

$$g_\nu = b_\nu f_\nu \quad \text{for} \quad \nu = 1, 2, \cdots .$$

Since K is compact, we may assume that the sequence $(g_1, f_1), (g_2, f_2), \cdots$ converges to $(g, f) \in K$. By Lemma 3.4.9, there is $b \in \mathrm{Aut}(W)$ such that $g = bf$ and such that $\{b_\nu\}_{\nu=1,2,\cdots}$ converges to b. Then $(b, f) \in \psi^{-1}(K)$ and $\{(b_\nu, f_\nu)\}_{\nu=1,2,\cdots}$ converges to (b, f). Q.E.D.

By Proposition 3.4.7 and Holmann [30, Satz 21], we get

Theorem 3.4.10. The quotient space $\mathfrak{Open}/\mathrm{Aut}(W)$ admits a complex space structure such that the canonical projection

$$\mathfrak{Open} \longrightarrow \mathfrak{Open}/\mathrm{Aut}(W)$$

is holomorphic and is a principal Aut(W)-bundle.

Corollary 3.4.11. (Namba [64]). Let V and W be compact complex manifolds. Then the quotient space $\mathrm{Open}(V,W)/\mathrm{Aut}(W)$ admits a complex space structure such that the projection

$$\mathrm{Open}(V,W) \longrightarrow \mathrm{Open}(V,W)/\mathrm{Aut}(W)$$

is holomorphic and is a principal Aut(W)-bundle.

Let V and W be compact complex manifolds. Then $\mathrm{Aut}(W)\times\mathrm{Aut}(V)$ acts on Open(V,W) as follows:

$$(b,a,f)\in \mathrm{Aut}(W)\times \mathrm{Aut}(V)\times \mathrm{Open}(V,W) \longmapsto bfa^{-1}\in \mathrm{Open}(V,W)\,.$$

The following proposition is an easy consequence of Proposition 3.4.7.

Proposition 3.4.12. Assume that Aut(V) is compact. Then the action of $\mathrm{Aut}(W)\times\mathrm{Aut}(V)$ on Open(V,W) is proper.

Hence, by Holmann [31, Satz 19],

Theorem 3.4.13. (Namba [64]). Let V and W be compact complex manifolds. Assume that Aut(V) is compact. Then the quotient space $\mathrm{Open}(V,W)/(\mathrm{Aut}(W)\times\mathrm{Aut}(V))$ admits a complex space structure.

The complex space $\mathrm{Open}(V,W)/(\mathrm{Aut}(W)\times\mathrm{Aut}(V))$ in the theorem can be considered as the moduli space of open holomorphic maps of V onto W.

Remark 3.4.14. Aut(V) naturally acts on Open(V,W)/Aut(W). The action is proper if Aut(V) is compact. Hence, in this case, (Open(V,W)/Aut(W))/Aut(V) is a complex space. We can easily show that it is naturally biholomorphic to $\mathrm{Open}(V,W)/(\mathrm{Aut}(W)\times\mathrm{Aut}(V))$.

Remark 3.4.15. (Namba [64, p.75]). The assumption that 'Aut(V) is compact' in the theorem can not be dropped. In fact, we show that $\mathrm{Open}(\mathbb{P}^1,\mathbb{P}^1)/(\mathrm{Aut}(\mathbb{P}^1)\times\mathrm{Aut}(\mathbb{P}^1))$ is not Hausdorff. For this purpose, it suffices to show that $R_3(\mathbb{P}^1)/(\mathrm{Aut}(\mathbb{P}^1)\times\mathrm{Aut}(\mathbb{P}^1))$ is not Hausdorff. Consider the following 1-parameter family of rational functions:

$$f_t(x) = x^3 + tx, \quad t \in \mathbb{C},$$

where $x = Z_1/Z_0$ is the imhomogeneous coordinate in $\mathbb{P}^1 - \infty$. Note that

$$f_0 \not\equiv f_1 \pmod{\mathrm{Aut}(\mathbb{P}^1) \times \mathrm{Aut}(\mathbb{P}^1)}.$$

In fact, arranging the ramification exponents, the types of the points of ramification of f_0 and f_1 are (3,3) and (3,2,2), respectively. On the other hand, if $t \neq 0$, then

$$f_t \equiv f_1 \pmod{\mathrm{Aut}(\mathbb{P}^1) \times \mathrm{Aut}(\mathbb{P}^1)}.$$

In fact, we define a_t, $b_t \in \mathrm{Aut}(\mathbb{P}^1)$ by

$$a_t(x) = (1/\sqrt{t})x,$$

$$b_t(x) = (1/\sqrt{t})^3 x.$$

Then we have $b_t f_t a_t^{-1} = f_1$. Thus $f_0 \pmod{\mathrm{Aut}(\mathbb{P}^1) \times \mathrm{Aut}(\mathbb{P}^1)}$ and $f_1 \pmod{\mathrm{Aut}(\mathbb{P}^1) \times \mathrm{Aut}(\mathbb{P}^1)}$ can not be separated.

Finally, let $\{V_t\}_{t \in T_g}$ be the family of compact Riemann surfaces of genus $g \geqq 2$ in Example 0.1.2 and let R_n, $(n \geqq 2)$, be the complex space defined in §3.3. Then $\mathrm{Aut}(\mathbb{P}^1)$ acts on R_n. By Theorem 3.3.5 and Theorem 3.4.10,

<u>Theorem 3.4.17.</u> The quotient space $R_n/\mathrm{Aut}(\mathbb{P}^1)$ is a non-singular complex space of dimension $2n+2g-5$. The projection

$$R_n \longrightarrow R_n/\mathrm{Aut}(\mathbb{P}^1)$$

is holomorphic and is a principal $\mathrm{Aut}(\mathbb{P}^1)$-bundle.

3.5. Two examples of Hol(V,V).

In this section, we compute Hol(V,V) for two simple V, mainly for their own interest.

(a). Hol(V,V), where V is a complex 1-torus.

This example is familiar to number theorists. Let ω be a complex number in the upper half plane $\mathbb{H}$. Put

$$V = V_\omega = \mathbb{C}/\Delta ,$$

$$\Delta = \mathbb{Z} + \mathbb{Z}\omega ,$$

$$\pi : \mathbb{C} \longrightarrow V_\omega, \quad \text{the projection} .$$

As in §1.4, for any point $y \in V_\omega$, we denote by t_y, the translation

$$t_y : z \in V_\omega \longmapsto z + y \in V_\omega .$$

It is an automorphism of V_ω. Let f be a holomorphic map of V_ω into itself. We put $y_o = f(0)$. Then the composition $t_{(-y_o)}f$ maps 0 to 0. Since $\mathbb{C}$ is simply connected, there is a holomorphic map $\tilde{f} : \mathbb{C} \longrightarrow \mathbb{C}$ such that $\tilde{f}(0) = 0$ and the diagram:

$$\begin{array}{ccc} \mathbb{C} & \xrightarrow{\tilde{f}} & \mathbb{C} \\ \pi \downarrow & & \downarrow \pi \\ V_\omega & \xrightarrow[t_{(-y_o)}f]{} & V_\omega \end{array}$$

commutes. Then, $(d\tilde{f}/dz)$ is regarded as a holomorphic function on V_ω. Hence it is a constant λ, i.e.,

$$\tilde{f} : z \in \mathbb{C} \longmapsto \lambda z \in \mathbb{C}$$

The complex number λ is uniquely determined by f. Put $s_\lambda = t_{(-y_o)}f$.

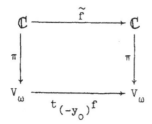

Then $f = t_{y_o} s_\lambda$. λ must satisfy the condition:

(1) $$\lambda\Delta \subset \Delta .$$

Converely, if $\lambda \in \mathbb{C}$ satisfies (1), then it is clear that the map

$$\tilde{s}_\lambda : z \in \mathbb{C} \longmapsto \lambda z \in \mathbb{C}$$

induces a holomorphic map $s_\lambda : V_\omega \longrightarrow V_\omega$ such that $s_\lambda(0) = 0$.

Now, (1) implies that there are integers p, q, r, s such that

$$\lambda = p + q\omega ,$$

$$\lambda\omega = r + s\omega .$$

Hence

(2) $$q\omega^2 + (p-s)\omega - r = 0 .$$

<u>Case 1</u>. When ω is not a quadratic imaginary.

In this case, we have $p = s$, $q = r = 0$ and $\lambda = p$. It is clear that, for $p > 0$, $f_\lambda = f_p$ is an unramified covering map of order p^2. Thus we conclude

$$Hol(V_\omega,V_\omega) = Const \cup Aut(V_\omega) \cup S_4(V_\omega,V_\omega) \cup \cdots \cup S_{n^2}(V_\omega,V_\omega) \cup \cdots ,$$

where

$$Const = \{all\ constant\ maps\} \cong V_\omega ,$$

$$Aut(V_\omega) = Aut_o(V_\omega) \cup Aut_o(V_\omega)s_{-1} ,$$

$$Aut_o(V_\omega) = \{t_y \mid y \in V_\omega\} \cong V_\omega ,$$

$$S_{n^2}(V_\omega,V_\omega) = Aut(V_\omega)s_n = Aut_o(V_\omega)s_n \cup Aut_o(V_\omega)s_{-n} .$$

(The unions are all disjoint unions.) Note that $S_{n^2}(V_\omega,V_\omega)$ is the set of all unramified covering maps of V_ω onto itself of order n^2.

Note also that $t_y s_n = s_n t_{y/n}$, so that

$$\mathrm{Aut}_o(V_\omega)s_n = s_n \mathrm{Aut}_o(V_\omega) .$$

($y/n = x \in V_\omega$ is a solution of the equation $nx = y$. n^2 solutions exist.)

<u>Case 2</u>. When ω is a quadratic imaginary.

In this case, there are uniquely determined integers $a, b, c \in$ with $a > 0$, $(a,b,c) = 1$ (relatively prime) and $b^2 - 4ac < 0$ such that

$$a\omega^2 + b\omega + c = 0 .$$

Then (2) implies that there is an integer m such that

$$q = ma, \quad p-s = mb, \quad -r = mc.$$

Hence, q, r and s are expressed by p and m as

$$q = ma, \quad r = -mc, \quad s = p-mb.$$

Thus we conclude

$$\mathrm{Hol}(V_\omega, V_\omega) = \cup_{(p,m)\in \mathbb{Z}^2} \mathrm{Aut}_o(V_\omega)s_{p+ma\omega}, \quad \text{(disjoint union)} .$$

Note that

$$\mathrm{Aut}_o(V_\omega)s_{p+ma\omega} = s_{p+ma\omega}\mathrm{Aut}_o(V_\omega) ,$$

$$\mathrm{Const} \doteq \mathrm{Aut}_o(V_\omega)s_0 \cong V_\omega .$$

It is easy to see that, if $\lambda = p+q\omega = p+ma\omega$ is not zero, then s_λ is an unramified covering map of V_ω onto itself of order

$$ps - qr = p^2 - bpm + acm^2.$$

Since $D = b^2 - 4ac < 0$, it is well known that, for a given integer n, the quadratic Diophantus equation

(3) $$p^2 - bpm + acm^2 = n$$

has at most finite solutions (p,m). (It may have no solution.) Put

$$S_n(V_\omega, V_\omega) = \{\text{all unramified covering maps of order } n \text{ of } V_\omega \text{ onto itself}\}.$$

We conclude

Proposition 3.5.1. For any positive integer n,

$$S_n(V_\omega, V_\omega) = \cup_{(p,m)} \mathrm{Aut}_o(V_\omega) s_{p+ma\omega},$$

where the union is a finite disjoint union extended over all solutions (p,m) of the Diophantus equation (3). ($S_n(V_\omega, V_\omega)$ is empty, if (3) has no solution.)

In particular, if we put $n = 1$, then

$$\mathrm{Aut}(V_\omega) = \cup_{(p,m)} \mathrm{Aut}_o(V_\omega) s_{p+ma\omega},$$

where the union is extended over all solutions (p,m) of the Diophantus equation:

(4) $$p^2 - bpm + acm^2 = 1.$$

We solve (4) as follows: (4) implies that

(5) $$\omega = \frac{r + s\omega}{p + q\omega}, \quad ps - rq = 1.$$

A modular transformation M is, by definition, an automorphism of the upper half plane $\mathbb{H}$ of the following type:

$$M(z) = \frac{r + sz}{p + qz}, \quad p, q, r, s \in \mathbb{Z}, \quad ps - rq = 1.$$

Two points $\omega, \omega' \in \mathbb{H}$ are said to be equivalent under modular transformations if there is a modular transformation M such that $M(\omega) = \omega'$. (In this case, M induces a biholomorphic map $f_{p+q\omega} : V_{\omega'} \longrightarrow V_\omega$.)

Now, it is well known that, if ω is neither equivalent to $\sqrt{-1}$ nor to $\zeta = \frac{1+\sqrt{-3}}{2}$, under modular transformations, then the relation (5) implies that either $p = s = 1$, $r = q = 0$ or $p = q = -1$, $r = q = 0$. Hence

<u>Proposition 3.5.2</u>. If $\omega \in \mathbb{H}$ is neither equivalent to $\sqrt{-1}$ nor to $\zeta = \frac{1+\sqrt{-3}}{2}$ under modular transformations, then

$$\mathrm{Aut}(V_\omega) = \mathrm{Aut}_o(V_\omega) \cup \mathrm{Aut}_o(V_\omega)s_{-1} .$$

If $\omega = \sqrt{-1}$ or $\omega = \zeta$, then we can solve (4) directly and get

<u>Proposition 3.5.3</u>. Put $i = \sqrt{-1}$ and $\zeta = \frac{1+\sqrt{-3}}{2}$. Then

$$\mathrm{Aut}(V_i) = \bigcup_{k=0}^{3} \mathrm{Aut}_o(V_i)s_{i^k}$$

$$\mathrm{Aut}(V_\zeta) = \bigcup_{k=0}^{5} \mathrm{Aut}_o(V_\zeta)s_{\zeta^k} .$$

<u>(b). Hol(V,V), where V is a (primary) Hopf surface</u>.

Among various Hopf surfaces, we consider the simplest one. Put $X = \mathbb{C}^2 - 0$, $0 = (0,0)$. A non-zero complex number β defines an automorphism of X as follows:

$$\beta : (z,w) \in X \longmapsto (\beta z, \beta w) \in X .$$

For a complex number α with $0 < |\alpha| < 1$, the group $G = \{\alpha^n \mid n \in \mathbb{Z}\}$ of automorphisms of X is a properly discontinuous group acting without fixed point. Hence we get a complex manifold of dimension 2:

$$V = V_\alpha = X/G_\alpha .$$

V_α is compact. In fact, it is diffeomorphic to $S^1 \times S^3$. It is called a <u>(primary) Hopf surface</u>. (In general, a <u>Hopf surface</u> is a 2-dimensional compact complex manifold whose universal covering space is X.)

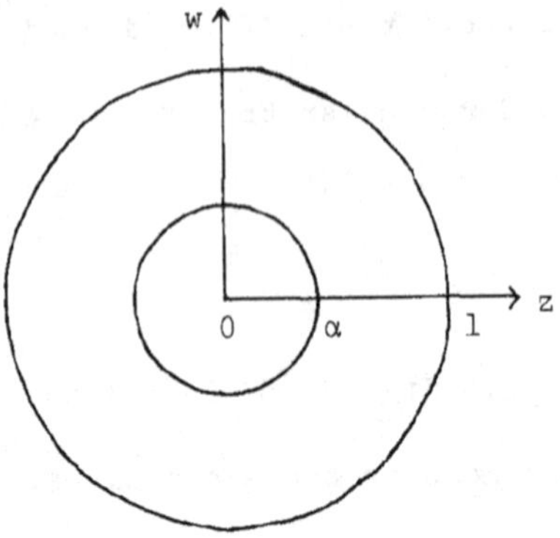

Let

$$\pi : X \longrightarrow V_\alpha$$

be the projection map.

Now, let $f : V_\alpha \longrightarrow V_\alpha$ be a holomorphic map. Then there is a holomorphic map $\tilde{f} : X \longrightarrow X$ such that the diagram:

$$\begin{array}{ccc} X & \xrightarrow{\tilde{f}} & X \\ {\scriptstyle\pi}\downarrow & & \downarrow{\scriptstyle\pi} \\ V_\alpha & \xrightarrow{f} & V_\alpha \end{array}$$

commutes. Hence

$$\pi(\tilde{f}(\alpha z, \alpha w)) = \pi(\tilde{f}(z,w)) \quad \text{for} \quad (z,w) \in X .$$

This means that there is an integer $k = k(z,w)$ such that

(1) $$\tilde{f}(\alpha z, \alpha w) = \alpha^k \tilde{f}(z,w) \quad \text{for} \quad (z,w) \in X .$$

Lemma 3.5.4. $k = k(z,w)$ is independent of $(z,w) \in X$.

Proof. $\tilde{f}$ is written as

(2) $$\tilde{f}(z,w) = (g(z,w), h(z,w)) \quad \text{for} \quad (z,w) \in X .$$

Then (1) is written as

$$(3) \qquad \begin{cases} g(\alpha z, \alpha w) = \alpha^k g(z,w) , \\ h(\alpha z, \alpha w) = \alpha^k h(z,w) , \end{cases}$$

for $(z,w) \in X$. Hence we have

$$k \log|\alpha| = \log(|g(\alpha z,\alpha w)| + |h(\alpha z,\alpha w)|) - \log(|g(z,w)| + |h(z,w)|) .$$

The right hand side is a continuous function of $(z,w) \in X$, while k is an integer. Hence k is independent of $(z,w) \in X$. <u>Q.E.D</u>.

By Hartogs' theorem (see, e.g., Gunning-Rossi [29]), the map $\tilde{f}$ is extended uniquely to a holomorphic map

$$\tilde{f} : \mathbb{C}^2 \longrightarrow \mathbb{C}^2 .$$

<u>Lemma 3.5.5</u>. The integer k in (1) is a non-negative integer. If $k = 0$, then f is a constant map. If $k > 0$, then f is not a constant map and $\tilde{f}(0) = 0$.

<u>Proof</u>. Assume that k is a negative integer. We have

$$\alpha^{-nk}\tilde{f}(\alpha^n z, \alpha^n w) = \tilde{f}(z,w) \quad \text{for} \quad (z,w) \in X \quad \text{and} \quad n \in \mathbb{Z} .$$

Fix $(z,w) \in X$. Then $f(z,w) \in X$. The left hand side converges to 0 as $n \longrightarrow +\infty$, for $\tilde{f}(\alpha^n z, \alpha^n w)$ converges to $\tilde{f}(0)$ as $n \longrightarrow +\infty$. This is a contradiction.

Assume that $k = 0$. If we write $\tilde{f}$ as in (2), then (3) implies that g and h induce holomorphic functions on compact V_α, so that they are constant. Hence f is a constant map.

Assume that $k > 0$. We have

$$\tilde{f}(\alpha^n z, \alpha^n w) = \alpha^{nk}\tilde{f}(z,w) \quad \text{for} \quad (z,w) \in X \quad \text{and} \quad n \in \mathbb{Z}.$$

Fix $(z,w) \in X$. Then the left hand side converges to $\tilde{f}(0)$ as $n \longrightarrow +\infty$, while the right hand side converges to 0. Hence $\tilde{f}(0) = 0$. In particular, $\tilde{f}$ is not a constant map. Hence f is not a constant map either. Q.E.D.

Lemma 3.5.6. The integer k in (1) does not depend on the choice of $\tilde{f}$.

Proof. Let $\tilde{\tilde{f}}$ be another holomorphic map of X into X such that $\pi\tilde{\tilde{f}} = \tilde{f}\pi$. Then

$$\pi(\tilde{\tilde{f}}(z,w)) = \pi(\tilde{f}(z,w)) \quad \text{for} \quad (z,w) \in X .$$

This means that there is an integer $m = m(z,w)$ such that

$$\tilde{\tilde{f}}(z,w) = \alpha^m \tilde{f}(z,w) .$$

A similar argument to the proof of Lemma 3.5.4 shows that m is independent of $(z,w) \in X$. Thus

$$\tilde{\tilde{f}}(\alpha z,\alpha w) = \alpha^m \tilde{f}(\alpha z,\alpha w) = \alpha^m \alpha^k \tilde{f}(z,w) = \alpha^k \tilde{\tilde{f}}(z,w) \quad \text{for} \quad (z,w) \in X .$$

Hence k does not depend on the choice of $\tilde{f}$. Q.E.D.

Now, assume that k is a positive integer. We expand the functions g and h in (2) in the power series of z and w at the origin:

$$g(z,w) = \sum_{p+q>0} c_{pq} z^p w^q ,$$

$$h(z,w) = \sum_{p+q>0} d_{pq} z^p w^q .$$

Then the condition (3) is written as

$$\sum_{p+q>0} \alpha^{p+q} c_{pq} z^p w^q = \sum_{p+q>0} \alpha^k c_{pq} z^p w^q ,$$

$$\sum_{p+q>0} \alpha^{p+q} d_{pq} z^p w^q = \sum_{p+q>0} \alpha^k d_{pq} z^p w^q .$$

Since $0 < |\alpha| < 1$, we have

$$c_{pq} = d_{pq} = 0, \quad \text{unless} \quad p+q = k\,.$$

Hence

$$g(z,w) = \sum_{p+q=k} c_{pq} z^p w^q\,,$$

$$h(z,w) = \sum_{p+q=k} d_{pq} z^p w^q\,.$$

Since $\tilde{f}$ maps $X = \mathbb{C}^2 - 0$ into X, the homogeneous polynomials g and h have no common non-constant factor. Hence, the resultant $R(g,h)$ is not zero.

Conversely, it is easy to see that, for homogeneous polynomials g and h of order k, if $R(g,h) \neq 0$, then $\tilde{f} = (g,h)$ induces a non-constant holomorphic map f of V_α into itself.

<u>Lemma 3.5.7</u>. If the integer k in (1) is positive, then f is a finite map of V_α onto itself of mapping order k^3.

<u>Proof</u>. First, we show that f is surjective. It suffices to show that $\tilde{f}$ is surjective. Since $R(g,h) \neq 0$, $\tilde{f}$ induces a surjective holomorphic map

$$(4) \qquad \hat{f} : (z:w) \in \mathbb{P}^1 \longrightarrow (g(z,w):h(z,w)) \in \mathbb{P}^1$$

of mapping order k, which makes the diagram

$$\begin{array}{ccc} X & \xrightarrow{\tilde{f}} & X \\ \downarrow{\scriptstyle\psi} & & \downarrow{\scriptstyle\psi} \\ \mathbb{P}^1 & \xrightarrow{\hat{f}} & \mathbb{P}^1 \end{array}$$

commutative, where ψ is the canonical projection. For a given point $(z,w) \in X$, let $(z',w') \in X$ be such that $\hat{f}\psi(z',w') = \psi(z,w)$.

Then $\psi(\tilde{f}(z',w')) = \psi(z,w)$, so that there is a non-zero complex number λ such that $\tilde{f}(z',w') = (\lambda z, \lambda w)$. Let ξ be a complex number such that $\xi^k = \lambda$. Then

$$\tilde{f}(\xi^{-1}z', \xi^{-1}w') = \xi^{-k}\tilde{f}(z',w') = \lambda^{-1}(\lambda z, \lambda w) = (z,w) .$$

Hence $\tilde{f}$ is surjective.

For a general point $(z,w) \in X$, there are k distinct points $P_1, \cdots, P_k \in \mathbb{P}^1$ such that

$$\hat{f}(P_i) = \psi(z,w), \quad i = 1, \cdots, k .$$

The above argument shows that there is a point $(z_i, w_i) \in X$ such that

$$\psi(z_i, w_i) = P_i ,$$

$$\tilde{f}(z_i, w_i) = (z,w)$$

for $i = 1, \cdots, k$. Since $P_i \neq P_j$ for $i \neq j$, we have

$$\pi(z_i, w_i) \neq \pi(z_j, w_j) \quad \text{for} \quad i \neq j .$$

Now, if $(z',w') \in X$ satisfies $f\pi(z',w') = \pi(z,w)$, then $\pi\tilde{f}(z',w') = \pi(z,w)$. Hence

$$\hat{f}\psi(z',w') = \psi(\tilde{f}(z',w')) = \psi(z,w) ,$$

so that $\psi(z',w') = P_i$ for some i. Thus there is a non-zero complex number λ such that $(z',w') = \lambda(z_i, w_i)$. Then

$$\tilde{f}(z',w') = \lambda^k\tilde{f}(z_i, w_i) = \lambda^k(z,w) .$$

On the other hand, $\pi\tilde{f}(z',w') = \pi(z,w)$ implies that there is an integer r such that $\tilde{f}(z',w') = \alpha^r(z,w)$. Hence, we have $\lambda^k = \alpha^r$. Thus λ must be one of the following numbers:

$$\beta^r \exp(2\pi\sqrt{-1}s/k), \quad r,s \in \mathbb{Z} ,$$

where $\beta \in \mathbb{C}$ satisfies $\beta^k = \alpha$. It is easy to see that

$$\pi(\beta^r \exp(2\pi\sqrt{-1}s/k)z_i, \beta^r \exp(2\pi\sqrt{-1}s/k)w_i) = \pi(z_i, w_i)$$

if and only if $r \equiv s \equiv 0 \pmod{k}$. Hence, we conclude that

$$f^{-1}(\pi(z,w)) = \{\pi(\beta^r \exp(2\pi\sqrt{-1}s/k)z_i, \beta^r \exp(2\pi\sqrt{-1}s/k)w_i) \mid r,s = 0, 1, \cdots, k-1, \ i = 1, \cdots, k\}.$$ Q.E.D.

Now, we put

$$\widetilde{M}_k = \{(c_0, c_1, \cdots, c_k, d_0, d_1, \cdots, d_k) \in \mathbb{C}^{2k+2} \mid R(g,h) \neq 0\},$$

where $R(g,h)$ is the resultant of the homogeneous polynomials:

$$g = c_0 w^k + c_1 z w^{k-1} + \cdots + c_k z^k,$$

$$h = d_0 w^k + d_1 z w^{k-1} + \cdots + d_k z^k.$$

Then $\widetilde{M}_k$ is a complex manifold of dimension $2k+2$. A non-zero complex number γ defines an automorphism of $\widetilde{M}_k$ as follows:

$$\gamma : (c_0, \cdots, c_k, d_0, \cdots, d_k) \longmapsto (\gamma c_0, \cdots, \gamma c_k, \gamma d_0, \cdots, \gamma d_k).$$

Then the group $G_\alpha = \{\alpha^n \mid n \in \mathbb{Z}\}$ of automorphisms of $\widetilde{M}_k$ is a properly discontinuous group acting without fixed point. Hence, we have a complex manifold

$$M_k = \widetilde{M}_k / G_\alpha.$$

Now, put

$$S_n(V_\alpha, V_\alpha) = \{\text{all finite holomorphic maps of } V_\alpha \text{ onto itself of mapping order } n\}.$$

Then, we conclude that

$$\mathrm{Hol}(V_\alpha, V_\alpha) = \mathrm{Const} \cup \mathrm{Aut}(V_\alpha) \cup S_8(V_\alpha, V_\alpha) \cup S_{27}(V_\alpha, V_\alpha) \cup S_{64}(V_\alpha, V_\alpha) \cdots$$

(disjoint union), where

$$\text{Const} = \{\text{constant maps}\} \cong V_\alpha ,$$

$$\text{Aut}(V_\alpha) = S_1(V_\alpha,V_\alpha) \cong GL(2,\mathbb{C})/G_\alpha ,$$

$$S_k;(V_\alpha,V_\alpha) \cong M_k, \quad k = 1, 2, \cdots .$$

Note that $\widetilde{M}_k/\mathbb{C}^*$ is biholomorphic to $R_k(\mathbb{P}^1) = \{$all rational functions of order $k\}$. Hence a holomorphic map $\Psi : S_k;(V_\alpha,V_\alpha) \longrightarrow R_k(\mathbb{P}^1)$ is induced. In fact, Ψ maps f to $\hat{f}$ (see (4)). Thus we have a holomorphic map

$$\Psi : \text{Hol}(V_\alpha,V_\alpha) \longrightarrow \text{Hol}(\mathbb{P}^1, \mathbb{P}^1) .$$

This is a principal $(\mathbb{C}^*/G_\alpha)$-bundle and is a "homomorphism" in the sense that

$$\Psi(f_1f_2) = \Psi(f_1)\Psi(f_2) \quad \text{for} \quad f_1, f_2 \in \text{Hol}(V_\alpha,V_\alpha) .$$

(f_1f_2 is the composition of f_2 and f_1.)

<u>Definition 3.5.8</u>. A semigroup S is said to be a <u>complex</u> (resp. <u>real</u>) <u>Lie semigroup</u> if (1) S is a complex (resp. real analytic) space and (2) the product $(a,b) \in S \times S \longmapsto ab \in S$ is a holomorphic (resp. real analytic) map.

For a compact complex manifold (space) V, $\text{Hol}(V,V)$ is an example of complex Lie semigroups. For a connected complex (resp. real) Lie group G, the set $\text{End}(G)$ of all holomorphic (resp. real analytic) homomorphisms of G into itself forms a complex (resp. real) Lie semigroup.

Some properties of complex Lie semigroups were given in Namba [63].

3.6. Deformation theory of holomorphic maps.

There are several definitions for "equivalence" of holomorphic maps of compact complex manifolds. We state here 2 typical definitions:

(1) When V varies and W is fixed:

In this case, two holomorphic maps

$$f : V \longrightarrow W \quad \text{and} \quad g : V' \longrightarrow W$$

are equivalent if there is a biholomorphic map $a : V \longrightarrow V'$ such that the following diagram commutes:

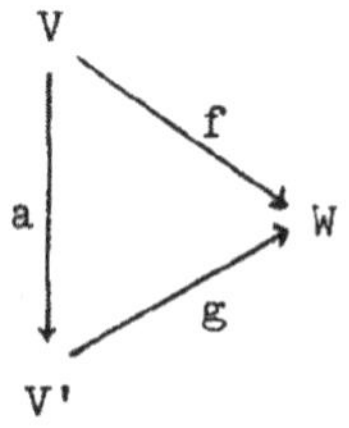

(2) When both V and W vary:

In this case, two holomorphic maps

$$f : V \longrightarrow W \quad \text{and} \quad g : V' \longrightarrow W'$$

are equivalent if there are biholomorphic maps $a : V \longrightarrow V'$ and $b : W \longrightarrow W'$ such that the following diagram commutes:

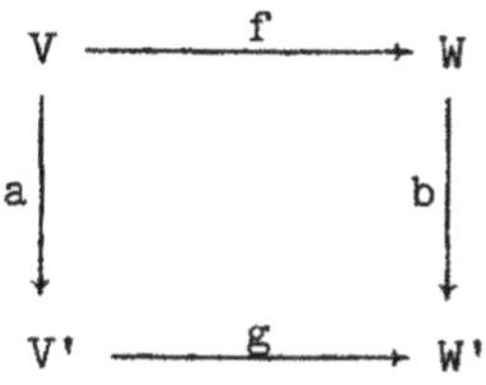

Corresponding to these cases, it is possible to develop deformation theories of holomorphic maps including Kuranishi-type theorems.

We first discuss the deformation theory corresponding to the case

(1) developped by Horikawa [32], Miyajima [54] and Kouchiyama [47].

Let W be a compact complex manifold, fixed once and for all. (In fact, the compactness of W is unnecessary in this case. But we assume it for simplicity.) A *family of holomorphic maps into* W is, by definition, a family $(X,\pi,S) = \{V_s\}_{s\in S}$ of compact complex manifolds, together with a holomorphic map $\mathfrak{F} : X \longrightarrow W$. We denote it by $(X,\pi,S,\mathfrak{F})$. We put $f_s = \mathfrak{F}|V_s : V_s \longrightarrow W$ and sometimes denote this family by $\{V_s, f_s\}_{s\in S}$.

A *morphism of* $(X',\pi',S',\mathfrak{F}') = \{V'_{s'}, f'_{s'}\}_{s'\in S'}$ *to* $(X,\pi,S,\mathfrak{F}) = \{V_s, f_s\}_{s\in S}$ is, by definition, a morphism $(h,\tilde{h})$ of (X',π',S') to (X,π,S) (see §0.1) which makes the diagram:

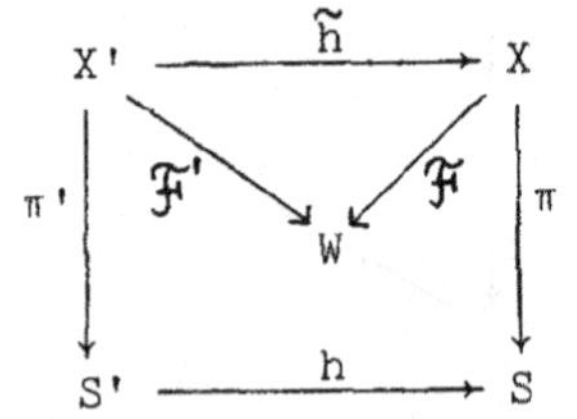

commutative. Thus, we can define an *isomorphism of families*.

A family $\{V_s, f_s\}_{s\in S}$ is said to be *complete at* $o \in S$ if, for any family $\{V'_{s'}, f'_{s'}\}_{s'\in S'}$ with a point $o' \in S'$ and a biholomorphic map $i : V'_{o'} \longrightarrow V_o$ which makes the diagram:

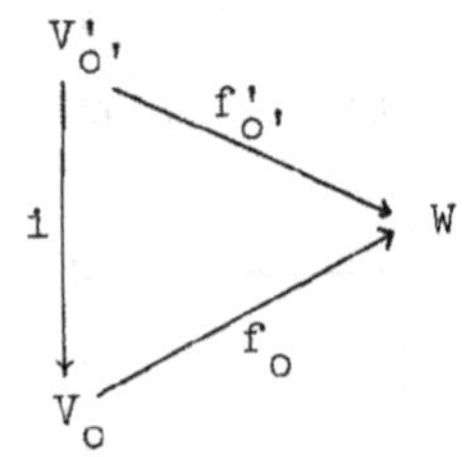

commutative, there are an open neighborhood U' of o' in S' and a morphism $(h,\tilde{h})$ of $\{V'_{s'}, f'_{s'}\}_{s'\in U'}$ to $\{V_s, f_s\}_{s\in S}$ such that (1) $h(o') = o$ and (2) $\tilde{h}_{o'} = i : V'_{o'} \longrightarrow V_o$.

$\{V_s, f_s\}_{s\in S}$ is said to be *complete* if it is complete at every point of S.

$\{V_s, f_s\}_{s\in S}$ is said to be <u>versal at</u> $o \in S$ if (1) it is complete at $o \in S$ and (2) the differential $(dh)_{o'}$ of the above map h at o' is uniquely determined, i.e., if $(h_1, \tilde{h}_1)$ is another such morphism, then $(dh)_{o'} = (dh_1)_{o'}$.

The existence of complete families is easy to show:

<u>Proposition 3.6.1</u>. For any holomorphic map $f : V \longrightarrow W$, there is a complete family $\{V_s, f_s\}_{s\in S}$ of holomorphic maps into W with a point $o \in S$ such that (1) $V_o = V$ and (2) $f_o = f$.

<u>Proof</u>. Let $\{V_t\}_{t\in T}$ be the Kuranishi family of V. Then, it is easy to see that the maximal family $\{f_s\}_{s\in S}$ of holomorphic maps of $\{V_t\}_{t\in T}$ into the trivial family $\{W\}_{t\in T}$ constructed in Theorem 3.2.3 with respect to $f : V = V_o \longrightarrow W$ is a complete family. <u>Q.E.D</u>.

The family constructed here is too big. We want to get the "smallest" complete families, i.e., versal families.

We define an infinitesimal deformation of a family $(X, \pi, S, \mathfrak{F}) = \{V_s, f_s\}_{s\in S}$ at $o \in S$. Put $f = f_o$. We use the same notations as in §0.1 and §3.1. $\mathfrak{F}$ is locally expressed by the equations:

$$w_i = f_i(z_i, s) \quad \text{for} \quad (z_i, s) \in U_i \times S .$$

They satisfy the compatibility conditions:

$$h_{ik}(f_k(z_k, s)) = f_i(g_{ik}(z_k, s), s) ,$$

where h_{ik} are the coordinate transformations in W. Hence

$$\left(\frac{\partial h_{ik}}{\partial w_k}\right)_{f_k(z_k,o)} \left(\frac{\partial f_k}{\partial s}\right)_{(z_k,o)} = \left(\frac{\partial f_i}{\partial s}\right)_{(z_i,o)} + \left(\frac{\partial f_i}{\partial z_i}\right)_{(z_i,o)} \left(\frac{\partial g_{ik}}{\partial s}\right)_{(z_k,o)}$$

$(z_i = g_{ik}(z_k, o).)$ This means that

$$\delta\eta = f_*\theta ,$$

where

$$\theta = \{\theta_{ik}\} \in Z^1(\mathcal{U}, \Theta_V), \quad \theta_{ik} = \left(\frac{\partial g_{ik}}{\partial s}\right)(z_k, o),$$

$$\eta = \{\eta_i\} \in C^0(\mathcal{U}, f^*\Theta_W), \quad \eta_i = \left(\frac{\partial f_i}{\partial s}\right)(z_i, o),$$

($\mathcal{U} = \{U_i\}_{i\in I}$.) Thus the pair (θ,η) is an element of the <u>hypercohomology group</u> $\check{\mathbb{H}}^1(\mathcal{U},\mathcal{L}) \cong \mathbb{H}^1(V,\mathcal{L})$ of the complex:

$$\mathcal{L} : 0 \longrightarrow \mathcal{L}^0 = \Theta_V \xrightarrow{f_*} \mathcal{L}^1 = f^*\Theta_W \longrightarrow 0 \longrightarrow 0 \longrightarrow \cdots .$$

We can easily show that $(\theta,\eta) \in \mathbb{H}^1(V,\mathcal{L})$ does not depend on the choice of the Stein covering $\{U_i\}$, etc.. We call it the <u>infinitesimal deformation of the family</u> $\{V_s, f_s\}_{s\in S}$ <u>at</u> $o \in S$ <u>to the direction</u> $\frac{\partial}{\partial s} \in T_oS$ and denote it by $\alpha_o(\frac{\partial}{\partial s})$. α_o is a linear map of T_oS into $\mathbb{H}^1(V,\mathcal{L})$, called the <u>characteristic map of the family at</u> o.

The family $\{V_s, f_s\}_{s\in S}$ is said to be <u>effectively parametrized at</u> $o \in S$ if α_o is injective. It is easy to see that, if the family is complete at $o \in S$ and effectively parametrized at o, then it is versal at o.

For the proof of the following proposition, see Horikawa [32] or Miyajima [54].

<u>Proposition 3.6.2</u>. There is the following exact sequence:

$$\begin{aligned} 0 &\longrightarrow \mathbb{H}^0(V,\mathcal{L}) \longrightarrow H^0(V,\Theta_V) \xrightarrow{f_*} H^0(V,f^*\Theta_W) \\ &\longrightarrow \mathbb{H}^1(V,\mathcal{L}) \longrightarrow H^1(V,\Theta_V) \xrightarrow{f_*} H^1(V,f^*\Theta_W) \\ &\longrightarrow \mathbb{H}^2(V,\mathcal{L}) \longrightarrow H^2(V,\Theta_V) \xrightarrow{f_*} H^2(V,f^*\Theta_W) \longrightarrow \cdots \end{aligned}$$

Now, we state

Theorem 3.6.3. (Miyajima [54]). For any holomorphic map $f : V \longrightarrow W$, there is a complete family $\{V_s, f_s\}_{s \in S}$ of holomorphic maps into W with a point $o \in S$ such that (1) $V_o = V$, (2) $f_o = f$ and (3) it is effectively parametrized at o, (hence is versal at o). Moreover, the parameter space S is given as follows: There are an open neighborhood U of 0 in $\mathbb{H}^1(V, \mathcal{L})$ and a holomorphic map $m : U \longrightarrow \mathbb{H}^2(V, \mathcal{L})$ such that (4) $S = \{\xi \in U \mid m(\xi) = 0\}$, (5) $o = 0$, (6) $(dm)_o = 0$ and (7) the characteristic map α_o is equal to $(di)_o$, where $i : S \hookrightarrow U$ is the inclusion map.

By Proposition 3.6.2,

Corollary 3.6.4. (Horikawa [32]). For a holomorphic map $f : V \longrightarrow W$, assume that

(a) $H^1(V, \Theta_V) \xrightarrow{f_*} H^1(V, f^* \Theta_W)$ is surjective,

(b) $H^2(V, \Theta_V) \xrightarrow{f_*} H^2(V, f^* \Theta_W)$ is injective.

Then, there is a complete family $\{V_s, f_s\}_{s \in S}$ of holomorphic maps into W with a point $o \in S$ such that (1) $V_o = V$, (2) $f_o = f$, (3) it is effectively parametrized at o and (4) o is a non-singular point of S and $\dim_o S = \dim \mathbb{H}^1(V, \mathcal{L})$.

Remark 3.6.5. Thus, Miyajima's theorem corresponds completely to Kuranishi's theorem (Theorem 0.1.3). His proof is also an analogy of Kuranishi [48].

Remark 3.6.6. In his master's thesis, N. Kouchiyama [47] proved the existence of versal families of holomorphic maps into W using a method similar to Commichau [10]. This method does not (to my eyes) give a concrete information of the parameter space S as in Miyajima's theorem. It does not therefore imply Horikawa's result (Corollary 3.6.4) as its corollary. However, his method is expected to be extended to the case of holomorphic maps of compact complex <u>spaces</u>, just like the relation between Commichau [10] and Grauert [24].

Now, we talk about deformation theory corresponding to the case (2). Many things go parallel to the case (1). But we state them.

A <u>family of holomorphic maps</u> is, by definition, two families $(X,\pi,S) = \{V_s\}_{s\in S}$ and $(Y,\mu,S) = \{W_s\}_{s\in S}$ of compact complex manifolds with the same parameter space S, together with a holomorphic map $\mathfrak{F} : X \longrightarrow Y$ which makes the diagram:

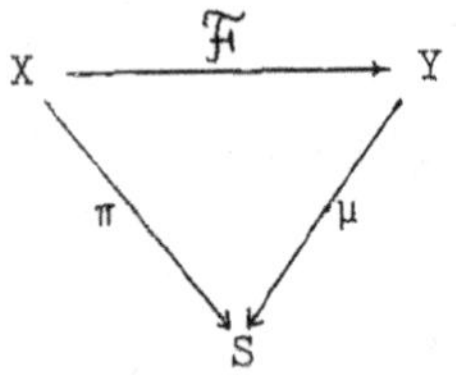

commutative. We denote it by $(X,\pi,Y,\mu,S,\mathfrak{F})$. We put $f_s = \mathfrak{F} | V_s : V_s \longrightarrow W_s$ and sometimes denote this family by $\{V_s,f_s,W_s\}_{s\in S}$.

A <u>morphism of</u> $(X',\pi',Y',\mu',S',\mathfrak{F}') = \{V'_{s'},f'_{s'},W'_{s'}\}_{s'\in S'}$ <u>to</u> $(X,\pi,Y,\mu,S,\mathfrak{F}) = \{V_s,f_s,W_s\}_{s\in S}$ is, by definition, a triple $(h,\tilde{h},\tilde{\tilde{h}})$, where $(h,\tilde{h})$ is a morphism of $\{V'_{s'}\}_{s'\in S'}$ to $\{V_s\}_{s\in S}$ and $(h,\tilde{\tilde{h}})$ is a morphism of $\{W'_{s'}\}_{s'\in S'}$ to $\{W_s\}_{s\in S}$ which make the diagram

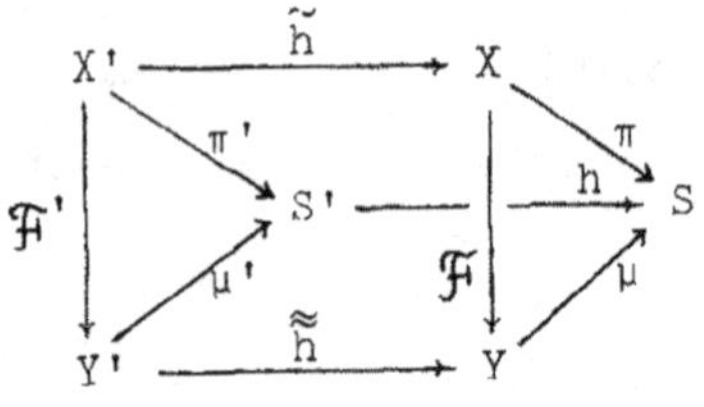

commutative. Thus, we can define an <u>isomorphism of families</u>.

A family $\{V_s,f_s,W_s\}_{s\in S}$ is said to be <u>complete at</u> $o\in S$ if, for any family $\{V'_{s'},f'_{s'},W'_{s'}\}_{s'\in S'}$ with a point $o'\in S'$ and biholomorphic maps $i : V'_{o'} \longrightarrow V_o$ and $j : W'_{o'} \longrightarrow W_o$ which make the diagram:

$$\begin{array}{ccc} V'_{o'} & \xrightarrow{\ i\ } & V_o \\ {\scriptstyle f'_{o'}}\downarrow & & \downarrow{\scriptstyle f_o} \\ W'_{o'} & \xrightarrow{\ j\ } & W_o \end{array}$$

commutative, there are an open neighborhood U' of o' is S' and a morphism $(h,\tilde{h},\approx{h})$ of $\{V'_{s'},f'_{s'},W'_{s'}\}_{s'\in U'}$ to $\{V_s,f_s,W_s\}_{s\in S}$ such that (1) $h(o') = o$, (2) $\tilde{h}_{o'} = i$ and (3) $\approx{h}_{o'} = j$.

$\{V_s,f_s,W_s\}_{s\in S}$ is said to be complete if it is complete at every point of S.

$\{V_s,f_s,W_s\}_{s\in S}$ is said to be versal at $o\in S$ if (1) it is complete at o and (2) the differential $(dh)_{o'}$ of the above map h at o' is uniquely determined.

Proposition 3.6.7. For any holomorphic map $f : V \longrightarrow W$, there is a complete family $\{V_s,f_s,W_s\}_{s\in S}$ with a point $o\in S$ such that (1) $V_o = V$, (2) $W_o = W$ and (3) $f_o = f$.

Proof. Let $\{V_t\}_{t\in T}$ and $\{W_u\}_{u\in U}$ be the Kuranishi families of V and W, respectively. Put $S = T\times U$. For $s = (t,u)\in S$, put $V_s = V_t$ and $W_s = W_u$. Then $\{V_s\}_{s\in S}$ and $\{W_s\}_{s\in S}$ are complete families of compact complex manifolds. Let $(\mathfrak{F},R,b) = \{f_r\}_{r\in R}$ be the maximal family of holomorphic maps of $\{V_s\}_{s\in S}$ into $\{W_s\}_{s\in S}$ constructed in Theorem 3.2.3 with respect to f. Then $\{V_{b(r)},f_r,W_{b(r)}\}_{r\in R}$ is a complete family which satisfies the requirement. Q.E.D.

Let $(X,\pi,Y,\mu,S,\mathfrak{F}) = \{V_s,f_s,W_s\}_{s\in S}$ be a family of holomorphic maps. For $o\in S$, put $V_o = V$, $W_o = W$ and $f_o = f$. We shrink S if necessary. We use the same notations as in §3.1. $\{Y_i\}_{i\in I}$ covers $f(V)$ $(\subset W)$. Let $A = \{\alpha\}$ be a set of indices with $I\cap A$ = empty. Let $\{Y_\alpha\}_{\alpha\in A}$ be a finite collection of open subsets of Y such that

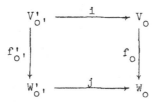

(1) $\overline{Y}_\alpha \cap f(V)$ = empty, for all $\alpha \in A$ and (2) $\{Y_i\} \cup \{Y_\alpha\}$ covers W. (A is empty if f is surjective.) Put $W_\alpha = W \cap Y_\alpha$. We may assume that each W_α is Stein and there is a holomorphic isomorphism $\zeta_\alpha : Y_\alpha \longrightarrow W_\alpha \times S$ such that the diagram

$$\begin{array}{ccc} Y_\alpha & \xrightarrow{\ \zeta_\alpha\ } & W_\alpha \times S \\ & \mu \searrow \quad \swarrow \text{proj} & \\ & S & \end{array}$$

commutes. We may assume that the map $\mathcal{F}$ satisfies $\mathcal{F}(X_i) \Subset Y_i$. Then, it is expressed by the equations:

$$w_i = f_i(z_i, s) \quad \text{for} \quad (z_i, s) \in U_i \times S .$$

They satisfy the comatibility conditions:

$$h_{ik}(f_k(z_k, s), s) = f_i(g_{ik}(z_k, s), s) .$$

Hence

$$\left(\frac{\partial h_{ik}}{\partial w_k}\right)(f_k(z_k,o),o)\left(\frac{\partial f_k}{\partial s}\right)(z_k,o) + \left(\frac{\partial h_{ik}}{\partial s}\right)(f_k(z_k,o),o)$$

$$= \left(\frac{\partial f_i}{\partial z_i}\right)(z_i,o)\left(\frac{\partial g_{ik}}{\partial s}\right)(z_k,o) + \left(\frac{\partial f_i}{\partial s}\right)(z_i,o)$$

$(z_i = g_{ik}(z_k, o))$. This means that

$$(*) \qquad \delta\eta = f_*\theta_1 - f^*\theta_2 ,$$

where

$$\theta_1 = \{\theta^1_{ik}\}_{i,k\in I} \in Z^1(\mathcal{U}, \Theta_V), \quad \theta^1_{ik} = \left(\frac{\partial g_{ik}}{\partial s}\right)(z_k, o) ,$$

$$\theta_2 = \{\theta^2_{ik}\}_{i,k\in I\cup A} \in Z^1(\mathcal{W}, \Theta_W), \quad \theta^2_{ik} = \left(\frac{\partial h_{ik}}{\partial s}\right)(w_k, o) ,$$

$$\eta = \{\eta_i\}_{i\in I} \in C^0(\mathcal{U}, f^*\Theta_W), \quad \eta_i = \left(\frac{\partial f_i}{\partial s}\right)(z_i, o) .$$

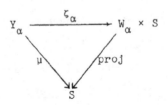

Here, we put $\mathcal{U} = \{U_i\}_{i\in I}$ and $\mathcal{W} = \{W_i\}_{i\in I} \cup \{W_\alpha\}_{\alpha\in A}$.

Now, for any integer $p \geq 0$, we put

$$C^p = C^p(\mathcal{U}, \mathcal{W}, \Theta_V, \Theta_W, f^*\Theta_W)$$

$$= C^p(\mathcal{U}, \Theta_V) \oplus C^p(\mathcal{W}, \Theta_W) \oplus C^{p-1}(\mathcal{U}, f^*\Theta_W) .$$

$(C^{-1}(\mathcal{U}, f^*\Theta_W) = 0.)$ We define a linear map

$$\Delta^p : C^p \longrightarrow C^{p+1}$$

by

$$\Delta^p(\xi, \zeta, \eta) = (\delta^p\xi, \delta^p\zeta, \delta^{p-1}\eta + (-1)^p(f_*\xi - f^*\zeta))$$

for $(\xi, \zeta, \eta) \in C^p$. Then, using $f_*\delta = \delta f_*$ and $f^*\delta = \delta f^*$, we get $\Delta^{p+1}\Delta^p = 0$. Thus we get a complex of $\mathbb{C}$-vector spaces. We denote its cohomology group by $\check{H}^p(\mathcal{U}, \mathcal{W}, \Theta_V, \Theta_W, f^*\Theta_W)$.

For $\xi \in Z^p(\mathcal{U}, \Theta_V)$, we denote by $\{\xi\}$ the cohomology class in $\check{H}^p(\mathcal{U}, \Theta_V)$ determined by ξ. Similar notations are used for $\zeta \in Z^p(\mathcal{W}, \Theta_W)$, etc..

It is easy to check that the following linear maps are well defined:

(1) $$\check{H}^p(\mathcal{U}, \Theta_V) \oplus \check{H}^p(\mathcal{W}, \Theta_W) \longrightarrow \check{H}^p(\mathcal{U}, f^*\Theta_W)$$

$$(\{\xi\}, \{\zeta\}) \longmapsto \{f_*\xi - f^*\zeta\} ,$$

(2) $$\check{H}^{p-1}(\mathcal{U}, f^*\Theta_W) \longrightarrow \check{H}^p(\mathcal{U}, \mathcal{W}, \Theta_V, \Theta_W, f^*\Theta_W)$$

$$\{\eta\} \longmapsto \{(0, 0, \eta)\} ,$$

(3) $$\check{H}^p(\mathcal{U}, \mathcal{W}, \Theta_V, \Theta_W, f^*\Theta_W) \longrightarrow \check{H}^p(\mathcal{U}, \Theta_V) \oplus \check{H}^p(\mathcal{W}, \Theta_W)$$

$$\{(\xi, \zeta, \eta)\} \longmapsto (\{\xi\}, \{\zeta\}) .$$

Then, the following lemma is easy to see.

<u>Lemma 3.6.8</u>. The following sequence is exact:

$$\cdots \longrightarrow \check{H}^{p-1}(\mathcal{U},\Theta_V) \oplus \check{H}^{p-1}(\mathcal{W},\Theta_W) \longrightarrow \check{H}^{p-1}(\mathcal{U},f^*\Theta_W) \longrightarrow$$

$$\check{H}^p(\mathcal{U},\mathcal{W},\Theta_V,\Theta_W,f^*\Theta_W) \longrightarrow \check{H}^p(\mathcal{U},\Theta_V)\oplus \check{H}^p(\mathcal{W},\Theta_W) \longrightarrow \check{H}^p(\mathcal{U},f^*\Theta_W) \longrightarrow \cdots .$$

Note that $\check{H}^p(\mathcal{U},\Theta_V)$, etc., does not depend on the choice of Stein coverings $\mathcal{U}$ and $\mathcal{W}$ and is isomorphic to $H^p(V,\Theta_V)$, etc.. Hence, by the five lemma, $\check{H}^p(\mathcal{U},\mathcal{W},\Theta_V,\Theta_W,f^*\Theta_W)$ does not depend on the choice of $\mathcal{U}$ and $\mathcal{W}$. We denote it by $H^p(\Theta_V,\Theta_W,f^*\Theta_W)$. Then, Lemma 3.6.8 is rewrittwn as

Proposition 3.6.9. The following sequence is exact:

$$0 \longrightarrow H^0(\Theta_V,\Theta_W,f^*\Theta_W) \longrightarrow H^0(V,\Theta_V)\oplus H^0(W,\Theta_W) \xrightarrow{f_*-f^*} H^0(V,f^*\Theta_W)$$

$$\longrightarrow H^1(\Theta_V,\Theta_W,f^*\Theta_W) \longrightarrow H^1(V,\Theta_V)\oplus H^1(W,\Theta_W) \xrightarrow{f_*-f^*} H^1(V,f^*\Theta_W)$$

$$\longrightarrow H^2(\Theta_V,\Theta_W,f^*\Theta_W) \longrightarrow H^2(V,\Theta_V)\oplus H^2(W,\Theta_W) \xrightarrow{f_*-f^*} H^2(V,f^*\Theta_W) \longrightarrow \cdots .$$

Returning back to (*), the triple (θ_1,θ_2,η) is an element of $Z^1(\mathcal{U},\mathcal{W},\Theta_V,\Theta_W,f^*\Theta_W)$ and hence gives an element of $H^1(\Theta_V,\Theta_W,f^*\Theta_W)$. We can easily show that this element is independent of the choice of the coverings $\mathcal{U}$ and $\mathcal{W}$. We denote it by $\beta_o(\frac{\partial}{\partial s})$ and call it the infinitesimal deformation of the family $\{V_s,f_s,W_s\}_{s\in S}$ at $o\in S$ to the direction $\frac{\partial}{\partial s}\in T_oS$. β_o is a linear map of T_oS into $H^1(\Theta_V,\Theta_W,f^*\Theta_W)$, called the characteristic map of the family at o.

The family $\{V_s,f_s,W_s\}_{s\in S}$ is said to be effectively parametrized at $o\in S$ if β_o is injective.

If $\{V_s,f_s,W_s\}_{s\in S}$ is complete at $o\in S$ and is effectively parametrized at o, then it is versal at o.

Now, we state

Theorem 3.6.10. For any holomorphic map $f : V \longrightarrow W$, there is a complete family $\{V_s,f_s,W_s\}_{s\in S}$ of holomorphic maps with a point $o\in S$

such that (1) $V_o = V$, (2) $W_o = W$, (3) $f_o = f$ and (4) it is effectively parametrized at o, (hence is versal at o).

Remark 3.6.11. In order to prove the theorem, I first tried to go parallel to Miyajima's argument. But, I had some difficulties to do so, (though I do not know if they are essential). Next, I tried to go parallel to Commichau [10] and Kouchiyama [47] and found it to get through. Hence, the parameter space S in the theorem is not so concrete as in Miyajima's Theorem 3.6.3. I can not therefore have a corollary corresponding to Corollary 3.6.4.

Now, we give a very short sketch of its proof. First note that

$$H^0(\Theta_V, \Theta_W, f^*\Theta_W) = \{(\xi,\zeta) \in H^0(V,\Theta_V) \oplus H^0(W,\Theta_W) \mid f_*\xi = f^*\zeta\}.$$

Put $r = \dim H^0(\Theta_V, \Theta_W, f^*\Theta_W)$. By an inductive process, we have

Lemma 3.6.12. There are mutually distinct points $P_1, \cdots P_{r_o}$ $(0 \leqq r_o \leqq r)$ on V and $Q_{r_o+1}, \cdots, Q_r$ on $W-f(V)$ with the following property: If $(\xi,\zeta) \in H^0(\Theta_V, \Theta_W, f^*\Theta_W)$ satisfies

$$\xi(P_\nu) = 0 \quad \text{for} \quad 1 \leqq \nu \leqq r_o,$$

$$\zeta(Q_\nu) = 0 \quad \text{for} \quad r_o+1 \leqq \nu \leqq r,$$

then $\xi = 0$ and $\zeta = 0$.

We fix these points once and for all. We denote by Θ_V^* and Θ_W^* the subsheaves of Θ_V and Θ_W consisting of germs of vector fields vanishing at every P_ν, $1 \leqq \nu \leqq r_o$, and at every Q_ν, $r_o+1 \leqq \nu \leqq r$, respectively.

In a similar way to define $H^p(\Theta_V, \Theta_W, f^*\Theta_W)$, we can define the cohomology group $H^p(\Theta_V^*, \Theta_W^*, f^*\Theta_W)$. By Lemma 3.6.12,

(**) $$H^0(\Theta_V^*, \Theta_W^*, f^*\Theta_W) = 0.$$

Moreover, we have

$$\dim H^1(\Theta_V^*, \Theta_W^*, f^*\Theta_W) = \dim H^1(\Theta_V, \Theta_W, f^*\Theta_W) + r_o(d-1) + (r-r_o)(m-1) .$$

($d = \dim V$ and $m = \dim W$.)

Now, we can construct the <u>normalizing operator</u> Ψ and the <u>smoothing operator</u> Ω in similar ways to Commichau [10]. Thus, by a parallel argument to Commichau [10], we can construct a family $(X, \pi, Y, \mu, T, \mathcal{F}) = \{V_t, f_t, W_t\}_{t \in T}$ with a point $o \in T$ such that $V_o = V$, $W_o = W$ and $f_o = f$, together with holomorphic sections

$$a_1, \cdots, a_{r_o} : T \longrightarrow X ,$$

$$b_{r_o+1}, \cdots, b_r : T \longrightarrow Y ,$$

such that $a_\nu(o) = P_\nu$, $1 \leqq \nu \leqq r_o$, and $b_\nu(o) = Q_\nu$, $r_o+1 \leqq \nu \leqq r$. This family is versal in the category of families of "holomorphic maps together with holomorphic sections". (In order to prove the completeness, we use (**).)

If we forget these sections, we get a complete family of holomorphic maps. Taking a suitable closed complex subspace S of T through o with $\dim T_o S \leqq \dim H^1(\Theta_V, \Theta_W, f^*\Theta_W)$, the subfamily $\{V_s, f_s, W_s\}_{s \in S}$ turns out to be complete and effectively parametrized at o.

<u>Example 3.6.13</u>. Let $V = V_\omega = \mathbb{C}/(\mathbb{Z} + \mathbb{Z}\omega)$, $\omega \in \mathbb{H}$, be a complex 1-torus. Let $f \in R_n(V)$ be an elliptic function of order n on V. First, note that

$$(1) \qquad H^0(\Theta_V, \Theta_{\mathbb{P}^1}, f^*\Theta_{\mathbb{P}^1}) = 0 .$$

In fact, let $\pi : \mathbb{C} \longrightarrow V$ be the projection map and let z be a coordinate on V induced by π. We have

$$H^0(V, \Theta_V) = \{a\frac{\partial}{\partial z} \mid a \in \mathbb{C}\} .$$

On the other hand, using the inhomogeneous coordinate $x = Z_1/Z_0$ in $\mathbb{P}^1 - \infty$, $H^0(\mathbb{P}^1, \Theta_{\mathbb{P}^1})$ can be written as

$$H^0(\mathbb{P}^1, \Theta_{\mathbb{P}^1}) = \{(bx^2+cx+d)\frac{\partial}{\partial x} \;\Big|\; b,c,d \in \mathbb{C}\} .$$

Hence, we have

$$f_*(a\frac{\partial}{\partial z}) = a(\frac{df}{dz})(z) ,$$

$$f^*((bx^2+cx+d)\frac{\partial}{\partial x}) = bf(z)^2+cf(z)+d ,$$

on $V - f^{-1}(\infty)$. Assume that they are equal. Then

$$af'(z) = bf(z)^2+cf(z)+d \quad \text{on} \quad V - f^{-1}(\infty) .$$

Hence, as elliptic functions on V,

$$af' = bf^2 + cf + d .$$

This is impossible unless $a = b = c = d = 0$. Thus we get (1). It is clear that

(2) $$H^1(\mathbb{P}^1, \Theta_{\mathbb{P}^1}) = 0 \quad \text{and} \quad H^1(V, f^*\Theta_{\mathbb{P}^1}) = 0 .$$

Hence, the exact sequence in Proposition 3.6.9 is written in this case as

$$0 \longrightarrow H^0(V,\Theta_V) \oplus H^0(\mathbb{P}^1, \Theta_{\mathbb{P}^1}) \xrightarrow{f_*-f^*} H^0(V,f^*\Theta_{\mathbb{P}^1})$$

$$\longrightarrow H^1(\Theta_V, \Theta_{\mathbb{P}^1}, f^*\Theta_{\mathbb{P}^1}) \longrightarrow H^1(V,\Theta_V) \longrightarrow 0 .$$

Hence

$$\dim H^1(\Theta_V, \Theta_{\mathbb{P}^1}, f^*\Theta_{\mathbb{P}^1}) = 2n - 3 .$$

Now, we construct the <u>global moduli space of elliptic functions of order</u> n $(n \geqq 2)$ as follows:

Let $\mathbb{H}$ be the upper half plane and let $(X,\pi,\mathbb{H}) = \{V_\omega\}_{\omega\in\mathbb{H}}$ be the family of complex 1-tori (see Example 0.1.1). For two points

$\omega, \omega' \in \mathbb{H}$, the tori V_ω and $V_{\omega'}$ are biholomorphic if and only if there is $s = \begin{pmatrix} a & b \\ c & d \end{pmatrix} \in SL(2, \mathbb{Z})$ such $\omega' = (a\omega+b)/(c\omega+d)$. In fact, s induces a holomorphic isomorphism

$$\tilde{s} : P \in V_\omega \longmapsto \frac{1}{c\omega+d} P \in V_{\omega'} .$$

Thus $s \in SL(2, \mathbb{Z})$ induces automorphisms $s : \mathbb{H} \longrightarrow \mathbb{H}$ and $\tilde{s} : X \longrightarrow X$ such that the diagram

$$\begin{array}{ccc} X & \xrightarrow{\ \tilde{s}\ } & X \\ {\scriptstyle \pi}\downarrow & & \downarrow{\scriptstyle \pi} \\ \mathbb{H} & \xrightarrow{\ s\ } & \mathbb{H} \end{array}$$

commutes, i.e., $(s, \tilde{s})$ is an automorphism of the family $\{V_\omega\}_{\omega \in \mathbb{H}}$ (see §0.1).

For $j, k = 0, 1, \cdots, n-1 \pmod n$, let

$$t_{(j,k)} : X \longrightarrow X$$

be an automorphism defined by

$$t_{(j,k)}(P) = P + \left(\frac{j}{n} + \frac{k}{n}\omega\right) \quad \text{for} \quad P \in V_\omega .$$

Then $(id, t_{(j,k)})$ is an automorphism of the family. ($id : \mathbb{H} \longrightarrow \mathbb{H}$ is the identity map.) Note that

$$t_{(j,k)} \tilde{s} = \tilde{s} t_{(jd+kb,\ jc+ka)} \quad \text{for} \quad s = \begin{pmatrix} a & b \\ c & d \end{pmatrix} \in SL(2, \mathbb{Z}) .$$

We denote by G the group of automorphisms of X generated by $\tilde{s}$, $s \in SL(2, \mathbb{Z})$, and $t_{(j,k)}$, $j, k = 0, 1, \cdots, n-1 \pmod n$.

Let 0_ω be the zero of V_ω as a group. We denote by $[n0_\omega]$ the line bundle determined by the divisor $n0_\omega = 0_\omega + \cdots + 0_\omega$ and by $G^1(|n0_\omega|)$ the Grassmann variety of all linear pencils contained in the complete linear system $|n0_\omega|$. Then $\{[n0_\omega]\}_{\omega \in \mathbb{H}}$ is a family of line bundles and hence the disjoint union

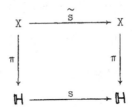

$$\mathbb{G}^1 = \cup_{\omega\in\mathbb{H}} G^1(|nO_\omega|)$$

is a complex space such that the projection $\mathbb{G}^1 \longrightarrow \mathbb{H}$ is a Grassmann bundle (see Theorem 4.1.9). Put

$$\mathcal{S}_\omega = \{l \in G^1(|nO_\omega|) \;\Big|\; l \text{ has no fixed point}\},$$

$$\mathcal{S} = \cup_{\omega\in\mathbb{H}} \mathcal{S}_\omega \quad \text{(disjoint union)}.$$

Then $\mathcal{S}_\omega$ and $\mathcal{S}$ are Zariski open in $G^1(|nO_\omega|)$ and in $\mathbb{G}^1$, respectively.

Now, the group G acts on $\mathbb{G}^1$ and $\mathcal{S}$ as follows: For $\gamma = t_{(j,k)}\tilde{s} \in G$ and $l = \{\lambda_0 D_0 + \lambda_1 D_1\}_{(\lambda_0:\lambda_1)\in\mathbb{P}^1} \in \mathbb{G}^1$,

$$\gamma(l) = \{\lambda_0\gamma(D_0) + \lambda_1\gamma(D_1)\}_{(\lambda_0:\lambda_1)\in\mathbb{P}^1},$$

where, if $D = P_1+\cdots+P_n \in |nO_\omega|$, then $\gamma(D) \in |nO_{\omega'}|$ $(\omega' = s(\omega))$ is defined by

$$\gamma(D) = \gamma(P_1) + \cdots + \gamma(P_n).$$

Note that the actions of G on $\mathbb{G}^1$ and $\mathcal{S}$ are proper. This follows easily from the fact that the action of $SL(2,\mathbb{Z})$ on $\mathbb{H}$ is proper.

Now the quotient space $\mathcal{S}/G$ can be considered as the <u>global moduli space of elliptic functions of order</u> n. It is a normal complex space of dimension 2n-3.

<u>Remark 3.6.14</u>. In general, for an integer r, $1 \leqq r \leqq n-1$, put

$$G^r(|nO_\omega|) = \{r\text{-dimensional linear subsystems of } |nO_\omega|\},$$

$$\mathbb{G}^r = \cup_{\omega\in\mathbb{H}} G^r(|nO_\omega|),$$

$$\mathcal{S}^r_\omega = \{L \in G^r(|nO_\omega|) \;\Big|\; L \text{ has no fixed point}\},$$

$$\mathcal{S}^r = \cup_{\omega\in\mathbb{H}} \mathcal{S}^r_\omega.$$

Then G acts properly on $\mathbb{G}^r$ and $\mathcal{S}^r$. The quotient space $\mathcal{S}^r/G$ can be considered as the <u>global moduli space of non-degenerate holomorphic maps</u> f <u>of complex 1-tori</u> V <u>into</u> $\mathbb{P}^r$ <u>such that</u>

$$n = (\text{ord } f)(\deg f(V)) ,$$

where $\deg f(V)$ is the degree of the image curve $f(V)$ in $\mathbb{P}^r$ and $\text{ord } f$ is the mapping order of $f : V \longrightarrow f(V)$. (Here, a holomorphic map $V \longrightarrow \mathbb{P}^r$ is said to be <u>non-degenerate</u> if $f(V)$ is <u>not</u> contained in any hyperplane in $\mathbb{P}^r$.) $\mathcal{S}^r/G$ is a normal complex space of dimension $(r+1)n - (r^2+2r)$.

Chapter 4. Families of effective divisors and linear systems on projective manifolds.

4.1. Linear fiber space of global sections.

Let $(X,\pi,S) = \{V_s\}_{s\in S}$ be a family of compact complex manifolds. A *family of holomorphic vector bundles* is, by definition, a quadruplet $(\mathbb{F},X,\pi,S)$, where $\mathbb{F}$ is a holomorphic vector bundle over X. For each $s\in S$, the restriction $F_s = \mathbb{F}|V_s$ of $\mathbb{F}$ to V_s is a holomorphic vector bundle over V_s. We often write $\{F_s,V_s\}_{s\in S}$ instead of $(\mathbb{F},X,S)$. As before, we denote by $\mathcal{O}(F_s)$ the sheaf over V_s of germs of holomorphic sections of F_s.

Fix a point $o\in S$. Put $V_o = V$ and $F_o = F$. We use the same notations as in §3.0 and §3.1. We may assume that $\mathbb{F}$ is trivial on X_i. Let $\{f_{ik}(z_k,s)\}$ be the transition matrices of $\mathbb{F}$ on $X_i\cap X_k$. For $\phi = \{\phi_i(z_i)\}\in C^0(F,|\ |)$, we define a continuous linear map

$$\tilde{\tau}_\phi : T_oS \longrightarrow C^1(F,|\ |)$$

as follows:

$$(1)\qquad (\tilde{\tau}_\phi(\tfrac{\partial}{\partial s}))_{ik}(z_i)$$

$$= (\partial f_{ik}/\partial s)_{(z_k,o)}\phi_k(z_k) - (\partial\phi_i/\partial z_i)_{z_i}(\partial g_{ik}/\partial s)_{(z_k,o)},$$

where $\frac{\partial}{\partial s}\in T_oS$ and $z_k = g_{ki}(z_i,o)\in U_i^e\cap U_k$.

Lemma 4.1.1. If $\xi\in H^0(V,\mathcal{O}(F))$, then $\tilde{\tau}_\xi(T_oS)\subset Z^1(F,|\ |)$.

Proof. Let $z_k = g_{ki}(z_i,o)$ be a point of $U_i^e\cap U_j^e\cap U_k$. Then, there are an open neighborhood Y in $(U_i^e\cap U_j^e\cap U_k)\times\Omega_\varepsilon$ and vector valued holomorphic functions $d^\lambda(z_k,s)$ on Y such that

$$(2)\qquad f_{ij}(g_{jk}(z_k',s),s)f_{jk}(z_k',s) = f_{ik}(z_k',s) + \Sigma_\lambda d^\lambda(z_k',s)v_\lambda(s),$$

for all $(z'_k,s)\in Y$, where v_λ, $1 \leqq \lambda \leqq p$, are holomorphic functions defining $\tilde{S}$ in $\tilde{\Omega}$. Taking the derivatives of (2) at (z_k,o) with respect to s and using (1) of §0.1, we get

$$(\partial f_{ij}/\partial z_j)_{(z_j,o)}(\partial g_{jk}/\partial s)_{(z_k,o)}f_{jk}(z_k,o)$$

$$+ (\partial f_{ij}/\partial s)_{(z_j,o)}f_{jk}(z_k,o) + f_{ij}(z_j,o)(\partial f_{jk}/\partial s)_{(z_k,o)}$$

$$= (\partial f_{ik}/\partial s)_{(z_k,o)},$$

where $z_j = g_{ji}(z_i,o)$. Multiplying $\xi_k(z_k)$ from the right and using the assumption that $\xi\in H^0(V, \Theta(F))$, we get

$$(\partial f_{ij}/\partial z_j)_{(z_j,o)}\xi_j(z_j)\theta_{jk} = -(\delta\psi)_{ijk}(z_i), \tag{3}$$

where $\theta_{jk} = (\partial g_{jk}/\partial s)_{(z_k,o)}$ and $\psi_{ik}(z_i) = (\partial f_{ik}/\partial s)_{(z_k,o)}\xi_k(z_k)$.

On the other hand, it holds

$$(\partial g_{ij}/\partial z_j)_{(z_j,o)}\theta_{jk} = \theta_{ik} - \theta_{ij}, \tag{4}$$

$$(\partial\xi_i/\partial z_i)_{z_i}(\partial g_{ij}/\partial z_j)_{(z_j,o)} = (\partial f_{ij}/\partial z_j)_{(z_j,o)}\xi_j(z_j) + f_{ij}(z_j,o)(\partial\xi_j/\partial z_j)_{z_j}. \tag{5}$$

Thus

$$\begin{aligned}(\delta\{(\partial\xi_i/\partial z_i)_{z_i}\theta_{ik}\})_{ijk}(z_i) &= f_{ij}(z_j,o)(\partial\xi_j/\partial z_j)_{(z_j,o)}\theta_{jk} + (\partial\xi_i/\partial z_i)_{(z_i,o)}\theta_{ij} \\ &\quad - (\partial\xi_i/\partial z_i)_{(z_i,o)}\theta_{ik} \\ &= f_{ij}(z_j,o)(\partial\xi_j/\partial z_j)_{(z_j,o)}\theta_{jk} \\ &\quad - (\partial\xi_i/\partial z_i)_{z_i}(\partial g_{ij}/\partial z_j)_{(z_j,o)}\theta_{jk}, \quad \text{(by (4))}, \\ &= -(\partial f_{ij}/\partial z_j)_{(z_j,o)}\xi_j(z_j)\theta_{jk}, \quad \text{(by (5))},\end{aligned}$$

$= (\delta\psi)_{ijk}(z_i)$, (by (3)).

Hence $\delta(\psi-\zeta) = 0$, where $\zeta = \{\zeta_{ik}(z_i)\}$ and $\zeta_{ik}(z_i) = (\partial\xi_i/\partial z_i)_{z_i}\theta_{ik}$. Thus $\tilde{\tau}_\xi(\frac{\partial}{\partial s})$ is a cocycle. <u>Q.E.D.</u>

Let $H : Z^1(F,|\ |) \longrightarrow H^1(V, \mathcal{O}(F))$ be the projection map defined in §3.0. Put $\tau_\xi = H\tilde{\tau}_\xi$. Then, for every $\xi \in H^0(V, \mathcal{O}(F))$,

$$\tau_\xi : T_oS \longrightarrow H^1(V, \mathcal{O}(F))$$

is a linear map.

The following theorem is again an analogy of Kuranishi's Theorem 0.1.3.

<u>Theorem 4.1.2</u>. Let $\{F_s,V_s\}_{s\in S}$ be a family of holomorphic vector bundles. Then, for any point $o \in S$, there are an open neighborhood S' of o in S and a vector bundle homomorphism

$$u : H^0(V_o, \mathcal{O}(F_o)) \times S' \longrightarrow H^1(V_o, \mathcal{O}(F_o)) \times S'$$

such that the disjoint union $\bigcup_{s\in S'} H^0(V_s, \mathcal{O}(F_s))$ is identified with the kernel of u, i.e., $\{(\xi,s) \mid u(\xi,s) = (0,s)\}$. Moreover, it holds

$$(du)_{(\xi,o)} = \begin{pmatrix} 0 & \tau_\xi \\ 0 & 1 \end{pmatrix} \quad \text{for} \quad \xi \in H^0(V_o, \mathcal{O}(F_o)) .$$

In the sequel, we give a proof of the theorem. It is quite similar to the proof of Theorem 3.2.3. In fact, a holomorphic section $\phi \in H^0(V_s, \mathcal{O}(F_s))$ is regarded as a holomorphic map

$$\phi : V_s \longrightarrow F_s \quad (= \text{the bundle space}) .$$

Hence $\bigcup_{s\in S'} H^0(V_s, \mathcal{O}(F_s))$ is regarded as a (closed) complex subspace of the relative Douady space $\bigcup_{s\in S'} \mathrm{Hol}(V_s,F_s)$. Thus the theorem may be reduced to Theorem 3.2.3. However, this special case has its own feature, so let us give a proof of the theorem.

Put $V_o = V$ and $F_o = F$. We use the same notations as in §3.0 and §3.1. Let ε be a sufficiently small positive number. For a (fixed) point $s \in S_\varepsilon$, we have a holomorphic isomorphism

$$(z_i, s) \in U_i \times s \longmapsto z_i \in U_i .$$

Combining this isomorphism with the holomorphic isomorphism

$$V_s \cap X_i = \eta_i^{-1}(U_i \times s) \xrightarrow{\eta_i} U_i \times s ,$$

we have a holomorphic isomorphism

$$h(i,s) : V_s \cap X_i \longrightarrow U_i .$$

Let ϕ be a global holomorphic section of F_s. Put $\phi_i = (\phi \mid V_s \cap X_i)h(i,s)^{-1}$. Then ϕ_i is a vector valued holomorphic function on U_i. The collection $\{\phi_i\}_{i\in I}$ must satisfy the following compatibility conditions:

(6) $f_{ik}(z_k,s)\phi_k(z_k) = \phi_i(g_{ik}(z_k,s))$ for $(z_k,s) \in \eta_k(X_i \cap X_k)$.

Conversely, it is clear that, if the collection of vector valued functions ϕ_i on U_i satisfies (6), then it defines a global holomorphic section of F_s.

Now, we define a map

$$K : C^0(F,|\ |) \times \Omega_\varepsilon \longrightarrow C^1(F,|\ |)$$

by

$$K(\phi,s)_{ik}(z_i) = f_{ik}(z_k,s)\phi_k(z_k) - \phi_i(g_{ik}(z_k,s)) ,$$

for $z_i \in U_i^e \cap U_k$ and $s \in \Omega_\varepsilon$, where $z_k = g_{ki}(z_i,o)$. It is easy to see that K is well defined. Put

$$M = \{(\phi,s) \in C^0(F,|\ |) \times S_\varepsilon \mid K(\phi,s) = 0\} .$$

Fix $s \in S_\varepsilon$. Let ϕ be a global holomorphic section of F_s over V_s. Put $\phi_i = (\phi \mid V_s \cap X_i)h(i,s)^{-1}$ as above and associate to ϕ an element $\phi = \{\phi_i\} \in C^0(F,|\ |)$. For any fixed point $z_i \in U_i^e \cap U_k$, $\eta_k^{-1}(z_k,s) \in X_i \cap X_k$ by Lemma 3.1.4. ($z_k = g_{ki}(z_i,o)$.) Hence $f_{ik}(z_k,s)\phi_k(z_k) = \phi_i(g_{ik}(z_k,s))$, by (6). Hence $K(\phi,s)_{ik}(z_i) = 0$. This shows that $K(\phi,s) = 0$, i.e., $(\phi,s) \in M$.

Conversely, let $(\phi,s) \in M$. Assume that $(z_k,s) \in X_i^{e'} \cap X_k^{e'}$. Then, by Lemma 3.1.5, $z_k \in U_i^e \cap U_k$. Since $K(\phi,s) = 0$,

$$f_{ik}(z_k,s)\phi_k(z_k) = \phi_i(g_{ik}(z_k,s)) .$$

Hence $\phi_i(z_i)$, $z_i \in U_i^{e'}$, $i \in I$, define a global holomorphic section ϕ of F_s over V_s.

Thus the problem is reduced to analyze the set M.

For a fixed point $s \in \Omega_\varepsilon$, $K(\ ,s)$ is a continuous linear map of $C^0(F,|\ |)$ into $C^1(F,|\ |)$. Hence, we can define a map

$$\tilde{K} : \Omega_\varepsilon \longrightarrow \mathcal{L}(C^0(F,|\ |),C^1(F,|\ |))$$

by

$$\tilde{K}(s)(\phi) = K(\phi,s), \quad \text{for} \quad \phi \in C^0(F,|\ |) ,$$

where $\mathcal{L}(C^0(F,|\ |),C^1(F,|\ |))$ is the Banach space of all continuous linear maps of $C^0(F,|\ |)$ into $C^1(F,|\ |)$ with the norm

$$|A| = \sup \{|A\phi|/|\phi| \mid \phi \in C^0(F,|\ |) - 0\} .$$

Note that $\tilde{K}(0) = \delta$. The proof of the following lemma is similar (and simpler) to Lemma 3.2.4, so we omit it.

<u>Lemma 4.1.3</u>. Take ε sufficiently small. Then

(1) $\tilde{K}$ is an analytic map.

(2) K is an analytic map and $(dK)_{(\phi,o)} = (\delta,\tilde{\tau}_\phi)$.

Now, we define an analytic map

$$L : C^0(F,|\ |) \times \Omega_\varepsilon \longrightarrow C^0(F,|\ |) \times \Omega_\varepsilon$$

by

$$L(\phi,s) = (\phi + E_0 B\Lambda K(\phi,s) - E_0\delta\phi,\ s)\ ,$$

where E_0, B and Λ are continuous linear maps defined in §3.0. For a fixed point $s \in \Omega_\varepsilon$, $L(\ ,s)$ is a continuous linear map of $C^0(F,|\ |)$ into itself. We regard $C^0(F,|\ |) \times \Omega_\varepsilon$ as a Banach vector bundle over Ω_ε. Then L is an analytic bundle map. We define a map

$$\tilde{L} : \Omega_\varepsilon \longrightarrow \mathcal{L}(C^0(F,|\ |),\ C^0(F,|\ |))$$

by

$$(\tilde{L}(s)(\phi),s) = L(\phi,s), \quad \text{for} \quad \phi \in C^0(F,|\ |)\ .$$

Then

$$\tilde{L}(s) = 1 + E_0 B\Lambda\tilde{K}(s) - E_0\delta\ .$$

Hence $\tilde{L}$ is an analytic map by Lemma 4.1.3. Since $\tilde{K}(o) = \delta$, we have $\tilde{L}(o) = 1$. By Proposition 1 of III of Douady [14], there exists a small positive number ε such that

$$L : C^0(F,|\ |) \times \Omega_\varepsilon \longrightarrow C^0(F,|\ |) \times \Omega_\varepsilon$$

is a vector bundle isomorphism. Put $\Psi = L^{-1}$. Put

$$T = \{(\xi,s) \in C^0(F,|\ |) \times S_\varepsilon \ \big|\ K\Psi(\xi,s) = 0\}\ .$$

Then it is clear that $M = \Psi(T)$.

Lemma 4.1.4. $T \subset H^0(V, \mathcal{O}(F)) \times S_\varepsilon$.

Proof. Take any $(\xi,s) \in T$. Put $\Psi(\xi,s) = (\phi,s)$. Then, since

$K(\phi,s) = 0$, we have

$$\xi = \phi + E_0 B \Lambda K(\phi,s) - E_0\delta\phi = \phi - E_0\delta\phi .$$

Hence

$$\delta\xi = \delta\phi - \delta E_0\delta\phi = \delta\phi - \delta\phi = 0 .$$

<u>Q.E.D.</u>

By this lemma,

$$T = \{(\xi,s) \in H^0(V, \mathcal{O}(F)) \times S_\varepsilon \mid K\Psi(\xi,s) = 0\} .$$

<u>Lemma 4.1.5</u>. If ε is sufficiently small, then

$$T = \{(\xi,s) \in H^0(V, \mathcal{O}(F)) \times S_\varepsilon \mid H\Lambda K\Psi(\xi,s) = 0\} ,$$

where $H : Z^1(F,|\ |) \longrightarrow H^1(V, \mathcal{O}(F))$ is the projection map defined in §3.0.

<u>Proof</u>. (Sketch.) By some calculations, we have the following estimate: If ε is sufficiently small, then

$$|\delta K(\phi,s)| \leqq c|K(\phi,s)|\cdot|s| \quad \text{for} \quad (\phi,s) \in C^0(F,|\ |) \times S_\varepsilon ,$$

where $c > 0$ is a constant. The rest of the proof is similar to that of Lemma 3.2.6.

<u>Q.E.D.</u>

The following lemma is proved in a similar way to Lemma 3.2.7.

<u>Lemma 4.1.6</u>. $d(H\Lambda K\Psi)_{(\xi,o)} = (0,\tau_\xi)$ for $\xi \in H^0(V, \mathcal{O}(F))$.

For a fixed point $s \in \Omega_\varepsilon$, $H\Lambda K\Psi(\ ,s)$ is a linear map of $H^0(V, \mathcal{O}(F))$ into $H^1(V, \mathcal{O}(F))$. We define a map

$$\tilde{u} : \Omega_\varepsilon \longrightarrow \mathcal{L}(H^0(V, \mathcal{O}(F)), H^1(V, \mathcal{O}(F)))$$

by

$$\tilde{u}(s)\xi = H\Lambda K\Psi(\xi, s) .$$

Then it is easy to see that $\tilde{u}$ is a holomorphic map. Finally, we define a vector bundle homomorphism

$$u : H^0(V, \mathcal{O}(F)) \times \Omega_\varepsilon \longrightarrow H^1(V, \mathcal{O}(F)) \times \Omega_\varepsilon$$

by

$$u(\xi,s) = (\tilde{u}(s)\xi, s) = (H\Lambda K\Psi(\xi,s), s) .$$

Now, it is clear that u satisfies the requirement. This proves Theorem 4.1.2.

From the definition of $\tilde{u}$, we get

Proposition 4.1.7. $((d\tilde{u})_o(\frac{\partial}{\partial s}))(\xi) = \tau_\xi(\frac{\partial}{\partial s})$ for $(\xi, \frac{\partial}{\partial s}) \in H^0(V, \mathcal{O}(F)) \times T_oS$.

Patching up the local data given in Theorem 4.1.2, we get the following theorem, which is considered as a special form of Schuster [77]. (It is possible to patch them up. We can do it directly. Or, see Proposition 4.1.13 below.)

Theorem 4.1.8. Let $\{F_s, V_s\}_{s\in S}$ be a family of holomorphic vector bundles. Then the disjoint union

$$\mathbb{H}^0 = \bigcup_{s\in S} H^0(V_s, \mathcal{O}(F_s))$$

admits a complex space structure so that $(\mathbb{H}^0, \lambda, s)$ is a linear fiber space in the sense of Grauert [23], where λ is the canonical projection of $\mathbb{H}^0$ onto S.

For a complex vector space A and a non-negative interger r, we denote by $G^r(A)$ the Grassmann variety of all $(r+1)$-dimensional linear subspaces of A. (If $\dim A \leqq r$, then $G^r(A)$ is empty.)

<u>Theorem 4.1.9</u>. Let $\{F_s, V_s\}_{s\in S}$ be a family of holomorphic vector bundles. Then, for any integer $r \geqq 0$, the disjoint union

$$\mathbb{G}^r = \bigcup_{s\in S} G^r(H^0(V_s, \mathcal{O}(F_s)))$$

admits a complex space structure so that the canonical projection $\mu : \mathbb{G}^r \longrightarrow S$ is a proper holomorphic map.

<u>Proof</u>. Take any $o \in S$ and $x_o \in G^r(H^0(V_o, \mathcal{O}(F_o)))$. Then, there are an open neighborhood U of x_o in $G^r(H^0(V_o, \mathcal{O}(F_o)))$ and linearly independent $(r+1)$-elements $\xi_0(x), \cdots, \xi_r(x)$ of $H^0(V_o, \mathcal{O}(F_o))$, depending holomorphically on $x \in U$, such that they span the linear subspace of $H^0(V_o, \mathcal{O}(F_o))$ corresponding to $x \in U$. Then, locally, $\mathbb{G}^r$ is given as the closed complex subspace

$$\{(x,s) \in U \times S' \mid \tilde{u}(s)\xi_\alpha(x) = 0, \quad 0 \leqq \alpha \leqq r\}$$

of $U \times S'$, where S' is an open neighborhood of o in S and $\tilde{u}$ is the map defined in the proof of Theorem 4.1.2 above. Patching up the local data, we get the complex space structure on $\mathbb{G}^r$. It is clear that μ is a proper holomorphic map. <u>Q.E.D.</u>

<u>Theorem 4.1.10</u>. Let $\{F_s, V_s\}_{s\in S}$ be a family of holomorphic vector bundles. Let o be a <u>non-singular</u> point of S.

(1) For $\xi \in H^0(V_o, \mathcal{O}(F_o))$, assume that the linear map

$$\tau_\xi : T_oS \longrightarrow H^1(V_o, \mathcal{O}(F_o))$$

is surjective. Then the complex space $\mathbb{H}^0$ in Theorem 4.1.8 is non-singular at ξ and

$$\dim_\xi \mathbb{H}^0 = \dim H^0(V_o, \mathcal{O}(F_o)) - \dim H^1(V_o, \mathcal{O}(F_o)) + \dim_o S.$$

(2) For $L \in G^r(H^0(V_o, \mathcal{O}(F_o)))$, let $\xi_0, \cdots, \xi_r \in H^0(V_o, \mathcal{O}(F_o))$ be a basis of L. Assume that the linear map

$$\frac{\partial}{\partial s} \in T_o S \longmapsto (\tau_{\xi_0}(\frac{\partial}{\partial s}), \cdots, \tau_{\xi_r}(\frac{\partial}{\partial s})) \in H^1(V_o, \mathcal{O}(F_o))^{r+1}$$

is surjective. Then the complex space $\mathbb{G}^r$ in Theorem 4.1.9 is non-singular at L and

$$\dim_L \mathbb{G}^r = (r+1)(h^0(F_o) - h^1(F_o) - r - 1) + \dim_o S ,$$

where $h^i(F_o) = \dim H^i(V_o, \mathcal{O}(F_o))$.

Proof. The complex space $\mathbb{H}^0$ is locally defined by

$$\{(\xi, s) \in H^0(V_o, \mathcal{O}(F_o)) \times S' \mid \tilde{u}(s)\xi = H\Lambda K\Psi(\xi, s) = 0\} .$$

(See the proof of Theorem 4.1.2 above.) The differential of $H\Lambda K\Psi$ at (ξ, o) is $(0, \tau_\xi)$ by Lemma 4.1.6. It is of maximal rank, if τ_ξ is surjective. If it is so and if o is a non-singular point of S, then $\mathbb{H}^0$ is non-singular at ξ and $\dim_\xi \mathbb{H}^0$ is given as above. (2) is proved in a similar way. Q.E.D.

We show that the condition in (2) of the theorem does not depend on the choice of the basis $\{\xi_o, \cdots, \xi_r\}$ of L. First, note that $\tau_\xi(\frac{\partial}{\partial s})$ depends bilinearly on $(\xi, \frac{\partial}{\partial s})$, i.e.,

$$\tau : (\xi, \frac{\partial}{\partial s}) \in H^0(V_o, \mathcal{O}(F_o)) \times T_o S \longmapsto \tau_\xi(\frac{\partial}{\partial s}) \in H^1(V_o, \mathcal{O}(F_o))$$

is a bilinear map. The restriction map

$$p_L : \alpha \in \mathrm{Hom}(H^0(V_o, \mathcal{O}(F_o)), H^1(V_o, \mathcal{O}(F_o)))$$
$$\longmapsto \alpha|L \in \mathrm{Hom}(L, H^1(V_o, \mathcal{O}(F_o)))$$

is a surjective linear map. On the other hand, the bilinear map τ is regarded as a linear map

$$\tau : T_oS \longrightarrow \mathrm{Hom}(H^0(V_o, \mathcal{O}(F_o)), H^1(V_o, \mathcal{O}(F_o))) .$$

It is now easy to see

<u>Lemma 4.1.11</u>. The condition in (2) of Theorem 4.1.10 is satisfied if and only if $p_L \cdot \tau$ is surjective. In particular, the condition does not depend on the choice of the basis $\{\xi_0, \cdots, \xi_r\}$ of L.

Finally, we give some functional properties of the complex spaces $\mathbb{H}^0$ and $\mathbb{G}^r$ (without proof).

From the local construction of the space $\mathbb{H}^0$, it is clear that $\mathbb{H}^0$ has the following property:

<u>Lemma 4.1.12</u>. The map

$$(\phi, P) \in \mathbb{H}^0 \times_S X \longmapsto \phi(P) \in \mathbb{F}$$

is holomorphic, where the vector bundle $\mathbb{F}$ is identified with its bundle space.

The following proposition also follows from the local construction of the space $\mathbb{H}^0$.

<u>Proposition 4.1.13</u>. The linear fiber space $\mathbb{H}^0$ has the following universal property: Assume that $\mathbb{H}'$ is another linear fiber space which is equal to $\mathbb{H}^0$ as sets and satisfies the condition that the map

$$(\phi, P) \in \mathbb{H}' \times {}_S X \longmapsto \phi(P) \in \mathbb{F}$$

is holomorphic. Then, the identification map $\mathbb{H}' \longrightarrow \mathbb{H}^0$ is holomorphic and is a morphism of linear fiber spaces in the sense of Grauert

[23].

Corollary 4.1.14. Let $h : T \longrightarrow S$ be a holomorphic map of a complex space T into S and let $(h^*\mathbb{F}, h^*X, \pi^*, T) = \{F_t, V_t\}_{t\in T}$ be the induced family of vector bundles over h. ($h^*\mathbb{F} = \mathbb{F} \times_X h^*X = \mathbb{F} \times_S T$.) Then the complex space $\mathbb{H}^0_T = \cup_t H^0(V_t, \mathcal{O}(F_t))$ constructed with respect to $\{F_t, V_t\}_{t\in T}$ is canonically isomorphic to $\mathbb{H}^0 \times_S T$ as linear fiber spaces.

Corollary 4.1.15. Let T be a complex subspace of S. Let $\{F_t, V_t\}_{t\in T}$ be the restriction of $\{F_s, V_s\}_{s\in S}$ to T. Then the complex space $\mathbb{H}^0_T$ constructed with respect to $\{F_t, V_t\}_{t\in T}$ is canonically isomorphic to the restriction $\mathbb{H}^0|T$ of $\mathbb{H}^0$ to T, as linear fiber spaces.

Next, $\mathbb{G}^r$ is locally given by

$$W = \{(x,s) \in U \times S' \mid \tilde{u}(s)\xi_\nu(x) = H\Lambda K\Psi(\xi_\nu(x),s) = 0, \quad 0 \leqq \nu \leqq r\}$$

(see the proof of Theorem 4.1.9). For $w = (x,s) \in W$, we define $\phi_\nu(w)$ by

$$(\phi_\nu(w), s) = \Psi(\xi_\nu(x), s) .$$

Then $\phi_0(w), \cdots, \phi_r(w)$ are regarded as global holomorphic sections of F_s and span the $(r+1)$-dimensional linear subspace L_w of $H^0(V_s, \mathcal{O}(F_s))$ corresponding to $w \in W$. That is to say:

Lemma 4.1.16. For any point $w_0 \in \mathbb{G}^r$, there are an open neighborhood W of w_0 in $\mathbb{G}^r$ and $\phi_0(w), \cdots, \phi_r(w) \in H^0(V_s, \mathcal{O}(F_s))$, ($w \in W$ and $s = \mu(w)$), such that (1) they span L_w, the linear subspace corresponding to w and (2) the maps

$$(w,P) \in W \times_S X \longmapsto \phi_\nu(w)(P) \in \mathbb{F}, \quad 0 \leqq \nu \leqq r ,$$

are holomorphic.

Proposition 4.1.17. The complex space $\mathbb{G}^r$ has the following universal property: If $\hat{\mathbb{G}}^r$ is another complex space which is equal to $\mathbb{G}^r$ as sets and satisfies the similar condition to Lemma 4.1.16, then the identification map $\hat{\mathbb{G}}^r \longrightarrow \mathbb{G}^r$ is holomorphic.

Corollary 4.1.18. Let $h : T \longrightarrow S$ be a holomorphic map and let $\{F_t, V_t\}_{t \in T}$ be the induced family over h. Then the complex space $\mathbb{G}^r_T$ constructed with respect to $\{F_t, V_t\}_{t \in T}$ is canonically biholomorphic to $\mathbb{G}^r \times_S T$.

Corollary 4.1.19. Let T be a complex subspace of S. Let $\{F_t, V_t\}_{t \in T}$ be the restriction of $\{F_s, V_s\}_{s \in S}$ to T. Then the complex space $\mathbb{G}^r_T$ constructed with respect to $\{F_t, V_t\}_{t \in T}$ is canonically biholomorphic to the restriction $\mathbb{G}^r|T$ of $\mathbb{G}^r$ to T.

Note 4.1.20. Mr. Ishida informed me that it is possible to draw the results of this section in a more general algebraic setting, by analyzing Grothendieck's coherent sheaf $\mathcal{Q}$ (see Grothendieck [27, Theorem 7.7.6, p.69]).

4.2. $\mathbb{D}_c(V)$ and the Jacobi map.

By a projective manifold, we mean a compact complex manifold imbedded in a complex projective space. Let V be a projective manifold. Let $\mathcal{O}$ (resp. $\mathcal{O}^*$) be the sheaf over V of germs of (resp. non-vanishing) holomorphic functions. The exact sequence of sheaves:

$$0 \longrightarrow \mathbb{Z} \xrightarrow{\ i\ } \mathcal{O} \xrightarrow{\ e\ } \mathcal{O}^* \longrightarrow 0$$

($e = 2\pi\sqrt{-1}\exp$) induces the exact sequence of cohomology groups:

$$0 \longrightarrow H^1(V, \mathbb{Z}) \xrightarrow{i^*} H^1(V, \mathcal{O}) \xrightarrow{e^*} H^1(V, \mathcal{O}^*) \xrightarrow{c} H^2(V, \mathbb{Z}) \longrightarrow \cdots .$$

The cohomology group $H^1(V, \mathcal{O}^*)$ is the set of all (isomorphism classes of) holomorphic line bundles on V. The linear map c maps a holomorphic line bundle B to its Chern class $c(B)$.

The kernel of c is equal to the image of e^* which is isomorphic to $H^1(V, \mathcal{O})/i^*H^1(V, \mathbb{Z})$. We write

$$\mathrm{Pic}^0(V) = \ker(c) = \{\text{isomorphism classes of holomorphic line bundles on } V \text{ whose Chern classes vanish.}\}$$

It is well known that $\mathrm{Pic}^0(V)$ is an abelian variety of dimension q, the irregularity of V.

It is also well known (see Kodaira-Spencer [42]) that the image of the linear map c is

$$H^{1,1}(V, \mathbb{Z}) = \{c \in H^2(V, \mathbb{Z}) \mid c \text{ is of type } (1,1)\} .$$

This is, by definition, equal to $j^{*-1}(H^{1,1}(V))$, where $j^* : H^2(V, \mathbb{Z}) \longrightarrow H^2(V, \mathbb{C})$ is the homomorphism induced by the inclusion map $j : \mathbb{Z} \hookrightarrow \mathbb{C}$ and $H^{1,1}(V)$ is the subspace of $H^2(V, \mathbb{C})$ in the Hodge decomposition:

$$H^2(V, \mathbb{C}) = H^{2,0}(V) \oplus H^{1,1}(V) \oplus H^{0,2}(V) .$$

We take $c \in H^{1,1}(V, \mathbb{Z})$ and fix it. Put

$$\mathbb{D}_c(V) = \{D \mid D \text{ is an \underline{effective} divisor on } V \text{ such that } c([D]) = c\} ,$$

where $[D]$ is the line bundle determined by D.

Problem. For which $c \in H^{1,1}(V, \mathbb{Z})$, is $\mathbb{D}_c(V)$ non-empty?

Assume that $\mathbb{D}_c(V)$ is non-empty. Fix an effective divisor $D_o \in \mathbb{D}_c(V)$ once and for all. Then, according to Weil [85] (also Kodaira [40]), $\mathbb{D}_c(V)$ is a (possibly reducible) projective variety (i.e., a compact complex space imbedded in a complex projective space) and the Jacobi map

$$\Phi : D \in \mathbb{D}_c(V) \longmapsto [D - D_o] \in \mathrm{Pic}^o(V)$$

is a proper holomorphic map.

On the other hand, for $t \in \mathrm{Pic}^o(V)$, let B_t be the line bundle on V corresponding to t. Then, by Kodaira [40], $\{B_t, V\}_{t\in \mathrm{Pic}^o(V)}$ is a family of holomorphic line bundles with the parameter space $\mathrm{Pic}^o(V)$. (V is fixed in this case.) We put, for $t \in \mathrm{Pic}^o(V)$,

$$F_t = B_t \otimes [D_o] .$$

Then $\{F_t, V\}_{t\in \mathrm{Pic}^o(V)}$ is again a family of holomorphic line bundles with the parameter space $\mathrm{Pic}^o(V)$.

Proposition 4.2.1. Let $\mathbb{G}^r = \mathbb{G}^r_c(V)$ $(r \geqq 0)$ be the complex space given in Theorem 4.1.9 with respect to the family $\{F_t, V\}_{t\in \mathrm{Pic}^o(V)}$. Then, $\mathbb{G}^0_c(V)$ is canonically biholomorphic to $\mathbb{D}_c(V)$.

Proof. For $\xi \in H^0(V, \mathcal{O}(F_t)) - 0$, we denote by $\langle \xi \rangle$ the line through ξ and 0 in $H^0(V, \mathcal{O}(F_t))$ and by (ξ) the zero divisor of ξ. Then the map

$$(\xi) \in \mathbb{D}_c(V) \longmapsto \langle \xi \rangle \in \mathbb{G}^0_c(V)$$

is clearly bijective and makes the diagram

$$\begin{array}{ccc} \mathbb{D}_c(V) & \longrightarrow & \mathbb{G}^0_c(V) \\ & {\scriptstyle\Phi}\searrow \quad \swarrow{\scriptstyle\mu} & \\ & \mathrm{Pic}^o(V) & \end{array}$$

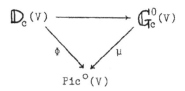

commutative. We show that it is biholomorphic.

Let us first recall the construction of the space $\mathbb{D}_c(V)$ by Weil [85]. (See also Kodaira [40].) Let E be a general hypersurface sectio on V in an ambient complex projective space such that the cohomology class $c([D_o+E]\otimes K_V^{-1})$ is positive, where K_V is the canonical bundle of V. Then

$$\Lambda = \{D \mid D \text{ is an effective divisor on } V \text{ with } c([D]) = c([D_o+E])\}$$

is a fiber bundle over $\mathrm{Pic}^o(V)$ whose standard fiber is a complex projective space. The projection of Λ onto $\mathrm{Pic}^o(V)$ is given by

$$\Phi' : D \in \Lambda \longmapsto [D - D_o - E] \in \mathrm{Pic}^o(V) .$$

Put

$$\mathbb{M} = \{D \in \Lambda \mid D - E \text{ is effective}\} .$$

Then $\mathbb{M}$ is a closed complex subspace of Λ. The map

$$D \in \mathbb{M} \longmapsto D - E \in \mathbb{D}_c(V)$$

is clearly bijective. Using this bijection, we introduce a complex space structure (in fact, an algebraic structure) on $\mathbb{D}_c(V)$. It can be proved that it does not depend on the choice of the auxiliary hypersurface section E of V.

Now, the projective fiber bundle Λ above is equal to $\mathbb{D}_{c'}(V)$, where $c' = c([D_o+E])$. Put

$$F'_t = B_t \otimes [D_o + E], \quad \text{for} \quad t \in \mathrm{Pic}^o(V) .$$

We first show that the complex space $\mathbb{G}^0 = \mathbb{G}^0_{c'}(V)$ given in Theorem 4.1.9 with respect to the family $\{F'_t, V\}_{t\in \mathrm{Pic}^o(V)}$ is canonically biholomorphic to Λ.

The assumption that $c([D]\otimes K_V^{-1})$ is positive implies that

$H^1(V, \mathcal{O}([D])) = 0$ (Kodaira vanishing theorem.) Hence, by Theorem 4.1.2, the linear fiber space $\hat{\mathbb{H}}^0$ given in Theorem 4.1.8 with respect to the family $\{F'_t, V\}_{t \in \mathrm{Pic}^0(V)}$ is a holomorphic vector bundle over $\mathrm{Pic}^0(V)$. Moreover, $\mathbb{G}^0_{c'}(V)$ is the projective bundle associated to $\hat{\mathbb{H}}^0$.

On the other hand, there is a holomorphic vector bundle $\mathbb{B}$ on $\mathrm{Pic}^0(V)$ such that Λ is the projective bundle associated to $\mathbb{B}$ (see Kodaira [40, p.730]). Both $\hat{\mathbb{H}}^0$ and $\mathbb{B}$ are identified with the disjoint union $\cup_t H^0(V, \mathcal{O}(F'_t))$. Thus, for our purpose, it is enough to show that the bijection $\beta : \mathbb{B} \longrightarrow \hat{\mathbb{H}}^0$ through the identification is biholomorphic. $\mathbb{B}$ satisfies the condition in Proposition 4.1.13, for $\mathbb{B}$ is locally trivial. Hence β is holomorphic. It is biholomorphic, for it is a bijective holomorphic map between complex manifolds.

Now, we prove that the map $\mathbb{D}_c(V) \longrightarrow \mathbb{G}^0_c(V)$ defined above is biholomorphic. Let $\{U_i\}$ be a finite Stein open covering of V. Assume that E is defined by a holomorphic function e_i on U_i, i.e.,

$$E \cap U_i = \{z \in U_i \mid e_i(z) = 0\} .$$

Then $e = \{e_i(z)\}$ is a holomorphic section of $[E]$ such that $(e) = E$. The map

$$i_e : \langle \xi \rangle \in \mathbb{G}^0_c(V) \longmapsto \langle \xi e \rangle \in \mathbb{G}^0_{c'}(V)$$

is clearly injective. By the local construction of $\mathbb{G}^0_c(V)$ in the proof of Theorem 4.1.2, we can easily show that i_e is a holomorphic imbedding. Now, the commutative diagram:

$$\begin{array}{ccc}
(\xi e) \in \mathbb{M}\mathbb{G}\Lambda & \overset{\cong}{\longmapsto} & \langle \xi e \rangle \in \mathbb{G}^0_{c'}(V) \\
\uparrow \wr\!\mid & & \uparrow i_e \\
(\xi) \in \mathbb{D}_c(V) & \longmapsto & \langle \xi \rangle \in \mathbb{G}^0_c(V)
\end{array}$$

implies that $\mathbb{D}_c(V) \longrightarrow \mathbb{G}^0_c(V)$ is biholomorphic. <u>Q.E.D.</u>

The following corollaries are easily proved by using the local construction of $\mathbb{G}_c^0(V)$ in the proof of Theorem 4.1.2.

Corollary 4.2.2. Let E be an effective divisor on V. Then the map

$$D \in \mathbb{D}_c(V) \longmapsto D + E \in \mathbb{D}_{c'}(V)$$

$(c = c([D_o])$ and $c' = c([D_o + E]))$ is a holomorphic imbedding.

Corollary 4.2.3. For $c, c' \in H^{1,1}(V, \mathbb{Z})$,

$$(D, D') \in \mathbb{D}_c(V) \times \mathbb{D}_{c'}(V) \longmapsto D + D' \in \mathbb{D}_{c+c'}(V)$$

is a holomorphic map.

For a divisor D on V, put $h^\nu(D) = \dim H^\nu(V, \mathcal{O}([D]))$ for $\nu = 0, 1, \cdots$. The following theorem was announced in Namba [65].

Theorem 4.2.4. Assume that there is $D \in \mathbb{D}_c(V)$ such that $h^0(D) > h^1(D)$. Then the Jacobi map $\Phi : \mathbb{D}_c(V) \longrightarrow \mathrm{Pic}^0(V)$ is surjective and each fiber of Φ has dimension at least $h^0(D) - h^1(D) - 1$.

Proof. By Proposition 4.2.1, we may identify $\mathbb{D}_c(V)$ with $\mathbb{G}_c^0(V)$. Put $o = \Phi(D)$. Then $F_o = [D]$. The proof of Theorem 4.1.2 implies that there are an open neighborhood U of o in $\mathrm{Pic}^0(V)$ and a $(h^1(D) \times h^0(D))$-matrix valued holomorphic funtion $\tilde{u}$ on U such that

$$\Phi^{-1}(U) = \{(\langle \xi \rangle, t) \in G^0(H^0(V, \mathcal{O}(F_o))) \times U \mid \tilde{u}(t)\xi = 0\}.$$

Take a point $t \in U$. If $h^0(D) > h^1(D)$, then $\operatorname{rank} \tilde{u}(t) \leqq h^1(D)$. Hence

$$\dim \ker \tilde{u}(t) \geqq h^0(D) - h^1(D) > 0 .$$

This proves that the restriction of Φ to $\Phi^{-1}(U)$ is a surjective holomorphic map onto U each of whose fiber has dimension at least $h^0(D)-h^1(D)-1$.

Since $\mathrm{Pic}^o(V)$ is irreducible, there must exist a (global) irreducible component X of $\mathbb{D}_c(V)$ such that $\Phi(X) = \mathrm{Pic}^o(V)$. In particular, Φ is surjective. Put

$$a_o = \min \{\dim H^0(V, \mathcal{O}(F_t)) \mid t \in \mathrm{Pic}^o(V)\} ,$$

$$W = \{t \in \mathrm{Pic}^o(V) \mid \dim H^0(V, \mathcal{O}(F_t)) = a_o\} .$$

Then W is a Zariski open subset of $\mathrm{Pic}^o(V)$ by Grauert [22] and Riemenschneider [74]. Since Φ is surjective, a_o is a positive integer. By the local construction of the space $\mathbb{G}_c^0(V)$, $\Phi^{-1}(W)$ is a $\mathbb{P}^{a_o-1}$-bundle over W. In particular, it is a complex manifold. Hence $\Phi^{-1}(W)$ is contained in a (global) irreducible component X' of $\mathbb{D}_c(V)$. We show that $X' = X$. Since $\Phi(X) = \mathrm{Pic}^o(V)$, there is $D_1 \in X$ such that $\Phi(D_1) \in W$. Then $D_1 \in X \cap \Phi^{-1}(W)$. Thus $X' = X$. (This shows that X' is the unique irreducible component such that $\Phi(X') = \mathrm{Pic}^o(V)$.)

Now, $W \cap U$ is a non-empty open subset of $\mathrm{Pic}^o(V)$. For a point $t \in W \cap U$,

$$a_o - 1 = \dim \Phi^{-1}(t) \geqq h^0(D) - h^1(D) - 1 .$$

By the definition of the integer a_o,

$$\dim \Phi^{-1}(t) \geqq h^0(D) - h^1(D) - 1 \quad \text{for all} \quad t \in \mathrm{Pic}^o(V) .$$

Q.E.D.

Remark 4.2.5. If V is a compact Riemann surface of genus g, then Theorem 4.2.4 reduces to Jacobi inversion. In fact, it is well known that $\mathbb{D}_c(V)$ is canonically identified with S^nV, where $n = \int_V c$. (We give a proof of this fact in §5.1.) By Riemann-Roch theorem, $h^0(D) - h^1(D) - 1 = n - g$ for $D \in S^nV$.

In general, we put $W_c = \Phi(\mathbb{D}_c(V))$. It is a closed complex subspace of $\mathrm{Pic}^0(V)$.

Theorem 4.2.6. For $D \in \mathbb{D}_c(V)$, assume that $h^0(D) \leqq h^1(D)$. Then, there are an open neighborhood U of $o = \Phi(D)$ and a $(h^1(D) \times h^0(D))$-matrix valued holomorphic function $\tilde{u}(t)$, $t \in U$, on U such $W_c \cap U$ is the set of zeros of all $(h^0(D) \times h^0(D))$-minors of $\tilde{u}(t)$.

Proof. We use the same notations as in the proof of Theorem 4.2.4. If $h^0(D) \leqq h^1(D)$, then

$$\{t \in U \mid \operatorname{rank} \tilde{u}(t) < h^0(D)\}$$
$$= \{t \in U \mid \text{all } (h^0(D) \times h^0(D))\text{-minors of } \tilde{u}(t) \text{ are zero}\}.$$

Q.E.D.

Remark 4.2.7. If V is a compact Riemann surface, then this theorem is known as Kempf's theorem (see Mumford [59, p.58]).

4.3. Semi-regularity theorem for linear systems.

For $c \in H^{1,1}(V, \mathbb{Z})$ and $r \geqq 0$, let $\mathbb{G}^r_c(V)$ be the complex space in Proposition 4.2.1 and let $\mu : \mathbb{G}^r_c(V) \longrightarrow \mathrm{Pic}^0(V)$ be the projection. The space $\mathbb{G}^r_c(V)$ is regarded as the set of all linear systems g^r_c on V of dimension r such that $c([D]) = c$ for any $D \in g^r_c$.

We want to apply Theorem 4.1.10 to the complex space $\mathbb{G}^r_c(V)$. We freely use the arguments in Kodaira [40].

As before, take $D_o \in \mathbb{D}_c(V)$ and fix it. For a suitable finite Stein open covering $\{U_i\}$ of V, D_o is defined by

$$D_o = \{f^o_i(z) = 0\},$$

where $f_i^o(z)$ is a holomorphic function of $z \in U_i$. Then the line bundle $[D_o]$ is given by the transition function $f_{ik}^o(z) = f_i^o(z)/f_k^o(z)$, $z \in U_i \cap U_k$.

Let $\{\omega_1, \cdots, \omega_q\}$ be a basis of $H^0(V, \Omega^1)$, the space of holomorphic 1-forms on V. Let $\overline{\omega_\nu}$ be the complex conjugate of ω_ν. Then $\{\overline{\omega_1}, \cdots, \overline{\omega_q}\}$ forms a basis of $H^0(V, \overline{\Omega^1})$, the space of anti-holomorphic 1-forms on V. We choose a point $z(i)$ in U_i. For a point $o \in \mathrm{Pic}^o(V)$, let U be a small open neighborhood of o in $\mathrm{Pic}^o(V)$. Then, the transition function $f_{ik}(z,t)$ of the line bundle $F_t = B_t \otimes [D_o]$, $t \in U$, is given by

$$f_{ik}(z,t) = f_{ik}^o(z)\tilde{f}_{ik}(t), \quad z \in U_i \cap U_k,$$

where

$$\tilde{f}_{ik}(t) = \exp \int_{z(i)}^{z(k)} (t_1\overline{\omega_1} + \cdots + t_q\overline{\omega_q}).$$

Here $\int_{z(i)}^{z(k)}$ denotes the integral along a smooth curve in $U_i \cup U_k$ combining $z(i)$ with $z(k)$ and $(t_1, \cdots, t_q)$ is the coordinate of the point $t \in U$ induced by the homomorphism:

$$H^0(V, \overline{\Omega^1}) \longrightarrow \mathrm{Pic}^o(V) \longrightarrow 0 .$$

Thus, for $\partial/\partial t = \Sigma_\nu c^\nu(\partial/\partial t^\nu) \in T_o\mathrm{pic}^o(V)$,

$$\begin{aligned}(\partial f_{ik}/\partial t)_{(z_k,o)} &= f_{ik}^o(z_k)\tilde{f}_{ik}(o) \int_{z(i)}^{z(k)} (\Sigma_\nu c^\nu\overline{\omega_\nu}) \\ &= f_{ik}(z_k,o) \int_{z(i)}^{z(k)} (\Sigma_\nu c^\nu\overline{\omega_\nu}) .\end{aligned}$$

Hence, for $\xi \in H^0(V, \mathcal{O}(F_o)) - 0$,

$$\begin{aligned}(\tilde{\tau}_\xi(\partial/\partial t))_{ik}(z_i) &= (\partial f_{ik}/\partial t)_{(z_k,o)}\xi_k(z_k) \\ &= f_{ik}(z_k,o)\xi_k(z_k) \int_{z(i)}^{z(k)} (\Sigma_\nu c^\nu\overline{\omega_\nu})\end{aligned}$$

$$= \xi_i(z_i) \int_{z(i)}^{z(k)} (\Sigma_\nu c^\nu \overline{\omega_\nu}) ,$$

where $z_i = g_{ik}(z_k)$ is the coordinate transformation in $U_i \cap U_k$.

We may regard $h = \{\int_{z(i)}^{z(k)} (\Sigma_\nu c^\nu \overline{\omega_\nu})\}$ as an element of $H^1(V, \mathcal{O})$. Then h corresponds to $\Sigma_\nu c^\nu \overline{\omega_\nu} \in H^{0,1}(V)$ under the Dolbeault isomorphism (see Kodaira-Spencer [42, p.871]). Thus we have linear isomorphisms:

$$h \in H^1(V, \mathcal{O}) \longmapsto \Sigma_\nu c^\nu \overline{\omega_\nu} \in H^{0,1}(V) \longmapsto \partial/\partial t = \Sigma_\nu c^\nu (\partial/\partial t^\nu) \in T_o \mathrm{Pic}^0(V) .$$

Define a bilinear map

$$\tau : H^0(V, \mathcal{O}(F_o)) \times H^1(V, \mathcal{O}) \longrightarrow H^1(V, \mathcal{O}(F_o))$$

by

$$\tau(\xi,h)_{ik}(z) = \xi_i(z) h_{ik}(z) ,$$

where $\xi = \{\xi_i(z)\} \in H^0(V, \mathcal{O}(F_o))$ and $h = \{h_{ik}(z)\} \in H^1(V, \mathcal{O})$. Put

$$\tau(\xi,h) = \tau_\xi(h) = \tau(h)(\xi)$$

by abuse of notation.

<u>Definition 4.3.1</u>. A linear system $g_c^r \in \mathbb{G}_c^r$ is said to be <u>semi-regular</u> if there are $D_0, \cdots, D_r \in g_c^r$ such that (1) they are independent as points of the projective space g_c^r and (2) the linear map

$$h \in H^1(V, \mathcal{O}) \longmapsto (\tau_{\xi_0}(h), \cdots, \tau_{\xi_r}(h)) \in H^1(V, \mathcal{O}(F))^{r+1}$$

is surjective, where $F = [D]$ for any $D \in g_c^r$ and $\xi_\alpha \in H^0(V, \mathcal{O}(F))$ satisfies $(\xi_\alpha) = D_\alpha$, $0 \leqq \alpha \leqq r$. ((ξ_α) is the zero divisor of ξ_α.)

Now, the above consideration and Theorem 4.1.10 imply

Theorem 4.3.2. (Semi-regularity theorem for linear systems.) If a linear system $g_c^r \in \mathbb{G}_c^r(V)$ is semi-regular, then it is a non-singular point of $\mathbb{G}_c^r(V)$ and

$$\dim_{g_c^r} \mathbb{G}_c^r(V) = (r+1)(h^0(D) - h^1(D) - r - 1) + q$$

for any $D \in g_c^r$.

If we put $r = 0$, then we get the usual Semi-regularity theorem by Kodaira-Spencer:

Theorem 4.3.3. (Semi-regularity theorem for divisors.) (Kodaira-Spencer [45], Mumford [57], Bloch [6]). If an effective divisor $D \in \mathbb{D}_c(V)$ is semi-regular, then it is a non-singular point of $\mathbb{D}_c(V)$ and

$$\dim_D \mathbb{D}_c(V) = h^0(D) - h^1(D) - 1 + q.$$

Next, we put

$$W_c^r = \{t \in \mathrm{Pic}^0(V) \mid h^0(F_t) \geq r+1\}.$$

Then, the following lemma is easy to see.

Lemma 4.3.4. W_c^r is the image of $\mu : \mathbb{G}_c^r(V) \longrightarrow \mathrm{Pic}^0(V)$. Hence it is a closed complex subspace of $\mathrm{Pic}^0(V)$. Moreover,

$$W_c = W_c^0 \supset W_c^1 \supset W_c^2 \supset \cdots .$$

The following theorem can be proved in a similar way to Theorem 4.2.6.

Theorem 4.3.5. For $o \in \mathrm{Pic}^0(V)$, there are an open neighborhood U of o in $\mathrm{Pic}^0(V)$ and a $(h^1(F_o) \times h^0(F_o))$-matrix valued holomorphic function $\tilde{u}(t)$, $t \in U$, on U such that $W_c^r \cap U$ is the set of zeros of all $(h^0(F_o)-r) \times (h^0(F_o)-r)$-minors of $\tilde{u}(t)$.

Note that the linear part of $\tilde{u}(t)$ is given by

$(d\tilde{u})_o(h) = \tau(h)$ for $h \in H^1(V, \mathcal{O})$, (see Proposition 4.1.7).

Proposition 4.3.6. Let Y be an irreducible component of W_c^r such that $Y \not\subset W_c^{r+1}$. Then,

$$\dim Y \geqq q - (r+1)h^1(F_o) \quad \text{for any point} \quad o \in Y \setminus W_c^{r+1} .$$

Proof. For any point $o \in Y \setminus W_c^{r+1}$, $\mu^{-1}(o)$ consists of a unique point $x_o = g_c^r = |F_o|$, which is a complete linear system. Hence, by the local construction of $\mathbb{G}_c^r(V)$ (see the proof of Theorem 4.1.9), an open neighborhood T of x_o in $\mathbb{G}_c^r(V)$ is given by:

$$T = \{(x_o,t) \in x_o \times U \mid \tilde{u}(t)\xi_\alpha(x_o) = 0, \quad 0 \leqq \alpha \leqq r\} ,$$

where U is an open neighborhood of o in $\mathrm{Pic}^0(V)$ and $\xi_\alpha(x_o)$, $0 \leqq \alpha \leqq r$, form a basis of $H^0(V, \mathcal{O}(F_o))$. Hence T is biholomorphic (under μ) to

$$T' = \{t \in U \mid \tilde{u}(t) = 0\} .$$

It is clear that T' is an open neighborhood of o in W_c^r. Note that $\tilde{u}$ is a $((r+1) \times h^1(F_o))$-matrix valued holomorphic function on U. Hence, for any irreducible component T'' of T',

$$\dim T'' \geqq q - (r+1)h^1(F_o) .$$

In particular,

$$\dim Y \geqq q - (r+1)h^1(F_o) . \qquad \text{Q.E.D.}$$

Theorem 4.3.7. For $D \in \mathbb{D}_c(V)$, let $\dim |D| = r$. If $|D|$ is semi-regular as a linear system, then W_c^r is non-singular at $\Phi(D)$

and $\dim_{\Phi(D)} W_c^r = q - (r+1)h^1(D)$.

<u>Proof</u>. Put $\Phi(D) = o$ and $|D| = x_o \in \mathbb{G}_c^r(V)$. We use the same notations as in the proof of the previous proposition. T is an open neighborhood of x_o in $\mathbb{G}_c^r(V)$. By the assumption that x_o is semi-regular, T is non-singular at x_o and

$$\dim_{x_o} T = (r+1)(h^0(D) - h^1(D) - r - 1) + q = q - (r+1)h^1(D) .$$

Since T' is an open neighborhood of $o = \Phi(D)$ in W_c^r which is biholomorphic to T under μ, T' is non-singular at o and

$$\dim_o T' = q - (r+1)h^1(D). \qquad \text{Q.E.D.}$$

For complex vector spaces A and B, we denote by $\mathrm{Hom}(A,B)$ the set of all linear maps of A into B. Let L be a linear subspace of $H^0(V, \mathcal{O}(F_o))$ of dimension $r+1$. Then the map

$$p_L : \alpha \in \mathrm{Hom}(H^0(V, \mathcal{O}(F_o)), H^1(V, \mathcal{O}(F_o))) \longmapsto \alpha \mid L \in \mathrm{Hom}(L, H^1(V, \mathcal{O}(F_o)))$$

is a surjective homomorphism. ($\alpha \mid L$ is the restriction of α to L.) Put

$$\tau_L = p_L \circ \tau : H^1(V, \mathcal{O}) \longrightarrow \mathrm{Hom}(L, H^1(V, \mathcal{O}(F_o))) .$$

Then, by Lemma 4.1.11,

<u>Lemma 4.3.8</u>. L is semi-regular if and only if τ_L is surjective.

<u>Corollary 4.3.9</u>. Let L and L' be linear subspaces of $H^0(V, \mathcal{O}(F_o))$ such that $L' \subset L$. If L is semi-regular, then L' is also semi-regular.

Note that, if L ($\dim L = r+1$) is semi-regular, then

$q \geqq (r+1)h^1(F_o)$.

The map $\mu : \mathbb{G}^r_c(V) \longrightarrow \mathrm{Pic}^o(V)$ is locally the restriction of the projection $G^r(H^0(V, \mathcal{O}(F_o))) \times \mathrm{Pic}^o(V) \longrightarrow \mathrm{Pic}^o(V)$. Hence

<u>Lemma 4.3.10</u>. If L is semi-regular, then the tangent space to $\mathbb{G}^r_c(V)$ at L is isomorphic to $T_L G^r(H^0(V, \mathcal{O}(F_o))) \times \ker \tau_L$.

This lemma clearly implies

<u>Proposition 4.3.11</u>. For a point $o \in \mathrm{Pic}^o(V)$, assume that <u>every</u> $L \in G^r(H^0(V, \mathcal{O}(F_o)))$ is semi-regular. Then, the normal bundle of $G^r(H^0(V, \mathcal{O}(F_o)))$ in $\mathbb{G}^r_c(V)$ is canonically identified with

$$N_o = \{(L,h) \in G^r(H^0(V, \mathcal{O}(F_o))) \times H^1(V, \mathcal{O}) \mid \tau_L(h) = 0\} .$$

Let $TC_o(W^r_c)$ be the <u>tangent cone of</u> W^r_c <u>at</u> o. By definition,

$$TC_o(W^r_c) = \mathrm{Spec}(\bigoplus_{i=0}^{\infty} m^i/m^{i+1}) ,$$

where m is the maximal ideal of the local ring $\mathcal{O}_{W^r_c, o}$. Under the assumption of the proposition, the holomorphic map $\mu : \mathbb{G}^r_c(V) \longrightarrow \mathrm{Pic}^o(V)$ induces a proper morphism of schemes:

$$\lambda_o : (L,h) \in N_o \longmapsto h \in TC_o(W^r_c) \subset H^1(V, \mathcal{O}) .$$

It is clear that its image is equal to

$$M_o = \{h \in H^1(V, \mathcal{O}) \mid \text{all } (h^0(F_o)-r) \times (h^0(F_o)-r)\text{-minors of } \tau(h) \text{ are zero}\} .$$

On the other hand, by Theorem 4.3.5, $TC_o(W^r_c)$ is, as a subset of $H^1(V, \mathcal{O})$, equal to the tangent cone $TC_o(\widetilde{M})$ at o of the closed comple subspace of $U \subset \mathrm{Pic}^o(V)$:

$$\widetilde{M} = \{t \in U \mid \text{all } (h^0(F_o)-r) \times (h^0(F_o)-r)\text{-minors of } \widetilde{u}(t) \text{ are zero}\}.$$

We know that the linear term of $\widetilde{u}$ is τ. Hence, every $(h^0(F_o)-r) \times (h^0(F_o)-r)$-minor of $\tau(h)$ appears as a leading term of a $(h^0(F_o)-r) \times (h^0(F_o)-r)$-minor of $\widetilde{u}(t)$ as a function of t. This means that $TC_o(\widetilde{M}) \subset M_o$. Hence, <u>as subsets of</u> $H^1(V, \mathcal{O})$,

$$\lambda_o(N_o) = TC_o(W_c^r) = TC_o(\widetilde{M}) = M_o .$$

Thus, we get the following lemma, which is a special case of Kempf's lemma, [37].

<u>Lemma 4.3.12</u>. Under the same assumption as Proposition 4.3.11, the proper morphism $\lambda_o : N_o \longrightarrow TC_o(W_c^r)$ is surjective.

Note that <u>as closed subschemes of</u> $H^1(V, \mathcal{O})$,

$$\lambda_o(N_o) \subset TC_o(W_c^r) \subset TC_o(\widetilde{M}) \subset M_o .$$

($\lambda_o(N_o)$ is reduced and irreducible.) Kempf [37] proved that they are all equal, if $r = 0$. His beautiful theorem is stated in our case as follows:

<u>Theorem 4.3.13. (Kempf's theorem for projective manifolds)</u>. For a point $o \in \mathrm{Pic}^0(V)$, assume that <u>every</u> divisor in $\Phi^{-1}(o)$ is semi-regular. Assume moreover that $h^0(F_o) \leqq h^1(F_o)+1$. Let N_o, $TC_o(W_c)$ and $\lambda_o : N_o \longrightarrow TC_o(W_c)$ be the normal bundle of $\Phi^{-1}(o)$ in $\mathbb{D}_c(V)$, the tangent cone of $W_c = \Phi(\mathbb{D}_c(V))$ at o and the morphism induced by the Jacobi map Φ, respectively. Then

(1) $\lambda_o : N_o \longrightarrow TC_o(W_c)$ is a rational resolution.

(2) The degree of $TC_o(W_c)$ is the binomial coefficient $\binom{h^1(F_o)}{h^0(F_o)-1}$.

(3) If $h^0(F_o) \leqq h^1(F_o)$, then the ideal defining $TC_o(W_c)$ is generated

by the maximal minors of the matrix $\tau(h)$ on $H^1(V, \Theta)$.

Remark 4.3.14. A proper birational morphism $h : X \longrightarrow Y$ between varieties is called a rational resolution if X is smooth, Y is normal and Cohen-Macaulay and the higher direct images $R^i h_* \mathcal{O}_X$ are zero for all $i > 0$ (see Kempf [37]).

Problem. Extend Kempf's theorem for $r > 0$.

4.4. Moduli of non-degenerate holomorphic maps into $\mathbb{P}^r$.

A holomorphic map f of a projective manifold V into $\mathbb{P}^r$ is said to be non-degenerate if the image $f(V)$ is not contained in any hyperplane of $\mathbb{P}^r$. (Note: In some literatures, this terminology has a different meaning. They say that a holomorphic map $f : V \longrightarrow W$ is said to be non-degenerate if there is a point $P \in V$ such that $\operatorname{rank}(df)_P = \dim V$.)

Let $f : V \longrightarrow \mathbb{P}^r$ be a non-degenerate holomorphic map. Then the hyperplane sections determine a linear system g_c^r on V of dimension r with neither fixed part nor base point. ($c = c([V \cap H])$, H = a hyperplane of $\mathbb{P}^r$.)

If $b \in \operatorname{Aut}(\mathbb{P}^r)$, then the composition bf is again a non-degenerate holomorphic map and determines the same linear system g_c^r as f. Conversely, if g_c^r is a linear system with neither fixed part nor base point, then g_c^r gives a non-degenerate holomorphic map determined up to $\operatorname{Aut}(\mathbb{P}^r)$.

Put

$$\mathbb{G}^r(V) = \cup_c \mathbb{G}_c^r(V), \quad \text{(the disjoint union).}$$

This is a complex space, (by considering every $\mathbb{G}_c^r(V)$ an open subspace of it). Put

$$\mathbb{F}^r(V) = \{g_c^r \in \mathbb{G}^r(V) \mid g_c^r \text{ has a fixed part or a base point}\},$$

$$\mathbb{F}_c^r(V) = \mathbb{F}^r(V) \cap \mathbb{G}_c^r(V).$$

<u>Lemma 4.4.1</u>. $\mathbb{F}^r(V)$ is a closed complex subspace of $\mathbb{G}^r(V)$.

<u>Proof</u>. Note that $\mathbb{F}^r(V) = \cup_c \mathbb{F}_c^r(V)$, the disjoint union. Hence it is enough to show that $\mathbb{F}_c^r(V)$ is a closed complex subspace of $\mathbb{G}_c^r(V)$. Consider the set

$$A = \{(P, g_c^r) \in V \times \mathbb{G}_c^r(V) \mid P \in D \text{ for all } D \in g_c^r\}.$$

If we show that the set A is a closed complex subspace of $V \times \mathbb{G}_c^r(V)$, then the lemma is proved. In fact, $\mathbb{F}_c^r(V) = p_2(A)$, where $p_2 : V \times \mathbb{G}_c^r(V) \longrightarrow \mathbb{G}_c^r(V)$ is the projection.

Now, take any points $o \in \mathrm{Pic}^0(V)$ and $x_o \in G^r(H^0(V, \mathcal{O}(F_o)))$. We use the same notations as in the proof of Theorem 4.1.9. $\mathbb{G}_c^r(V)$ is locally given by

$$\{(x,s) \in U \times S' \mid \tilde{u}(s)\xi_\alpha(x) = 0, \quad 0 \leqq \alpha \leqq r\}.$$

Put

$$\Psi(\xi_\alpha(x), s) = (\{\Psi_i(\xi_\alpha(x), s)\}, s).$$

(Ψ is the bundle map defined in the proof of Theorem 4.1.2.) Then $\Psi_i(\xi_\alpha(x), s)(z_i)$ depends holomorphically on $(z_i, x, s) \in U_i \times U \times S'$. Put

$$A_i = \{(z_i, x, t) \in U_i \times U \times S' \mid \tilde{u}(s)\xi_\alpha(x) = 0 \text{ and } \Psi_i(\xi_\alpha(x), s)(z_i) = 0 \text{ for } 0 \leqq \alpha \leqq r\}.$$

Then A_i is a closed complex subspace of $U_i \times U \times S'$. Since A is locally given by A_i, the lemma is proved. <u>Q.E.D.</u>

Put

$$\mathrm{Hol}_{\text{non-deg}}(V, \mathbb{P}^r) = \{\text{all non-degenerate holomorphic maps of } V \text{ into } \mathbb{P}^r\}.$$

It is an open subspace of $\mathrm{Hol}(V, \mathbb{P}^r)$. It is clear that $\mathrm{Aur}(\mathbb{P}^r)$ acts <u>freely</u> on $\mathrm{Hol}_{\text{non-deg}}(V, \mathbb{P}^r)$. As was noted above, there is a canonical bijection:

$$\alpha : \mathrm{Hol}_{\text{non-deg}}(V, \mathbb{P}^r)/\mathrm{Aut}(\mathbb{P}^r) \longrightarrow \mathbb{G}^r(V) - \mathbb{F}^r(V).$$

Then, as is expected, we get the following theorem.

<u>Theorem 4.4.2</u>. The orbit space $\mathrm{Hol}_{\text{non-deg}}(V, \mathbb{P}^r)/\mathrm{Aut}(\mathbb{P}^r)$ has a complex space structure such that (1) it is biholomorphic to $\mathbb{G}^r(V) - \mathbb{F}^r(V)$ under α and (2) the projection

$$\mathrm{Hol}_{\text{non-deg}}(V, \mathbb{P}^r) \longrightarrow \mathrm{Hol}_{\text{non-deg}}(V, \mathbb{P}^r)/\mathrm{Aut}(\mathbb{P}^r)$$

is a principal $\mathrm{Aut}(\mathbb{P}^r)$-bundle.

<u>Proof</u>. $\mathbb{G}^r_c(V)$ is, as a set, equal to the disjoint union of Grassmann varieties $G^r(H^0(V, \mathcal{O}(F_t)))$, $t \in \mathrm{Pic}^0(V)$. If we consider the disjoint union

$$S^r_c(V) = \cup_t S^r(H^0(V, \mathcal{O}(F_t)))$$

of Stiefel varieties $S^r(H^0(V, \mathcal{O}(F_t)))$ (= the set of all (r+1)-frames $(\phi_0, \cdots, \phi_r)$ of $H^0(V, \mathcal{O}(F_t))$), then $S^r_c(V)$ has a complex space structure. This is shown in a similar way to Theorem 4.1.9. Put

$$S^r(V) = \cup_c S^r_c(V), \quad \text{(the disjoint union)}.$$

It is a complex space, (by considering every $S^r_c(V)$ an open subspace of it). The projection

$$p : S^r(V) \longrightarrow \mathbb{G}^r(V)$$

is holomorphic and has a holomorphic local cross section through any point of $S^r(V)$. This follows from the local construction of the spaces $S^r(V)$ and $\mathbb{G}^r(V)$. It is clear that the general linear group $GL(r+1,\mathbb{C})$ acts freely on $S^r(V)$ and there is a canonical bijection

$$\gamma : S^r(V)/GL(r+1,\mathbb{C}) \longrightarrow \mathbb{G}^r(V) .$$

The orbit space $S^r(V)/GL(r+1,\mathbb{C})$ has a complex space structure such that (1) it is biholomorphic to $\mathbb{G}^r(V)$ through γ and (2) the projection $S^r(V) \longrightarrow S^r(V)/GL(r+1,\mathbb{C})$ is a principal $GL(r+1,\mathbb{C})$-bundle. This follows from the fact that $p : S^r(V) \longrightarrow \mathbb{G}^r(V)$ has holomorphic local cross sections. Put

$$\widetilde{F}^r(V) = \{(\phi_0,\cdots,\phi_r) \in S^r(V) \mid \text{there is } P \in V \text{ such that } \phi_\alpha(P) = 0,\ 0 \leqq \alpha \leqq r\} .$$

Then, as Lemma 4.4.1, we can show that $\widetilde{F}^r(V)$ is a closed complex subspace of $S^r(V)$. Note that

$$p(\widetilde{F}^r(V)) = \mathbb{F}^r(V) \quad \text{and} \quad p^{-1}(\mathbb{F}^r(V)) = \widetilde{F}^r(V) .$$

Note also that $GL(r+1,\mathbb{C})$ acts freely on $\widetilde{F}^r(V)$ and the bijection

$$\widetilde{F}^r(V)/GL(r+1,\mathbb{C}) \longrightarrow \mathbb{F}^r(V)$$

is biholomorphic.

Now, $(\phi_0,\cdots,\phi_r) \in S^r(V) - \widetilde{F}^r(V)$ determines a non-degenerate holomorphic map

$$f : P \in V \longmapsto (\phi_0(P):\cdots:\phi_r(P)) \in \mathbb{P}^r .$$

Conversely, a non-degenerate holomorphic map $f : V \longrightarrow \mathbb{P}^r$ determines an element of $S^r(V) - \widetilde{F}^r(V)$ up to non-zero constant. Hence there is a bijection

$$(S^r(V) - \widetilde{F}^r(V))/\mathbb{C}^* \longrightarrow \mathrm{Hol}_{\text{non-deg}}(V, \mathbb{P}^r) .$$

We show that this is biholomorphic. In fact, the map

$$(P,(\phi_0,\cdots,\phi_r)) \in V \times (S^r(V) - \tilde{F}^r(V)) \longmapsto (\phi_0(P):\cdots:\phi_r(P)) \in \mathbb{P}^r$$

is holomorphic. (c.f., Lemma 4.1.16). On the other hand, if Y is a complex space and $F : V \times Y \longrightarrow \mathbb{P}^r$ is a holomorphic map, then F is locally written as

$$F(P,y) = (\phi_0(y)(P):\cdots:\phi_r(y)(P)) ,$$

where $(\phi_0(y),\cdots,\phi_r(y)) \in S^r(V) - \tilde{F}^r(V)$. It is easy to see that ϕ_α can be chosen so that $\phi_\alpha(y)(P)$ depends holomorphically on (P,y). Hence, Proposition 4.1.13 implies that $\phi_\alpha(y)$ $(\in \mathbb{H}^0)$ depends holomorphically on y. This shows that

$$y \in Y \longmapsto (\phi_0(y),\cdots,\phi_r(Y)) \in S^r(V) - \tilde{F}^r(V)$$

is holomorphic. Thus $S^r(V) - \tilde{F}^r(V)$ has a maximal property (see §0.2 or §3.2). Thus the canonical map

$$S^r(V) - \tilde{F}^r(V) \longrightarrow \mathrm{Hol}_{\text{non-deg}}(V, \mathbb{P}^r)$$

is holomorphic and is a principal $\mathbb{C}^*$-bundle. Hence $\mathrm{Hol}_{\text{non-deg}}(V, \mathbb{P}^r)$ is biholomorphic to $(S^r(V) - \tilde{F}^r(V))/\mathbb{C}^*$.

Now, $\mathrm{Aut}(\mathbb{P}^r)$ acts freely on $(S^r(V) - \tilde{F}^r(V))/\mathbb{C}^*$ and the orbit space $((S^r(V) - \tilde{F}^r(V))/\mathbb{C}^*)/\mathrm{Aut}(\mathbb{P}^r)$ has a complex space structure such that (1) it is biholomorphic to $\mathbb{G}^r(V) - \mathbb{F}^r(V)$ and (2) the projection

$$(S^r(V) - \tilde{F}^r(V))/\mathbb{C}^* \longrightarrow ((S^r(V) - \tilde{F}^r(V))/\mathbb{C}^*)/\mathrm{Aut}(\mathbb{P}^r)$$

is a principal $\mathrm{Aut}(\mathbb{P}^r)$-bundle.

This completes the proof of the theorem. Q.E.D.

<u>Lemma 4.4.3</u>. $\mathrm{Aut}(V)$ acts on $\mathbb{G}^r(V)$ and on $\mathbb{F}^r(V)$.

Proof. For $a \in \mathrm{Aut}(V)$ and $D = (\xi) \in \mathbb{D}_c(V)$, $(\xi \in H^0(V, \mathcal{O}(F_o)))$, we put

$$(a^{-1})^*D = (\xi \cdot a^{-1}) \in \mathbb{D}_{(a^{-1})^*c}(V) ,$$

where $\xi \cdot a^{-1} \in H^0(V, \mathcal{O}((a^{-1})^*F_o))$ and $(a^{-1})^*c = c([(\xi \cdot a^{-1})])$. Thus $\mathrm{Aut}(V)$ acts on $\mathbb{D}(V) = \cup_c \mathbb{D}_c(V)$ as follows:

$$(a,D) \longmapsto (a^{-1})^*D .$$

It is easy to see that the action is holomorphic.

Next, for $g_c^r \in \mathbb{G}^r(V)$, put

$$(a^{-1})^*g_c^r = \{(a^{-1})^*D \mid D \in g_c^r\} .$$

Then $(a^{-1})^*g_c^r$ is an element of $\mathbb{G}^r(V)$ and

$$(a, g_c^r) \longmapsto (a^{-1})^*g_c^r$$

defines an action of $\mathrm{Aut}(V)$ on $\mathbb{G}^r(V)$ (and on $\mathbb{F}^r(V)$). It is easy to see that the action is holomorphic. Q.E.D.

Two non-degenerate holomorphic maps $f, h : V \longrightarrow \mathbb{P}^r$ are said to be equivalent, $f \sim h$, if there are automorphisms $a \in \mathrm{Aut}(V)$ and $b \in \mathrm{Aut}(\mathbb{P}^r)$ such that $ha = bf$. The set of all equivalence classes is nothing but the orbit space

$$M(V, \mathbb{P}^r) = \mathrm{Hol}_{\text{non-deg}}(V, \mathbb{P}^r)/(\mathrm{Aut}(\mathbb{P}^r) \times \mathrm{Aut}(V)) ,$$

where the action is given by

$$(b,a,f) \longmapsto bfa^{-1}.$$

In general, it may not even be a Hausdorff space. If there is a suitable complex space structure on it, we may call it the moduli space of non-degenerate holomorphic maps of V into $\mathbb{P}^r$.

Theorem 4.4.4. If $\mathrm{Aut}(V)$ is compact, then the orbit space $M(V, \mathbb{P}^r) = \mathrm{Hol}_{\text{non-deg}}(V, \mathbb{P}^r)/(\mathrm{Aut}(\mathbb{P}^r) \times \mathrm{Aut}(V))$ has a complex space structure such that (1) it is biholomorphic to $(\mathbb{G}^r(V) - \mathbb{F}^r(V))/\mathrm{Aut}(V)$ and (2) if U is open in $\mathrm{Hol}_{\text{non-deg}}(V, \mathbb{P}^r)$ and f is a holomorphic function on U which is stable under the action of $\mathrm{Aut}(\mathbb{P}^r) \times \mathrm{Aut}(V)$, then f induces a holomorphic function $\hat{f}$ on $p(U)$. (p : $\mathrm{Hol}_{\text{non-deg}}(V, \mathbb{P}^r) \longrightarrow M(V, \mathbb{P}^r)$ is the projection.)

Proof. By the assumption and by Holmann's theorem ([31, Satz 19]), $\mathbb{G}^r(V)/\mathrm{Aut}(V)$ has a complex space structure, which is characterized by the following property: If U is open in $\mathbb{G}^r(V)$ and if f is a holomorphic function on U which is stable under the action of $\mathrm{Aut}(V)$, then f induces a holomorphic function $\hat{f}$ on $\hat{p}(U)$, ($\hat{p}$: $\mathbb{G}^r(V) \longrightarrow \mathbb{G}^r(V)/\mathrm{Aut}(V)$ is the projection map).

Note that $\mathrm{Aut}(V)$ acts on $\mathbb{F}^r(V)$ and the action is compatible with the biholomorphic map

$$\alpha : \mathrm{Hol}_{\text{non-deg}}(V, \mathbb{P}^r)/\mathrm{Aut}(\mathbb{P}^r) \longrightarrow \mathbb{G}^r(V) - \mathbb{F}^r(V)$$

(see Theorem 4.4.2). Hence there is a bijection:

$$\hat{\alpha} : M(V, \mathbb{P}^r) \longrightarrow (\mathbb{G}^r(V) - \mathbb{F}^r(V))/\mathrm{Aut}(V) .$$

Through this bijection, we introduce a complex space structure on $M(V, \mathbb{P}^r)$. Then it has the property (2). This follows from the property of $\mathbb{G}^r(V)/\mathrm{Aut}(V)$ mensioned above and from Theorem 4.4.2. Q.E.D.

We give here one of the simplest examples.

Example 4.4.5. (c.f., §1.4 and Remark 3.6.14.) Let $V = \mathbb{C}/(\mathbb{Z}+\mathbb{Z}\omega)$ be a complex 1-torus. Let 0 be the zero of V as a group. Let n be an integer such that $n \geqq 3$. Let C_n be the image of the map

$$\Phi_{|n0|} : V \longrightarrow \mathbb{P}^{n-1} .$$

Then $\Phi_{|n0|}$ is a biholomorphic map of V onto C_n, a non-singular elliptic curve of degree n in $\mathbb{P}^{n-1}$. Let $\mathcal{S}_n^r$ be the open subspace of the Grassmann variety of all (n-2-r)-dimensional linear subspaces of $\mathbb{P}^{n-1}$ which do not intersect with C_n. Let G be the finite subgroup of $\mathrm{Aut}(V)$ defined in §1.4. Then, G acts on $\mathbb{P}^{n-1}$ as a finite group of projective transformations sending C_n onto itself. Thus G acts on $\mathcal{S}_n^r$.

Now, as in §1.4, $M(V, \mathbb{P}^r)$ is divided into connected components as follows:

$$M(V, \mathbb{P}^r) = M_{r+1}(V, \mathbb{P}^r) \cup M_{r+2}(V, \mathbb{P}^r) \cup \cdots ,$$

where

$$M_{r+1}(V, \mathbb{P}^r) = \text{one point} ,$$

$$M_n(V, \mathbb{P}^r) \cong \mathcal{S}_n^r / G \quad \text{for} \quad n \geqq r+2 .$$

The complex space $M_n(V, \mathbb{P}^r)$ can be considered as the <u>moduli space of non-degenerate holomorphic maps</u> $f : V \longrightarrow \mathbb{P}^r$ <u>such that</u>

$$n = (\mathrm{ord}\ f)\cdot(\deg f(V)) ,$$

where $\mathrm{ord}\ f$ is the mapping order of $f : V \longrightarrow f(V)$.

Note that every element of $M_n(V, \mathbb{P}^r)$ is obtained as follows: (1) $\Phi_{|(r+1)0|}$, if $n = r+1$. (2) the composition

$$V \xrightarrow{\Phi_{|n0|}} C_n \longrightarrow \mathbb{P}^r \ (\mathrm{mod}\ G), \quad \text{if} \quad n \geqq r+2 ,$$

where $C_n \longrightarrow \mathbb{P}^r$ is the projection with the center $L \in \mathcal{S}_n^r$.

Chapter 5. Families of linear systems on compact Riemann surfaces.

5.1. $\mathbb{G}_n^r(V)$ for a compact Riemann surface V.

Let V be a compact Riemann surface of genus g. Then $H^{1,1}(V, \mathbb{Z}) = H^2(V, \mathbb{Z})$ is isomorphic to $\mathbb{Z}$. The isomorphism is given by

$$c \in H^{1,1}(V, \mathbb{Z}) \longmapsto n = \int_V c \in \mathbb{Z}.$$

We identify $H^{1,1}(V, \mathbb{Z})$ with $\mathbb{Z}$ using this isomorphism. Then $\mathbb{G}_c^r(V)$ is written as $\mathbb{G}_n^r(V)$. Note that $\mathbb{G}_n^r(V)$ is <u>the set of all linear systems</u> g_n^r <u>on</u> V <u>of degree</u> n <u>and of dimension</u> r. It is empty for $n \leqq 0$. For $r \geqq 1$, put

$$\mathbb{F}_n^r(V) = \{g_n^r \in \mathbb{G}_n^r(V) \mid g_n^r \text{ has a fixed point}\}.$$

Then, by Lemma 4.4.1, $\mathbb{F}_n^r(V)$ is a closed complex subspace of $\mathbb{G}_n^r(V)$. Put

$$\mathrm{Hol}_{\text{non-deg}}(V, \mathbb{P}^r)_n = \{f \in \mathrm{Hol}_{\text{non-deg}}(V, \mathbb{P}^r) \mid n = (\mathrm{ord}\, f)(\deg f(V))\},$$

where $\mathrm{ord}\, f$ is the mapping order of $f : V \longrightarrow f(V)$. It is open and closed in $\mathrm{Hol}_{\text{non-deg}}(V, \mathbb{P}^r)$, which is divided into the disjoint union:

$$\mathrm{Hol}_{\text{non-deg}}(V, \mathbb{P}^r) = \bigcup_{n=1}^{\infty} \mathrm{Hol}_{\text{non-deg}}(V, \mathbb{P}^r)_n .$$

$\mathrm{Aut}(\mathbb{P}^r)$ acts freely on $\mathrm{Hol}_{\text{non-deg}}(V, \mathbb{P}^r)_n$. By Theorem 4.4.2,

$$\mathrm{Hol}_{\text{non-deg}}(V, \mathbb{P}^r)_n/\mathrm{Aut}(\mathbb{P}^r) \cong \mathbb{G}_n^r(V) - \mathbb{F}_n^r(V).$$

In particular, if $r = 1$, then

$$\mathrm{Hol}_{\text{non-deg}}(V, \mathbb{P}^1)_n = R_n(V) .$$

Hence

Theorem 5.1.1. $R_n(V)/\mathrm{Aut}(\mathbb{P}^1) \cong \mathbb{G}_n^1(V) - \mathbb{F}_n^1(V)$.

Thus we have constructed the complex space $R_n(V)/\mathrm{Aut}(\mathbb{P}^1)$ again.

Example 5.1.2. An element $f \in \mathrm{Hol}_{\text{non-deg}}(\mathbb{P}^1, \mathbb{P}^r)_n$ is given by $f = (F_0:\cdots:F_r)$, i.e.,

$$f(P) = (F_0(P):\cdots:F_r(P)) \in \mathbb{P}^r \quad \text{for} \quad P \in \mathbb{P}^1 ,$$

where $F_0,\cdots,F_r$ are linearly independent homogeneous polynomials of 2-variables (Z_0,Z_1) of degree n such that the set

$$\{(Z_0:Z_1) \in \mathbb{P}^1 \mid F_0(Z_0:Z_1) = \cdots = F_r(Z_0:Z_1) = 0\}$$

is empty. On the other hand, $\mathbb{G}_n^r(\mathbb{P}^1) - \mathbb{F}_n^r(\mathbb{P}^1)$ is the open subspace of Grassmann variety of all linear subspaces L of $\mathbb{P}^n$ of dimension $n-r-1$ which do not intersect with the rational curve

$$C_n = \Phi_{|n0|}(\mathbb{P}^1) = \{(Z_0:\cdots:Z_n) \mid Z_2Z_0 = Z_1^2,\ Z_3Z_0^2 = Z_1^3,\ \cdots,\ Z_nZ_0^{n-1} = Z_1^{n-1}\} .$$

$(0 = (1:0).)$ The element of $\mathrm{Hol}_{\text{non-deg}}(\mathbb{P}^1, \mathbb{P}^r)_n/\mathrm{Aut}(\mathbb{P}^r)$ corresponding to L is given by the projection $\pi_L : C_n \longrightarrow \mathbb{P}^r$ with the center L.

The following lemma is a generalization of Lemma 2.1.1.

Lemma 5.1.3. Let $n \leq g+r-1$. Then, for any $f \in \mathrm{Hol}_{\text{non-deg}}(V, \mathbb{P}^r)_n$, there are linearly independent $\omega_0,\cdots,\omega_r \in H^0(V, \mathcal{O}(K_V))$ such that $f = (\omega_0:\cdots:\omega_r)$, i.e.,

$$f(P) = (\omega_0(P):\cdots:\omega_r(P)) \in \mathbb{P}^r \quad \text{for} \quad P \in V .$$

Proof. We may write $f = (\xi_0:\cdots:\xi_r)$, where $\xi_0,\cdots,\xi_r$ are linearly independent elements of $H^0(V, \mathcal{O}(F))$ for some F with $\deg F = n$. Note that

$$h^0(K \otimes F^{-1}) = h^0(F) + g - 1 - n \geqq r + g - n \geqq 1$$

Thus there is a non-zero $\eta \in H^0(V, \mathcal{O}(K \otimes F^{-1}))$. Put

$$\omega_0 = \xi_0\eta, \cdots, \omega_r = \xi_r\eta .$$

Then they satisfy the requirement. Q.E.D.

Proposition 5.1.4. Let V be hyperelliptic and of genus g. Then, for $n \leqq g+r-1$,

$$\mathrm{Hol}_{\text{non-deg}}(V, \mathbb{P}^r)_n \begin{cases} = \text{empty}, & \text{if } n \text{ is odd} . \\ \cong \mathrm{Hol}_{\text{non-deg}}(\mathbb{P}^1, \mathbb{P}^r)_{n/2}, & \text{if } n \text{ is even} . \end{cases}$$

Proof. This follows from the previous lemma in a similar way to the proof of Proposition 2.1.2. Q.E.D.

Theorem 5.1.5. Let $V = C$ be a non-singular plane curve of degree d. Then

(1) If $d \geqq 3$, then $\mathbb{G}^1_{d-1}(C) \cong C$ and $\mathbb{F}^1_{d-1}(C)$ is empty.

(2) If $d \geqq 5$, then $\mathbb{G}^1_d(C) - \mathbb{F}^1_d(C) \cong \mathbb{P}^2 - C$, $\mathbb{G}^2_d(C)$ is one point and $\mathbb{F}^2_d(C)$ is empty.

Proof. (1) follows from Theorem 2.3.1. The first part of (2) follows from Proposition 2.3.6.

Let g^2_d be the linear system of all line sections of C. We first show that this is the only element of $\mathbb{G}^2_d(C) - \mathbb{F}^2_d(C)$. Let $\hat{g}^2_d$ be another element of $\mathbb{G}_d^2(V) - \mathbb{F}_d^2(C)$. Let $\hat{g}_d^1$ be a pencil in $\hat{g}_d^2$ without fixed point. (Such $\hat{g}_d^1$ exists. See the proof of Lemma 1.3.2.) By Proposition 2.3.6, $\hat{g}_d^1$ corresponds to a projection $\pi_P : C \longrightarrow \mathbb{P}^1$ with the center $P \in \mathbb{P}^2 - C$. In particular, every $D \in \hat{g}_d^1$ is a line section of C, i.e., $D \in g_d^2$. Hence $\hat{g}_d^2 \subset g_d^2$ so that $\hat{g}_d^2 = g_d^2$.

Finally, we show that $\mathbb{F}_d^2(C)$ is empty. Let $\hat{g}_d^2 \in \mathbb{F}_d^2(C)$. Let P_o be a fixed point of $\hat{g}_d^2$. Then, we may write

$$\hat{g}_d^2 = \hat{g}_{d-1}^2 + P_o .$$

But there is no $\hat{g}_{d-1}^2$, for otherwise, $\mathbb{G}_{d-1}^1(C)$ has the dimension at least 2, which contradicts to (1). Q.E.D.

Remark 5.1.6. In (2) of Theorem 5.1.5, $\mathbb{G}_d^1(C)$ has 2 components. One of them is $\mathbb{F}_d^1(C)$, which is biholomorphic to $C \times C$, and another is the closure of $\mathbb{G}_d^1(C) - \mathbb{F}_d^1(C)$ in $\mathbb{G}_d^1(C)$, which is biholomorphic to $\mathbb{P}^2$.

Now, we consider the case $r = 0$. It is well known that $\mathbb{D}_n(V)$ is canonically isomorphic to the n-th symmetric product S^nV. Here, we give a proof of this fact.

Lemma 5.1.7. $\mathbb{D}_n(V)$ is canonically biholomorphic to S^nV.

Proof. By Proposition 4.2.1, it is enough to show that $\mathbb{G}_n^0(V)$ is canonically biholomorphic to S^nV. The 0-th Picard variety $\mathrm{Pic}^0(V)$ is the Jacobi variety $J(V)$ in this case. Let $\{F_t\}_{t\in J(V)}$, $F_t = B_t \otimes [D_o]$, be the family of line bundles defined in §4.2. Let $\mathbb{H}^0$ be the linear fiber space in Theorem 4.1.8 with respect to $\{F_t\}_{t\in J(V)}$. Put

$$\mathbb{H}^* = \cup_{t\in J(V)} (H^0(V, \mathcal{O}(F_t)) - 0) .$$

Then, $\mathbb{H}^*$ is an open subspace of $\mathbb{H}^0$. We define a map (called the divisor map after Gunning [28])

$$\mathcal{D} : \mathbb{H}^* \longrightarrow S^nV$$

by $\mathcal{D}(\xi) = (\xi) \in S^nV$, where (ξ) is the zero divisor of $\xi \in \mathbb{H}^*$.

We first show that $\mathcal{D}$ is holomorphic. For a point $o \in J(V)$, let U be a small open neighborhood of o in $J(V)$ and let

$$\Psi : H^0(V, \mathcal{O}(F)) \times U \longrightarrow C^0(F, |\ |) \times U$$

be the analytic map defined in §4.1. $(F = F_o.)$ For $(\xi,t) \in H^0(V, \mathcal{O}(F)) \times U$, we write

$$\Psi(\xi,t) = (\{\phi_i(\xi,t)(z_i)\},t),$$

where $\phi_i(\xi,t)(z_i)$ is a holomorphic function of $(z_i,\xi,t) \in U_i \times H^0(V, \mathcal{O}(F)) \times U$. For a fixed $\xi^o \in H^0(V, \mathcal{O}(F)) - 0$, let $P^1 \in U_i$ be a zero point of ξ^0 with the multiplicity ν_1. Let B_ε be a small closed ε-ball in U_i with the center P^1. Then, the functions

$$\frac{1}{2\pi\sqrt{-1}} \int_{\partial B_\varepsilon} \frac{\partial(\phi_i(\xi,t)(z_i))/\partial z_i}{\phi_i(\xi,t)(z_i)} z_i^\alpha \, dz_i, \quad \alpha = 1,\cdots,\nu_1,$$

are holomorphic for (ξ,t). If $\phi = \{\phi_i(\xi,t)\} \in H^0(V, \mathcal{O}(F_t))$, then the above functions give local coordinates of the point $(\phi) = \mathcal{D}(\phi)$ in S^nV near $(\xi^o) = \mathcal{D}(\xi^o)$, (see Gunning [28]). Hence $\mathcal{D}$ is holomorphic

Now, it is clear that $\mathcal{D}$ induces a bijective holomorphic map $\mathbb{G}_n^0(V) \longrightarrow S^nV$. Since $\mathbb{G}_n^0(V)$ is compact and S^nV is non-singular, this map is biholomorphic by Zariski's Main Theorem (see, e.g., Ueno [82, p.9]). Q.E.D.

As in §1.3, we put

$$G_n^r = \{D \in S^nV \mid h^0(D) \geq r+1\},$$

$$W_n^r = \phi(G_n^r) = \{t \in J(V) \mid h^0(F_t) \geqq r+1\},$$

where $\phi : S^nV \longrightarrow J(V)$ is the Jacobi map. (Don't confuse $\mathbb{G}_n^r(V)$ and G_n^r!) Then, we get the following corollary, which is a modification of Gunning [28, Theorem 16(b)].

Corollary 5.1.8. $\phi : G_n^r - G_n^{r+1} \longrightarrow W_n^r \setminus W_n^{r+1}$ is a fiber bundle with the standard fiber $\mathbb{P}^r$.

Proof. This follows easily from Corollary 4.1.19. Q.E.D.

Lemma 5.1.7 implies in particular that $\mathbb{D}_n(V)$ is non-singular and of dimension n. But we can prove it also by showing that every divisor $D \in \mathbb{D}_n(V)$ is semi-regular. In order to show it, we first rewrite the condition of semi-regularity for a linear system g_n^r using Serre duality.

For $r \geqq 0$, let g_n^r be a linear system on V of dimension r and degree n. Let L be the linear subspace of $H^0(V, \mathcal{O}(F))$, ($F = [D]$, $D \in g_n^r$), corresponding to g_n^r. For a basis $\{\xi_0, \cdots, \xi_r\}$ of L, put $D_\nu = (\xi_\nu)$ and put

$$H^0(V, \mathcal{O}(K_V - D_\nu)) = \{\omega \in H^0(V, \mathcal{O}(K_V)) \mid (\omega) \geqq D_\nu\},$$

for $0 \leqq \nu \leqq r$. They are linear subspaces of $H^0(V, \mathcal{O}(K_V))$ of dimension $h^1(F) = h^0(F)+g-n-1$. We say that they are independent if the following condition is satisfied: If $\omega_\nu \in H^0(V, \mathcal{O}(K_V-D_\nu))$ satisfy $\sum_{\nu=0}^{r} \omega_\nu = 0$, then all $\omega_\nu = 0$, $0 \leqq \nu \leqq r$.

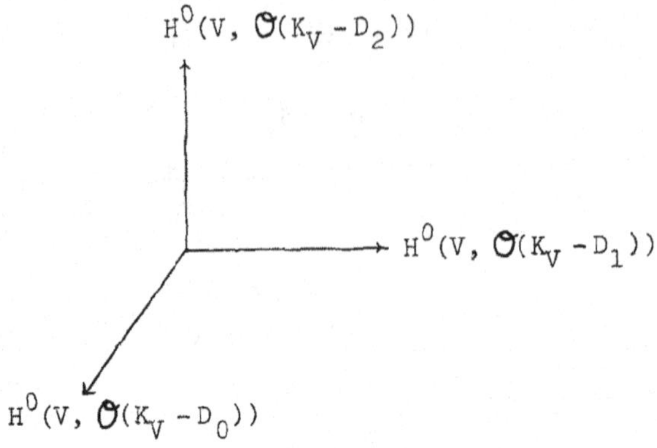

Theorem 5.1.9. $g_n^r \in \mathbb{G}_n^r(V)$ is semi-regular if and only if $H^0(V, \mathcal{O}(K_V - D_\nu))$, $0 \leqq \nu \leqq r$, are independent.

Proof. We use the same notations as above. Let $\{U_i\}$ be an open covering of V. Write $\xi_\nu = \{\xi_{\nu i}(z_i)\}$, where $\xi_{\nu i}(z_i)$ is a holomorphic function on U_i. (z_i is a coordinate in U_i.) We compute the dual of the linear map

$$\tau_{\xi_\nu} : h = \{h_{ik}\} \in H^1(V, \mathcal{O}) \longmapsto \{\xi_{\nu i} h_{ik}\} \in H^1(V, \mathcal{O}(F)).$$

Put $h_{ik} = g_k - g_i$, where g_i is a C^∞-function on U_i. Then $\bar{\partial} g_i = \bar{\partial} g_k$ is a global $\bar{\partial}$-closed (0,1)-form corresponding to $h = \{h_{ik}\}$ under the Dolbeault isomorphism

$$H^1(V, \mathcal{O}) \longrightarrow H^{0,1}(V) .$$

In a similar way, $\xi_{\nu i} \bar{\partial} g_i$ corresponds to $\tau_{\xi_\nu}(h)$ under the Dolbeault isomorphism

$$H^1(V, \mathcal{O}(F)) \longrightarrow H^{0,1}(V, F) .$$

On the other hand, the dual of $H^1(V, \mathcal{O})$ is

$$H^0(V, \Omega^1) = H^0(V, \mathcal{O}(K_V)) = H^{1,0}(V)$$

and the pairing which gives the duality is

$$\langle h,\omega\rangle = \int_V \omega\wedge\bar{\partial} g_i .$$

The duality between $H^1(V, \mathcal{O}(F))$ and $H^0(V, \mathcal{O}(K_V \otimes F^{-1}))$ is given in a similar way. Hence, for $\eta = \{\eta_i\} \in H^0(V, \mathcal{O}(K_V \otimes F^{-1}))$,

$$\begin{aligned}\langle h, \tau^*_{\xi_\nu}(\eta)\rangle &= \langle \tau_{\xi_\nu}(h), \eta\rangle \\ &= \int_V (\eta_i dz_i)\wedge(\xi_{\nu i}\bar{\partial} g_i) \\ &= \int_V \xi_{\nu i}\eta_i dz_i \wedge \bar{\partial} g_i .\end{aligned}$$

Hence

$$\tau^*_{\xi_\nu} : \eta = \{\eta_i\} \in H^0(V, \mathcal{O}(K_V \otimes F^{-1})) \longmapsto \xi_\nu\eta = \{\xi_{\nu i}\eta_i\} \in H^0(V, \mathcal{O}(K_V)).$$

It is clear that $\tau^*_{\xi_\nu}$ is injective. (Hence τ_{ξ_ν} is surjective.) Its image is just $H^0(V, \mathcal{O}(K_V - D_\nu))$.

Now, note that the dual space of $H^1(V, \mathcal{O}(F))^{r+1}$ is $H^0(V, \mathcal{O}(K_V \otimes F^{-1}))^{r+1}$ under the pairing

$$\langle(\zeta_0, \cdots, \zeta_r), (\eta_0, \cdots, \eta_r)\rangle = \Sigma_\nu \langle \zeta_\nu \ \eta_\nu\rangle .$$

Put

$$\tau_r : (h_0, \cdots, h_r) \in H^1(V, \mathcal{O})^{r+1}$$

$$\longmapsto (\tau_{\xi_0}(h_0), \cdots, \tau_{\xi_r}(h_r)) \in H^1(V, \mathcal{O}(F))^{r+1},$$

$$\Delta = \{(h, \cdots, h) \in H^0(V, \mathcal{O})^{r+1} \mid h \in H^1(V, \mathcal{O})\} .$$

Assume that $\tau_r(\Delta) \neq H^1(V, \mathcal{O}(F))^{r+1}$. Then, there is a non-zero $(\eta_0, \cdots, \eta_r) \in H^0(V, \mathcal{O}(K_V \otimes F^{-1}))^{r+1}$ such that $\langle \tau_r(\Delta), (\eta_0, \cdots, \eta_r)\rangle = 0$. Then

$$0 = \langle \Delta, \tau^*_r(\eta_0, \cdots, \eta_r)\rangle = \langle \Delta, (\tau^*_{\xi_0}(\eta_0), \cdots, \tau^*_{\xi_r}(\eta_r))\rangle .$$

This means that, for all $h \in H^1(V, \mathcal{O})$,

$$0 = \Sigma_\nu \langle h, \tau^*_{\xi_\nu}(\eta_\nu)\rangle = \langle h, \Sigma_\nu \xi_\nu \eta_\nu \rangle$$

Hence $\Sigma_\nu \xi_\nu \eta_\nu = 0$. Put $\omega_\nu = \xi_\nu \eta_\nu \in H^0(V, \mathcal{O}(K_V - D_\nu))$. Then at least one of ω_ν is non-zero and $\omega_0 + \cdots + \omega_r = 0$.

The converse is shown by tracing the converse of the above argument. Q.E.D.

Corollary 5.1.10.

(1) Every $D \in S^n V$ is semi-regular.

(2) $g^1_n \in \mathbb{G}^1_n(V)$ is semi-regular if and only if $H^1(V, \mathcal{O}([2D - \hat{D}])) = 0$, where $D \in g^1_n$ and $\hat{D}$ is the fixed part of the linear pencil g^1_n.

(2)' If $g^r_n \in \mathbb{G}^r_n(V)$ $(r \geqq 1)$ is semi-regular, then $H^1(V, \mathcal{O}([2D - \hat{D}])) = 0$, where $D \in g^r_n$ and $\hat{D}$ is the fixed part of g^r_n.

(3) Every linear subsystem of $|K_V|$ is semi-regular.

(4) If g^r_n is semi-regular and D is an effective divisor, then $g^r_n + D$ is also semi-regular.

(5) If $D \in g^r_n$ satisfies $h^1(D) \leqq 1$, then g^r_n is semi-regular.

(6) If g^r_n is semi-regular, then $g \geqq (r+1)h^1(D)$ for $D \in g^r_n$.

(7) If g^r_n is semi-regular, then any linear subsystem g^s_n of g^r_n is also semi-regular.

Proof. (1), (3), (4), (5), (6) and (7) are trivial. We show (2). Let D_0 and D_1 be linearly equivalent divisors in g^1_n such that G.C.D.$(D_0, D_1) = \hat{D}$, the fixed part of g^1_n. Then

"$H^0(V, \mathcal{O}(K_V - D_0))$ and $H^0(V, \mathcal{O}(K_V - D_1))$ are independent."

$$\Longleftrightarrow$$

"$H^0(V, \mathcal{O}(K_V - D_0)) \cap H^0(V, \mathcal{O}(K_V - D_1)) = 0$."

$$\Longleftrightarrow$$

"If $\omega \in H^0(V, \mathcal{O}(K_V))$ satisfies $(\omega) \geqq D_0$ and $(\omega) \geqq D_1$, then $\omega = 0$."

$\Longleftrightarrow$

"$H^0(V,\mathcal{O}(K_V-D_2)) = 0$, where $D_2 = \text{L.C.M.}(D_0,D_1) = D_0+D_1-\hat{D}$."

$\Longleftrightarrow$

$$\text{"}H^1(V, \mathcal{O}([2D_0-\hat{D}])) = 0.\text{"}$$

(2)' can be shown in a similar way. Q.E.D.

Semi-regularity theorem (Theorem 4.3.2) is rewritten in the present case as follows:

Theorem 5.1.11. If $g_n^r \in \mathbb{G}_n^r(V)$ is semi-regular, then $\mathbb{G}_n^r(V)$ is non-singular at g_n^r and

$$\dim_{g_n^r}\mathbb{G}_n^r(V) = (r+1)(n-r) - rg .$$

Remark 5.1.12. Severi [78, p.159], says that, for a general V, $\dim\mathbb{G}_n^r(V) = (r+1)(n-r) - rg$ (with respect to his complex space $\mathbb{G}_n^r(V)$), provided $(r+1)(n-r) - rg \geqq 0$.

Now, let

$$\tau : H^0(V, \mathcal{O}(F)) \times H^1(V, \Theta) \longrightarrow H^1(V, \mathcal{O}(F))$$

be the bilinear map defined in §4.3. Put as before

$$\tau(\xi,h) = \tau_\xi(h) = \tau(h)(\xi) \quad \text{for} \quad (\xi,h) \in H^0(V, \mathcal{O}(F)) \times H^1(V, \Theta) .$$

Let $\{\xi_0,\cdots,\xi_s\}$ and $\{\zeta_1,\cdots,\zeta_t\}$ be basis of $H^0(V, \mathcal{O}(F))$ and $H^1(V, \mathcal{O}(F))$, respectively. ($s = h^0(F)-1$, $t = h^1(F)$.) Let $\{\eta_1,\cdots,\eta_t\}$ be the basis of $H^0(V, \mathcal{O}(K_V \otimes F^{-1}))$ dual to $\{\zeta_1,\cdots,\zeta_t\}$. Put

$$\tau(\xi_\alpha,h) = \Sigma_\beta a_{\alpha\beta}(h)\zeta_\beta .$$

Then, as in the proof of Theorem 5.1.9,

$$a_{\alpha\beta}(h) = \langle \tau(\xi_\alpha, h), \eta_\beta \rangle = \langle h, \xi_\alpha \eta_\beta \rangle .$$

Hence

Lemma 5.1.13. The matrix-valued function $\tau(h)$ is given by

$$\tau(h)(\xi_\alpha) = \Sigma_\beta \langle h, \xi_\alpha \eta_\beta \rangle \zeta_\beta ,$$

$$\tau(h) = (\langle h, \xi_\alpha \eta_\beta \rangle) .$$

This lemma, combined with Lemma 4.3.8, proves Theorem 5.1.9, again. In fact, if L is a linear subspace of $H^0(V, \mathcal{O}(F))$ spanned by $\xi_0, \cdots, \xi_r$, then

"L is semi-regular."

$$\Longleftrightarrow$$

"For any $a_{\alpha\beta} \in \mathbb{C}$, $0 \leqq \alpha \leqq r$, $1 \leqq \beta \leqq h^1(F)$, there is $h \in H^1(V, \mathcal{O})$ such that $\langle h, \xi_\alpha \eta_\beta \rangle = a_{\alpha\beta}$."

$$\Longleftrightarrow$$

"$\xi_\alpha \eta_\beta$, $0 \leqq \alpha \leqq r$, $1 \leqq \beta \leqq h^1(F)$, are linearly independent in $H^0(V, \mathcal{O}(K_V))$."

Now,

$$W_n^r = \{t \in J(V) \mid h^0(F_t) \geqq r+1\}$$

is the image of $\mathbb{G}_n^r(V)$ under the projection $\mu : \mathbb{G}_n^r(V) \longrightarrow J(V)$ (see Lemma 4.3.4). In particular,

$$W_n = W_n^0 = \phi(S^n V),$$

where $\phi : S^n V \longrightarrow J(V)$ is the Jacobi map. Proposition 4.3.6 and Theorem 4.3.7 are rewritten in the present case as follows:

<u>Proposition 5.1.14</u>. Let Y be an irreducible component of W_n^r such that $Y \not\subset W_n^{r+1}$. Then

$$\dim Y \geqq (r+1)(n-r) - rg .$$

<u>Theorem 5.1.15</u>. For $D \in S^n V$, put $r = \dim |D|$. Assume that $|D|$ is semi-regular as a linear system. Then W_n^r is non-singular at $\phi(D)$ and $\dim_{\phi(D)} W_n^r = (r+1)(n-r) - rg$.

<u>Corollary 5.1.16</u>. For $D \in S^n V$, assume that $h^0(D) = 2$ and $h^1(2D-\hat{D}) = 0$, where $\hat{D}$ is the fixed part of the pencil $|D|$. Then W_n^1 is non-singular at $\phi(D)$ and $\dim_{\phi(D)} W_n^1 = 2n-2-g$.

<u>Remark 5.1.17</u>. Proposition 5.1.14 is Theorem 14 (b) of Gunning [28]. The assumption $Y \not\subset W_n^{r+1}$ is unnecessary if $n \leqq g$. In fact, if $n \leqq g$, then $W_n^r \setminus W_n^{r+1}$ is dense in W_n^r (see Proposition 1.3.15). Corollary 5.1.16 is found in Martens [52, II, p.98]. In this paper, he showed conversely that if V is non-hyperelliptic and $g \geqq 5$ and if $D \in S^{g-1}V$ satisfies $h^0(D) = 2$, $[2D] = K_V$ and $|D|$ has no fixed point, then W_{g-1}^1 is singular at $\phi(D)$.

Now the Jacobi map

$$\phi : P \in V \longmapsto \left(\int_{P_0}^{P} \omega_1, \cdots, \int_{P_0}^{P} \omega_g\right) \text{ (mod periods) } \in J(V)$$

is an imbedding. For $s \in J(V)$, let $G^0(T_s J(V))$ be the projective space of all lines through 0 in the tangent space $T_s J(V)$ to $J(V)$ at s. Let

$$\hat{t}_{-s} : G^0(T_s J(V)) \longrightarrow G^0(T_0 J(V))$$

be the biholomorphic map induced by the translation

$$t_{-s} : x \in J(V) \longmapsto x - s \in J(V) .$$

(0 = zero of the group $J(V)$.) Then

Lemma 5.1.18. (See, e.g., Andreotti-Mayer [3]). For $P \in V$,

$$\Phi_K(P) = \hat{t}_{-\phi(P)}((d\phi)_P(T_P V)),$$

where $\Phi_K : V \longrightarrow \mathbb{P}^{g-1}$ is the canonical map and $G^0(T_0 J(V))$ is identified with the dual space $\mathbb{P}^{g-1}$ of $|K_V|$.

Proof. This is clear from the fact:

$$(d\phi)_P = (\omega_1(P), \cdots, \omega_g(P)).$$

Q.E.D.

Now, consider the projection map

$$\mu : \mathbb{G}_n^r(V) \longrightarrow W_n^r \subset J(V).$$

For a point $s \in W_n^r$, assume that every element of $\mu^{-1}(s) = G^r(H^0(V, \mathcal{O}(F_s$ is semi-regular. Then, by Proposition 4.3.11, the normal bundle of $\mu^{-1}(s)$ in $\mathbb{G}_n^r(V)$ is identified with

$$N_s = \{(L,h) \in \mu^{-1}(s) \times H^1(V, \mathcal{O}) \mid \tau(h)(L) = 0\}.$$

The projection

$$\lambda_s : (L,h) \in N_s \longmapsto h \in H^1(V, \mathcal{O})$$

maps N_s onto the tangent cone $TC_s(W_n^r)$ of W_n^r at s, which is, as a reduced subscheme, equal to

$$M_s = \{h \in H^1(V, \mathcal{O}) \mid \text{all } (h^0(F_s)-r) \times (h^0(F_s)-r)\text{-minors of } \tau(h) \text{ are zero}\}$$

(see §4.3). The tangent cone $TC_s(W_n^r)$ is, by translation, regarded as a closed subscheme in $T_0 J(V)$ (= $H^1(V, \mathcal{O})$). We denote by $\widehat{TC}_s(W_n^r)$

the <u>projectivized</u> <u>tangent</u> <u>cone</u>. It is a (not necessarily reduced) closed subscheme of $\mathbb{P}^{g-1}$.

As in §2.5, for $\omega \in H^0(V, \mathcal{O}(K_V))$, let $\langle\omega\rangle$ be the hyperplane in $\mathbb{P}^{g-1}$ determined by ω. For an effective divisor D, put

$$S_D = \bigcap_{\langle\omega\rangle \geqq D}\langle\omega\rangle .$$

Then, S_D is a linear subspace of $\mathbb{P}^{g-1}$. It is the minimal linear subspace containing D. Note that

$$\dim S_D = g-1-h^1(D) = \deg D - h^0(D).$$

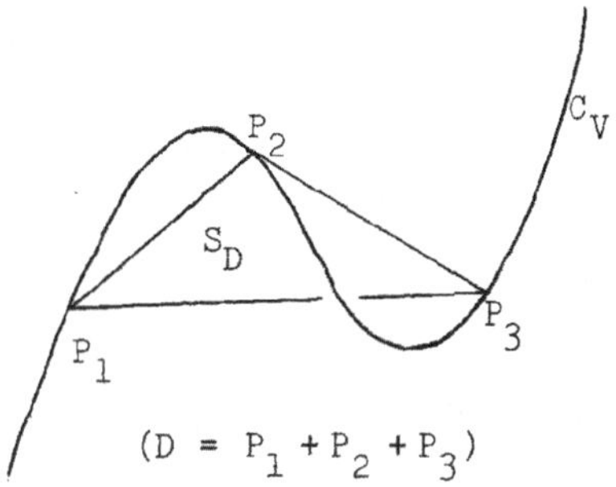

$(D = P_1 + P_2 + P_3)$

For a linear system L, put

$$S_L = \bigcap_{D \in L}S_D .$$

<u>Proposition 5.1.19</u>. For $s \in W_n^r$, assume that <u>every</u> element of $\mu^{-1}(s)$ is semi-regular. Then, as sets,

$$\widehat{TC}_s(W_n^r) = \bigcup_{L \in \mu^{-1}(s)}S_L .$$

Moreover, if $\dim |F_s| > r$, then $\widehat{TC}_s(W_n^r)$ contains the canonical curve $C_V = \Phi_K(V)$.

<u>Proof</u>. Let $\{\zeta_1, \cdots, \zeta_t\}$ and $\{\eta_1, \cdots, \eta_t\}$ be dual basis of $H^1(V, \mathcal{O}(F_s))$ and $H^0(V, \mathcal{O}(K_V \otimes F_s^{-1}))$. $(t = h^1(F_s).)$ Fix $L \in \mu^{-1}(s)$. Then, by Lemma 5.1.13,

$$\{h \in H^1(V, \mathcal{O}) \mid \tau(h)(L) = 0\}$$

$$= \{h \in H^1(V, \mathcal{O}) \mid \langle h, \xi\eta_\beta\rangle = 0, \ 1 \leqq \beta \leqq t, \ \text{for}$$

$$\text{all} \ \xi \in L\}$$

$$= \{h \in H^1(V, \mathcal{O}) \mid \langle h, \xi H^0(V, \mathcal{O}(K_V \otimes F_s^{-1}))\rangle = 0$$

$$\text{for all} \quad \xi \in L\} .$$

This shows the first part of the proposition.

If $P \in V$ is a fixed point of $|F_s|$, then $\Phi_K(P) \in S_D$ for all $D \in |F_s|$. Hence $\Phi_K(P) \in \widehat{TC}_s(W_n^r)$.

If $P \in V$ is not a fixed point of $|F_s|$, then the point $\Phi_{|F_s|}(P) \in \mathbb{P}^N$, $(N = \dim |F_s|)$, is well defined. Assume that $N = \dim |F_s| > r$. Then there is a linear subsystem L of $|F_s|$ of dimension r having P as a fixed point. In fact, if S is a $(N-r-1)$-dimensional linear subspace of $\mathbb{P}^N$ passing $\Phi_{|F_s|}(P)$, then the system of all hyperplanes of $\mathbb{P}^N$ through S determines such L. Then $\Phi_K(P) \in S_D$ for all $D \in L$. Hence

$$\Phi_K(P) \in S_L \subset \widehat{TC}_s(W_n^r) .$$

Q.E.D.

Corollary 5.1.20. For $s \in W_n$,

$$\widehat{TC}_s(W_n) = \cup_{D \in \phi^{-1}(s)} S_D .$$

Moreover, if $\dim |F_s| \geqq 1$, then $\widehat{TC}_s(W_n)$ contains the canonical curve $C_V = \Phi_K(V)$.

Note that, if we drop the assumption of semi-regularity in Proposition 5.1.19, then we only get $\widehat{TC}_s(W_n^r) \subset \cup_{L \in \mu^{-1}(s)} S_L$.

Theorem 5.1.21. Let $n \leqq g-1$. For $s \in J(V)$, assume that (1) $a = \dim |F_s| \geqq 1$ and (2) every element of $\mu^{-1}(s)$ is semi-regular, where $\mu : \mathbb{G}_n^{a-1}(V) \longrightarrow J(V)$ is the projection map. Then, the intersection $\bigcap_{l,l'} Q_{l,l'}$ of all quadrics $Q_{l,l'}$ of rank $\leqq 4$ in $\mathbb{P}^{g-1}$, determined by lines $l \subset |F_s|$ and $l' \subset |K_V \otimes F_s^{-1}|$ (see §2.5), is (as sets) equal to the projectivized tangent cone $\widehat{TC}_s(W_n^{a-1})$ of W_n^{a-1} at s.

Proof. By the assumption, we have $h^1(F_s) \geqq 2$. Note that the quadric $Q_{l,l'}$ in $\mathbb{P}^{g-1}$ determined by lines $l \subset |F_s|$ and $l' \subset |K_V \otimes F_s^{-1}|$ is given by

$$Q_{l,l'} = \cup_{D\in l}\langle D+l'\rangle = \cup_{D'\in l'}\langle l+D'\rangle ,$$

where

$$\langle D+l'\rangle = \cap_{D'\in l'}\langle D+D'\rangle$$

$$\langle D+D'\rangle = \langle \omega\rangle, \text{ where } (\omega) = D+D' .$$

Let $\widetilde{Q}_{l,l'}$ be the pull back of $Q_{l,l'}$ in $\mathbb{C}^g = H^1(V, \mathcal{O})$. Then, it is given by

$$\widetilde{Q}_{l,l'} = \cup_{\xi\in L-0}\{z \in \mathbb{C}^g \mid \langle z,\xi L'\rangle = 0\}$$

$$= \cup_{\eta\in L'-0}\{z \in \mathbb{C}^g \mid \langle z,L\eta\rangle = 0\} ,$$

where $L \subset H^0(V, \mathcal{O}(F_s))$ and $L' \subset H^0(V, \mathcal{O}(K_V \otimes F_s^{-1}))$ are 2-dimensional linear subspaces corredponding to l and l', respectively.

Let $\{\xi_1,\xi_2\}$ and $\{\zeta_1,\eta_2\}$ be basis of L and L', respectively. Then, it is clear that $z \in \widetilde{Q}_{l,l'}$ if and only if there is $(t_1,t_2) \in \mathbb{C}^2-0$ such that

$$\begin{cases} t_1\langle z,\xi_1\eta_1\rangle + t_2\langle z,\xi_2\eta_1\rangle = 0 , \\ t_1\langle z,\xi_1\eta_2\rangle + t_2\langle z,\xi_2\eta_2\rangle = 0 . \end{cases}$$

This holds if and only if

$$\det\begin{pmatrix} \langle z,\xi_1\eta_1\rangle & \langle z,\xi_1\eta_2\rangle \\ \langle z,\xi_2\eta_1\rangle & \langle z,\xi_2\eta_2\rangle \end{pmatrix} = 0 .$$

Hence, $z \in \cap_{l,l'}\widetilde{Q}_{l,l'}$ if and only if $\operatorname{rank}\,(\langle z,\xi_\alpha\eta_\beta\rangle) \leqq 1$, where $\{\xi_1,\cdots,\xi_{h^0(F_s)}\}$ and $\{\eta_1,\cdots,\eta_{h^1(F_s)}\}$ are basis of $H^0(V, \mathcal{O}(F_s))$ and $H^0(V, \mathcal{O}(K_V \otimes F_s^{-1}))$, respectively. This last statement holds if and only if $z \in TC_s(W_n^{a-1})$. Hence

$$\bigcap_{l,l'} Q_{l,l'} = \widehat{TC}_s(W_n^{a-1}) \quad .$$

Q.E.D.

Corollary 5.1.22. Let $n \leqq g-1$. For $s \in J(V)$, assume that $\dim |F_s| = 1$. Then the intersection $\bigcap_{l'} Q_{|F_s|,l'}$ of all quadrics $Q_{|F_s|,l'}$ of rank $\leqq 4$ in $\mathbb{P}^{g-1}$ determined by $|F_s|$ and lines $l' \subset |K_V \otimes F_s^{-1}|$ is equal to $\widehat{TC}_s(W_n)$.

Remark 5.1.23. If we drop the assumption (2) in Theorem 5.1.21, then we only get

$$\widehat{TC}_s(W_n^{a-1}) \subset \bigcap_{l,l'} Q_{l,l'} \; .$$

Finally, we prove

Theorem 5.1.24. Let V be a non-hyperelliptic compact Riemann surface of genus $g \geqq 4$. Then $\mathbb{G}^1_{g-1}(V)$ is pure $(g-4)$-dimensional.

Before giving our proof of the theorem, we give an example of $\mathbb{G}^1_{g-1}(V)$ in order to illustrate the idea of the proof.

Example 5.1.25. (c.f., Remark 5.1.6). Let C be a non-singular plane quintic. The genus of C is 6. The linear system g_5^2 of all line sections of C is complete. By Theorem 5.1.5, any linear pencil $g_5^1 \in \mathbb{G}^1_5(C) - \mathbb{F}^1_5(C)$ is obtained by the projection π_P with the center $P \in \mathbb{P}^2 - C$, i.e.,

$$g_5^1 = \{D_\lambda(\pi_P)\}_{\lambda \in \mathbb{P}^1} \, .$$

On the other hand, any $g_5^1 \in \mathbb{F}^1_5(C)$ is written as

$$g_5^1 = \{D_\lambda(\pi_P) + Q\}_{\lambda \in \mathbb{P}^1} \, .$$

where π_P is the projection with the center $P \in C$ and $Q \in C$ is the fixed point of g_5^1. Thus, $\mathbb{G}_5^1(C)$ is pure 2-dimensional and has 2 components, which are biholomorphic to $\mathbb{P}^2$ and $C \times C$, respectively. Note that every $g_5^1 \in \mathbb{G}_5^1(C) - \mathbb{F}_5^1(C)$ is not semi-regular, since $[2D] = K_C$ for any line section D of C. On the other hand, $\{D_\lambda(\pi_P)+Q\}_{\lambda \in \mathbb{P}^1} \in \mathbb{F}_5^1(C)$ is semi-regular, unless $P = Q$.

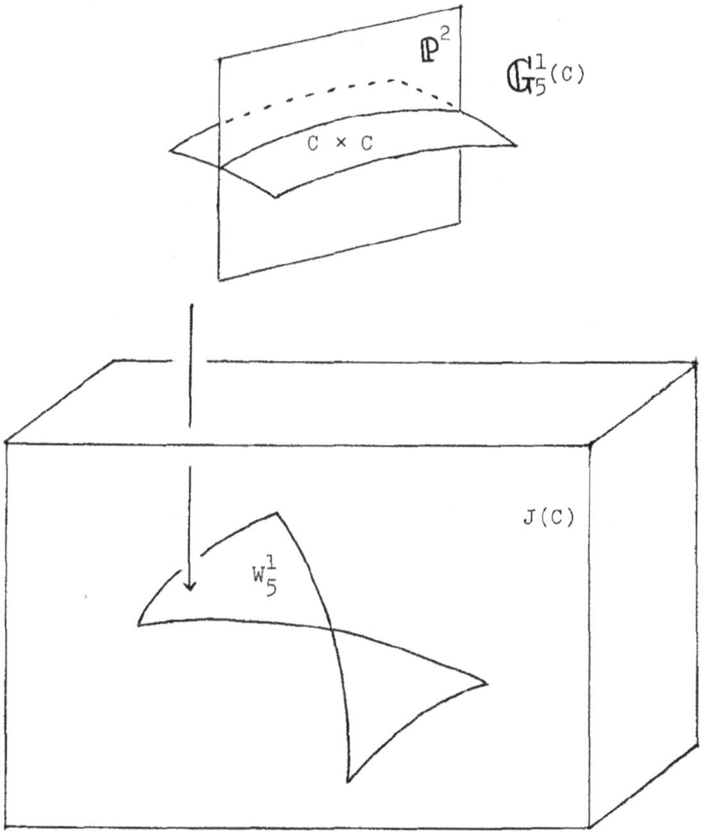

Proof of Theorem 5.1.24. By Corollary 2.1.8, $R_{g-1}(V)$ is pure (g-1)-dimensional. Hence the open subset

$$\mathbb{G}_{g-1}^1(V) - \mathbb{F}_{g-1}^1(V) \quad (\cong R_{g-1}(V)/\mathrm{Aut}(\mathbb{P}^1) \text{ by Theorem 5.1.1})$$

of $\mathbb{G}^1_{g-1}(V)$ is pure (g-4)-dimensional. Hence, it suffices to show that

(1) $\dim_l \mathbb{G}^1_{g-1}(V) = g-4$ for any $l \in \mathbb{F}^1_{g-1}(V)$.

To show (1), we may assume that l is a <u>non-singular</u> point of $\mathbb{G}^1_{g-1}(V)$. Hence, by Theorem 5.1.11, it suffices to show

(2) For any $l \in \mathbb{F}^1_{g-1}(V)$ and for any open neighborhood U of l in $\mathbb{G}^1_{g-1}(V)$, there is a <u>semi-regular</u> $l' \in U$.

Write

$$l = \{D_\lambda(f) + \hat{D}\}_{\lambda \in \mathbb{P}^1} ,$$

where f is a meromorphic function on V of order $n \leqq g-2$ and $\hat{D}$ ($\deg \hat{D} = d$) is the fixed part of l. Note that

$$n + d = g - 1 \quad \text{and} \quad d \geqq 1 .$$

Note that the map

$$D \in S^d V \longmapsto \{D_\lambda(f) + D\}_{\lambda \in \mathbb{P}^1} \in \mathbb{G}^1_{g-1}(V)$$

is holomorphic. Thus, in order to prove (2), it suffices by (2) of Corollary 5.1.10 to show

(3) For any open neighborhood U of $\hat{D}$ in $S^d V$, there is $D \in U$ such that $h^1(2D_\infty(f)+D) = 0$.

Since $S^d V$ is irreducible, it suffices to show

(4) There is a <u>proper</u> closed complex subspace Y of $S^d V$ such that

$$h^1(2D_\infty(f) + D) = 0 \quad \text{for all} \quad D \in S^d V - Y .$$

We define a holomorphic map σ by

$$\sigma : D \in S^d V \longmapsto 2D_\infty(f) + D \in S^{2n+d} V .$$

Put

$$Y = \sigma^{-1}(G^n_{2n+d}) .$$

Then, by Riemann-Roch theorem,

$$Y = \{D \in S^dV \mid h^1(2D_\infty(f) + D) \geqq 1\} .$$

In order to show that Y satisfies (4), we have to check $Y \neq S^dV$. First note that

(5) $h^1(2D_\infty(f)) \leqq d.$

In fact, note that $\deg(2D_\infty(f)) = 2n < 2g-2$. By the assumption, V is non-hyperelliptic. Hence, by Clifford's theorem (see, e.g., Gunning [28, p.60]),

$$h^0(2D_\infty(f)) \leqq [\frac{2n+1}{2}] = n .$$

Hence, by Riemann-Roch theorem, we get (5).

Next, we show that there is $D \in S^dV$ such that $h^1(2D_\infty(f)+D) = 0$. Take a point $Q_1 \in V$ which is <u>not</u> a fixed point of $|K_V-[2D_\infty(f)]|$. Then

$$h^1(2D_\infty(f) + Q_1) = h^1(2D_\infty(f)) - 1 .$$

Take $Q_2 \in V$ which is <u>not</u> a fixed point of $|K_V-[2D_\infty(f)+Q_1]|$. Then

$$h^1(2D_\infty(f) + Q_1 + Q_2) = h^1(2D_\infty(f)) - 2 .$$

Repeating this process and using (5), we find

$$D = Q_1 + \cdots + Q_d \in S^dV$$

such that $h^1(2D_\infty(f)+D) = 0$. <u>Q.E.D.</u>

<u>Corollary 5.1.26. (Martens [52, I], Andreotti-Mayer [3])</u>. If V is non-hyperelliptic and $g \geqq 4$, then W^1_{g-1} is pure $(g-4)$-dimensional.

Proof. By Proposition 1.3.15, it suffices to show that $W^1_{g-1} \setminus W^2_{g-1}$ is pure (g-4)-dimensional. Let $\mu : \mathbb{G}^1_{g-1}(V) \longrightarrow W^1_{g-1}$ be the projection map. Then, by the proof of Proposition 4.3.6,

$$\mu : \mu^{-1}(W^1_{g-1} \setminus W^2_{g-1}) \longrightarrow W^1_{g-1} \setminus W^2_{g-1}$$

is biholomorphic. Q.E.D.

Corollary 5.1.27. If V is non-hyperelliptic and $g \geqq 4$, then

$$2n - g - 2 \leqq \dim \mathbb{G}^1_n(V) \leqq n - 3 \quad \text{for} \quad 3 \leqq n \leqq g - 1 .$$

Proof. The first inequality is obtained by the local construction of $\mathbb{G}^1_n(V)$ (see the proof of Theorem 4.1.9).

Note that the map

$$\psi : (l, D) \in \mathbb{G}^1_n(V) \times S^dV \longmapsto l + D \in \mathbb{G}^1_{g-1}(V)$$

is holomorphic. ($d = g-1-n$.) The map ψ is finite-to-one, for a given $l' \in \mathbb{G}^1_{g-1}(V)$ can be written as

$$l' = l + D, \text{ where } l \in \mathbb{G}^1_n(V) \text{ and } D \in S^dV,$$

in only finitely many ways. Thus

$$\dim \mathbb{G}^1_n(V) + \dim S^dV \leqq \dim \mathbb{G}^1_{g-1}(V) = g - 4 .$$

Hence

$$\dim \mathbb{G}^1_n(V) \leqq n - 3 .$$ Q.E.D.

Corollary 5.1.28. (Martens [52, I]). If V is non-hyperelliptic and $g \geqq 4$, then

$$2n-g-2 \leqq \dim W^1_n \leqq n-3 \quad \text{for} \quad 3 \leqq n \leqq g-1 .$$

Remark 5.1.29. We can easily show that if V is hyperelliptic, then $\mathbb{G}_n^1(V)$ is pure (n-2)-dimensional for $2 \leqq n \leqq g$.

5.2. The space $\mathbb{G}_n^r$ and weak semi-regularity.

Let $g \geqq 2$ and T_g be the Teichmüller space of compact Riemann surfaces of genus g. For $t \in T_g$, we denote by V_t the compact Riemann surface corresponding to t. Then, $X_g = \cup_t V_t$, the disjoint union, is a complex manifold and $(X_g, \pi_g, T_g) = \{V_t\}_{t \in T_g}$ is a family of compact Riemann surfaces, where $\pi_g : X_g \longrightarrow T_g$ is the projection (see Example 0.1.2).

For an integer $n \geqq 0$, let $J_n(V_t)$ be the Jacobi variety of degree n, i.e., the set of all (isomorphism classes of) holomorphic line bundles on V_t of degree n. It is biholomorphic to $J(V_t) = J_0(V_t)$. It is well known (see Grothendieck [26, Exp. 16]) that $J_n = \cup_t J_n(V_t)$, the disjoint union, is a complex manifold and $(J_n, \tilde{\pi}, T_g) = \{J_n(V_t)\}_{t \in T_g}$ is a family of Abelian varieties. ($\tilde{\pi} : J_n \longrightarrow T_g$ is the projection.) (We can directly prove this fact using local holomorphic sections of $\pi_g : X_g \longrightarrow T_g$ and the space $J_0 = \cup_t J_0(V_t)$ constructed by Earle [16].)

For $s \in J_n$, put $V_s = V_{\tilde{\pi}(s)}$. Then $\{V_s\}_{s \in J_n}$ is a family of compact Riemann surfaces. In fact, it is the induced family $(\hat{X}_g, \hat{\pi}, J_n)$ of $\{V_t\}_{t \in T_g}$ over $\tilde{\pi}$. ($\hat{X}_g = \tilde{\pi}^* X_g$.)

Let T be an open subset of T_g such that $\pi_g : X_g \longrightarrow T_g$ has a holomorphic section on T. We fix such T for a while. Put $\pi_g^{-1}(T) = X_g'$ and $\tilde{\pi}^{-1}(T) = J_n'$. Let F_s be the line bundle of degree n on V_s corresponding to $s \in J_n'$. Then it is known (see Grothendieck [26, Exp. 16]) that $\{F_s, V_s\}_{s \in J_n'}$ is a family of line bundles. The transition functions $\{f_{ik}(z_k, s)\}$ of F_s is given as follows:

Fix $o \in T$. Let $\mathcal{U} = \{X_i\}$ be a finite covering of V_o by open subsets X_i of X_g' and let $\eta_i : X_i \longrightarrow U_i \times W$ be a biholomorphic map which makes the diagram

$$\begin{array}{ccc} X_i & \xrightarrow{\ \eta_i\ } & U_i \times W \\ & {\scriptstyle \pi_g}\searrow \qquad \swarrow{\scriptstyle \mathrm{proj}} & \\ & W & \end{array}$$

commutative, where U_i is an open subset in V_o with a coordinate z_i and W is an open neighborhood of o in T with a coordinate system $t = (t_1, \cdots, t_{3g-3})$.

By Kodaira-Spencer [43], $\cup_t H^1(V_t, \mathcal{O}_t)$ (the disjoint union) is a holomorphic vector bundle on T_g. Hence, there are cocycles

$$h^\alpha = \{h_{ik}^\alpha(z,t)\} \in Z^1(\mathcal{U}, \mathcal{O}_{\pi_g^{-1}(W)}), \quad 1 \leqq \alpha \leqq g,$$

whose restrictions to any V_t, $t \in W$, form a basis of $H^1(V_t, \mathcal{O}_t)$. Let $\{q_i\}$ be a partition of unity subordinate to the covering $\{U_i\}$ of V_o. Then $\Sigma_i q_i h_{ik}^\alpha$ is a C^∞-function on X_k. Note that

$$\Sigma_i q_i h_{ik}^\alpha - \Sigma_i q_i h_{ij}^\alpha = h_{jk}^\alpha .$$

Hence the (global) $\bar\partial$-closed (0,1)-form

$$\omega^\alpha = \bar\partial(\Sigma_i q_i h_{ik}^\alpha) = \Sigma_i h_{ik}^\alpha \bar\partial q_i = \Sigma_i h_{ik}^\alpha \left(\frac{dq_i}{d\bar z_i}\right) d\bar z_i$$

corresponds to h^α under Dolbeault's isomorphism. Note that ω^α, $1 \leqq \alpha \leqq g$, depend holomorphically on t and their restrictions $\omega^\alpha(t)$ to any V_t, $t \in W$, form a basis of $H^0(V_t, \bar\Omega_t^1)$, the space of anti-holomorphic 1-forms on V_t.

Let $t \in W \longmapsto P_t \in V_t$ be a holomorphic (local) cross section of $\pi_g : X_g \longrightarrow T_g$. Then $\{[nP_t]\}_{t \in W}$ is clealy a family of line bundles. We may write

$$[nP_t] = \{f_{ik}^o(z_k, t)\},$$

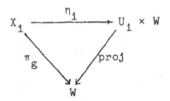

where $f_{ik}^{o}(z_k,t)$ are transition functions of $[nP_t]$ and depends holomorphically on $(z_k,t) \in U_k \times W$.

Now, fix $s_o \in J_n(V_o)$. The line bundle $F_{s_o} \otimes [nP_o]^{-1}$ is of degree 0, so that it corresponds to a point $\hat{s}_o \in J(V_o)$. Let $x = (x_1, \cdots, x_g)$ be a coordinate system in an open neighborhood D of $\hat{s}_o$ in $J(V_o)$. Put $\hat{s}_o = (x_1^o, \cdots, x_g^o)$. Note that $D \times W$ can be identified with an open neighborhood of s_o in J_n.

Choose a point $z(i)$ in U_i and an oriented smooth curve in $U_i \cup U_k$ combining $z(i)$ with $z(k)$.

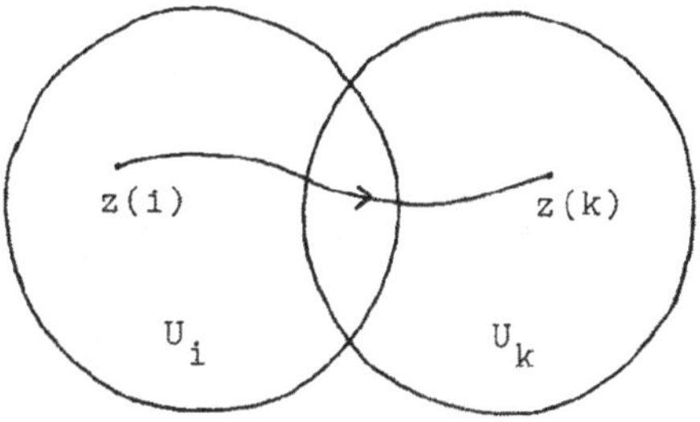

Now, by the argument in §4.3, the transition functions of F_s, $s = (x,t) \in D \times W$, is given by

$$f_{ik}(z_k,x,t) = f_{ik}^{o}(z_k,t) \cdot \exp \int_{z(i)}^{z(k)} (x_1 \omega^1(t) + \cdots + x_g \omega^g(t)) .$$

In particular,

$$F_{s_o} = \{f_{ik}(z_k,x^o,o)\} ,$$

$$f_{ik}(z_k,x^o,o) = f_{ik}^{o}(z_k,o) \cdot \exp \int_{z(i)}^{z(k)} (x_1^o \omega^1(o) + \cdots + x_g^o \omega^g(o)) .$$

Note that $f_{ik}(z_k,x,t)$ depends holomorphically on (z_k,x,t).

Now, we denote by $\mathbb{G}_{n,T}^{r}$ the complex space $\mathbb{G}^{r}$ in Theorem 4.1.9 with respect to the family $\{F_s, V_s\}_{s \in J_n'}$.

We define the set $\mathbb{G}_n^r$ by

$$\mathbb{G}_n^r = \cup_{t\in T_g}\mathbb{G}_n^r(V_t), \quad \text{(disjoint union)}.$$

Then, there is a canonical injection

$$\mathbb{G}_{n,T}^r \hookrightarrow \mathbb{G}_n^r .$$

We use this map as a "local coordinate system" in $\mathbb{G}_n^r$. Using the expression of the transition functions of F_s above, we can easily patch these data up and get a complex space structure on $\mathbb{G}_n^r$ such that (1) the projection

$$\mu : \mathbb{G}_n^r \longrightarrow J_n$$

is a proper holomorphic map and (2) each fiber of

$$\pi : \mathbb{G}_n^r \xrightarrow{\mu} J_n \xrightarrow{\tilde{\pi}} T_g$$

is biholomorphic to the complex space $\mathbb{G}_n^r(V_t)$ in §5.1.

In particular

$$\mathbb{G}_n^0 = \cup_{t\in T_g} S^n V_t, \quad \text{(disjoint union)}.$$

For $r \geqq 1$, put

$$\mathbb{F}_n^r = \{g_n^r \in \mathbb{G}_n^r \mid g_n^r \text{ has a fixed point}\}.$$

Then, in a similar way to the proof of Lemma 4.4.1, we get

<u>Lemma 5.2.1</u>. $\mathbb{F}_n^r$ is a closed complex subspace of $\mathbb{G}_n^r$. Moreover,

$$\mathbb{F}_n^r = \cup_t \mathbb{F}_n^r(V_t), \quad \text{(disjoint union)}.$$

Now, put

$$\mathbb{H}\mathrm{ol}_{\text{non-deg}}(n,r) = \cup_t \mathrm{Hol}_{\text{non-deg}}(V_t, \mathbb{P}^r)_n, \quad \text{(disjoint union)},$$

(see §5.1). Then it is an open subspace of

$$\mathbb{H}\mathrm{ol} = \cup_t \mathrm{Hol}(V_t, \mathbb{P}^r), \quad \text{(disjoint union)},$$

(see Theorem 3.2.9). Note that $\mathrm{Aut}(\mathbb{P}^r)$ acts freely on $\mathbb{H}\mathrm{ol}_{\text{non-deg}}(n,r)$. The proof of the following theorem is similar to that of Theorem 4.4.2, so we omit it.

Theorem 5.2.2. The orbit space $\mathbb{H}\mathrm{ol}_{\text{non-deg}}(n,r)/\mathrm{Aut}(\mathbb{P}^r)$ has a complex space structure which is biholomorphic to $\mathbb{G}_n^r - \mathbb{F}_n^r$ and makes $\mathbb{H}\mathrm{ol}_{\text{non-deg}}(n,r)$ a principal $\mathrm{Aut}(\mathbb{P}^r)$-bundle on it.

Let $R_n = \cup_t R_n(V_t)$ be the complex space defined in §3.3. It is clear that $R_n = \mathbb{H}\mathrm{ol}_{\text{non-deg}}(n,1)$.

Corollary 5.2.3. $R_n/\mathrm{Aut}(\mathbb{P}^1)$ is biholomorphic to $\mathbb{G}_n^1 - \mathbb{F}_n^1$.

Now, we rewrite the condition of Theorem 4.1.10 in the present case. First, note that

$$\left(\frac{\partial f_{ik}}{\partial x}\right)_{(z_k, x^o, o)} = f_{ik}(z_k, x^o, o)h_{ik},$$

where

$$h_{ik} = \int_{z(i)}^{z(k)} \left(\sum_\alpha a^\alpha \omega^\alpha(o)\right),$$

$$\left(\frac{\partial}{\partial x}\right)_{x^o} = \sum_\alpha a^\alpha \left(\frac{\partial}{\partial x^\alpha}\right)_{x^o} \in T_{x^o}J(V_o).$$

Note that $h = \{h_{ik}\} \in H^1(V_o, \mathcal{O})$ corresponds to $\sum_\alpha a^\alpha\omega^\alpha(o) \in H^0(V_o, \bar{\Omega}_o^1)$ under Dolbeault's isomorphism. We identify $T_{x^o}J(V_o)$ with $H^1(V_o, \mathcal{O})$ by the linear isomorphism:

$$\sum_\alpha a^\alpha \left(\frac{\partial}{\partial x^\alpha}\right)_{x^o} \in T_{x^o}J(V_o) \longmapsto \sum_\alpha a^\alpha\omega^\alpha(o) \in H^0(V_o, \bar{\Omega}_o^1) \longmapsto h \in H^1(V_o, \mathcal{O}).$$

Hence, for $\xi \in H^0(V_o, \mathcal{O}(F_{s_o}))$,

$$\tau_\xi(h)_{ik} = f_{ik}(z_k,x^o,o)h_{ik}\xi_k = \xi_i h_{ik} .$$

Hence

(1) $\tau_\xi(h) = \tau(\xi,h)$, the bilinear map defined in §4.3.

Next, put

$$f_{ik}(z_k,t) = f_{ik}(z_k,x^o,t) .$$

$\eta_i\eta_k^{-1}$ is written as

$$\eta_i\eta_k^{-1}(z_k,t) = (g_{ik}(z_k,t),t) .$$

Then, from the equality

$$f_{ik}(z_k,t) = f_{ij}(g_{jk}(z_k,t),t)f_{jk}(z_k,t) ,$$

we get the following (2) and (3).

(2) $$\left(\frac{\partial \log f_{ik}(z_k,t)}{\partial t}\right)_{t=o} =$$

$$\left(\frac{\partial \log f_{ij}(z_j,t)}{\partial t}\right)_{t=o} + \left(\frac{\partial \log f_{jk}(z_k,t)}{\partial t}\right)_{t=o} + \theta_{jk}\left(\frac{\partial \log f_{ij}(z_j,o)}{\partial z_j}\right) ,$$

where $\frac{\partial}{\partial t} = \sum_\nu b^\nu\left(\frac{\partial}{\partial t_\nu}\right)_o$, $\theta_{jk} = \sum_\nu b^\nu\left(\frac{\partial g_{jk}}{\partial t_\nu}\right)_{t=o}$ and $z_j = g_{jk}(z_k,o)$.

($\theta = \{\theta_{jk}\} \in H^1(V_o,\Theta)$, $\Theta = \mathcal{O}(TV_o)$.)

(3) $$\frac{\partial \log f_{ik}(z_k,o)}{\partial z_k} = \left(\frac{\partial z_j}{\partial z_k}\right)\frac{\partial \log f_{ij}(z_j,o)}{\partial z_j} + \frac{\partial \log f_{jk}(z_k,o)}{\partial z_k} .$$

Let θ_i be a C^∞-function on U_i such that

(4) $$\theta_{ik} = \left(\frac{\partial z_i}{\partial z_k}\right)\theta_k - \theta_i$$

Then

$$-\theta_{ik}\left(\frac{d\xi_i}{dz_i}\right) = -\left(\frac{\partial z_i}{\partial z_k}\right)\theta_k\left(\frac{d\xi_i}{dz_i}\right) + \theta_i\left(\frac{d\xi_i}{dz_i}\right)$$

$$= -\theta_k\left(\frac{d\xi_i}{dz_k}\right) + \theta_i\left(\frac{d\xi_i}{dz_i}\right)$$

$$= -\theta_k f_{ik}\left(\frac{d\xi_k}{dz_k}\right) - \theta_k \xi_k\left(\frac{\partial f_{ik}}{\partial z_k}\right) + \theta_i\left(\frac{d\xi_i}{dz_i}\right)$$

$$= -f_{ik}(z_k,o)\theta_k\left(\frac{d\xi_k}{dz_k}\right) + \theta_i\left(\frac{d\xi_i}{dz_i}\right) - \xi_i\theta_k\left(\frac{\partial \log f_{ik}(z_k,o)}{\partial z_k}\right).$$

$\left(\frac{\partial z_i}{\partial z_k} = \frac{\partial g_{ik}(z_k,o)}{\partial z_k}.\right)$ Hence

$$\tau_\xi\left(\frac{\partial}{\partial t}\right)_{ik}(z_i) = \frac{\partial f_{ik}}{\partial t}\xi_k - \theta_{ik}\frac{d\xi_i}{dz_i}$$

$$= \xi_i\left(\frac{\partial \log f_{ik}(z_k,t)}{\partial t}\right)_{t=0} - \theta_{ik}\frac{d\xi_i}{dz_i}$$

$$= \xi_i\left(\left(\frac{\partial \log f_{ik}(z_k,t)}{\partial t}\right)_{t=0} - \theta_k \frac{\partial \log f_{ik}(z_k,o)}{\partial z_k}\right)$$

$$- \left(f_{ik}(z_k,o)\theta_k\left(\frac{d\xi_k}{dz_k}\right) - \theta_i\left(\frac{d\xi_i}{dz_i}\right)\right).$$

Lemma 5.2.4. $\left\{\left(\frac{\partial \log f_{ik}(z_k,t)}{\partial t}\right)_{t=0} - \theta_k \frac{\partial \log f_{ik}(z_k,o)}{\partial z_k}\right\}$ is a 1-cocycle of the sheaf on V_o of C^∞-functions.

Proof. Put

$$a_{ik}(z_k) = \left(\frac{\log f_{ik}(z_k,t)}{t}\right)_{t=0},$$

$$u_{ik}(z_k) = \frac{\partial \log f_{ik}(z_k,o)}{\partial z_k}.$$

Then, by (2),

(5) $a_{ik} - a_{ij} - a_{jk} = \theta_{jk}u_{ij}$.

By (3) and (4),

(6) $u_{ik}\theta_k - u_{ij}\theta_j - u_{jk}\theta_k = \theta_{jk}u_{ij}$.

(5) and (6) imply that $\{a_{ik} - u_{ik}\theta_k\}$ is a 1-cocycle. Q.E.D.

By this lemma, there is C^∞-function $b_i(z_i)$ on U_i such that

$$b_k(z_k) - b_i(z_i) = \left(\frac{\partial \log f_{ik}(z_k,t)}{\partial t}\right)_{t=0} - \Theta_k \frac{\partial \log f_{ik}(z_k,o)}{\partial z_k}.$$

Then

$$\tau_\xi\left(\frac{\partial}{\partial t}\right)_{ik}(z_i) = f_{ik}(z_k,o)\left(\xi_k b_k - \left(\frac{d\xi_k}{dz_k}\right)\theta_k\right) - \left(\xi_i b_i - \left(\frac{d\xi_i}{dz_i}\right)\theta_i\right).$$

Put $v_i = \xi_i b_i - \left(\frac{d\xi_i}{dz_i}\right)\theta_i$. Then

$$\bar\partial v_i = f_{ik}(z_k,o)\bar\partial v_k$$

is a F_{s_o}-valued (global) $\bar\partial$-closed (0,1)-form corresponding to $\tau_\xi(\frac{\partial}{\partial t})$ under Dolbeault's isomorphism.

We identify T_oT_g with $H^1(V_o,\Theta)$ by the linear isomorphism

$$\frac{\partial}{\partial t} = \sum_\nu b^\nu\left(\frac{\partial}{\partial t_\nu}\right)_o \in T_oT_g \longmapsto \theta = \{\theta_{ik}\} \in H^1(V_o,\Theta)\ ,$$

where $\theta_{ik} = \sum_\nu b^\nu\left(\frac{\partial g_{ik}}{\partial t_\nu}\right)(z_k,o)$. Then $\tau_\xi(\frac{\partial}{\partial t})$ is written as $\tau_\xi(\theta)$.

We denote by $\hat\tau^*_\xi$ the dual map of

$$\tau_\xi : H^1(V_o,\Theta) \longrightarrow H^1(V_o,\mathcal{O}(F_{s_o}))\ .$$

Then, for $\eta = \{\eta_i\} \in H^0(V_o,\mathcal{O}(K_{V_o}\otimes F_{s_o}^{-1}))$,

(7) $$\langle\theta,\hat\tau^*_\xi(\eta)\rangle = \langle\tau_\xi\theta,\eta\rangle = \int_{V_o}\eta_i dz_i\wedge\bar\partial v_i$$

$$= \int_{V_o}\xi_i\eta_i dz_i\wedge\bar\partial b_i - \int_{V_o}\eta_i \frac{d\xi_i}{dz_i} dz_i\wedge\bar\partial\theta_i\ .$$

Now,

$$\tau_\xi(h,\theta) = \tau_\xi(h) + \tau_\xi(\theta) = \tau(\xi,h) + \tau_\xi(\theta)\ .$$

The dual τ^*_ξ of τ_ξ is the linear map

$$\tau^*_\xi : H^0(V_o,\mathcal{O}(K_{V_o}\otimes F_{s_o}^{-1})) \longrightarrow H^0(V_o,\mathcal{O}(K_{V_o})) \times H^0(V_o,\mathcal{O}(K_{V_o}^{\otimes 2}))$$

given by

$$\langle (h,\theta),\ \tau_\xi^* \eta\rangle = \langle h,\ \tau_\xi^*\eta\rangle + \langle \theta,\ \hat{\tau}_\xi^*\eta\rangle$$

$$= \langle h,\ \xi\eta\rangle + \langle \theta,\ \hat{\tau}_\xi^*\eta\rangle\ ,$$

where $\eta \in H^0(V_o, \mathcal{O}(K_{V_o} \otimes F_{s_o}^{-1}))$.

Let $\xi^0,\cdots,\xi^r \in H^0(V_o, \mathcal{O}(F_{s_o}))$ be linearly independent. Consider the linear map

$$\tau_r : ((h_0,\theta_0),\cdots,(h_r,\theta_r)) \in (H^1(V_o,\mathcal{O}) \times H^1(V_o,\Theta))^{r+1}$$

$$\longmapsto (\tau_{\xi^0}(h_0,\theta_0),\cdots,\tau_{\xi^r}(h_r,\theta_r)) \in H^1(V_o, \mathcal{O}(F_{s_o}))^{r+1}.$$

Let

$$\Delta = \{((h,\theta),\cdots,(h,\theta)) \in (H^1(V_o, \mathcal{O}) \times H^1(V_o, \Theta))^{r+1}\}$$

be the diagonal. Then

"$\tau_r(\Delta) = H^1(V_o, \mathcal{O}(F_{s_o}))^{r+1}$."

$\Longleftrightarrow$

"If $(\eta^0,\cdots,\eta^r) \in (H^0(V_o, \mathcal{O}(K_{V_o} \otimes F_{s_o}^{-1})))^{r+1}$ satisfies

$\langle \tau_r(\Delta),(\eta^0,\cdots,\eta^r)\rangle = 0$, then $\eta^0 = \cdots = \eta^r = 0$."

$\Longleftrightarrow$

"If $(\eta^0,\cdots,\eta^r) \in (H^0(V_o, \mathcal{O}(K_{V_o} \otimes F_{s_o}^{-1})))^{r+1}$ satisfies

$\langle \Delta,\ (\tau^*\eta^0,\cdots,\tau^*\eta^r)\rangle = 0$, then $\eta^0 = \cdots = \eta^r = 0$."

$\Longleftrightarrow$

"If $(\eta^0,\cdots,\eta^r) \in (H^0(V_o, \mathcal{O}(K_{V_o} \otimes F_{s_o}^{-1})))^{r+1}$ satisfies

$\langle h,\ \sum_\alpha \tau_{\xi^\alpha}^*\eta^\alpha\rangle + \langle \theta, \sum_\alpha \hat{\tau}_{\xi^\alpha}^*\eta^\alpha\rangle = 0$ for all $(h,\theta) \in$

$H^1(V_o, \mathcal{O}) \times H^1(V_o, \Theta)$, then $\eta^0 = \cdots = \eta^r = 0$."

$\Longleftrightarrow$

"If $(\eta^0, \cdots, \eta^r) \in (H^0(V_o, \mathcal{O}(K_{V_o} \otimes F_{s_o}^{-1})))^{r+1}$ satisfies

$$\begin{cases} \sum_\alpha \xi^\alpha \eta^\alpha = 0 , \\ \sum_\alpha \hat{\tau}^*_{\xi^\alpha} \eta^\alpha = 0 , \end{cases}$$

then $\eta^0 = \cdots = \eta^r = 0$."

If $\sum_\alpha \xi^\alpha \eta^\alpha = 0$, then, by (7),

(8) $$\langle \theta, \sum_\alpha \hat{\tau}^*_{\xi^\alpha} \eta^\alpha \rangle = -\int_{V_o} \left(\sum_\alpha \eta_i^\alpha \frac{d\xi_i^\alpha}{dz_i}\right) dz_i \wedge \bar{\partial} \theta_i ,$$

where $\xi^\alpha = \{\xi_i^\alpha\}$ and $\eta^\alpha = \{\eta_i^\alpha\}$.

Lemma 5.2.5. For $\xi^\alpha = \{\xi_i^\alpha\} \in H^0(V_o, \mathcal{O}(F_{s_o}))$ and $\eta^\alpha = \{\eta_i^\alpha\} \in H^0(V_o, \mathcal{O}(K_{V_o} \otimes F_{s_o}^{-1}))$, $0 \leqq \alpha \leqq r$, assume that $\sum_\alpha \xi^\alpha \eta^\alpha = 0$. Tnen

$$\left\{\sum_\alpha \eta_i^\alpha \left(\frac{d\xi_i^\alpha}{dz_i}\right)\right\} \in H^0(V_o, \mathcal{O}(K_{V_o}^{\otimes 2})) .$$

Proof. Note that $\xi_i^\alpha = f_{ik}(z_k, 0)\xi_k^\alpha$ and $\eta_i^\alpha = \left(\frac{\partial z_k}{\partial z_i}\right) f_{ki}(z_i, o)\eta_k^\alpha$. Hence

$$\frac{d\xi_i^\alpha}{dz_i} = \frac{\partial f_{ik}(z_k, o)}{\partial z_i} \xi_k^\alpha + f_{ik}(z_k, o)\left(\frac{\partial z_k}{\partial z_i}\right)\left(\frac{d\xi_k^\alpha}{dz_k}\right)$$

Thus

$$\begin{aligned} \sum_\alpha \eta_i^\alpha \left(\frac{d\xi_i^\alpha}{dz_i}\right) &= \sum_\alpha \left(\frac{\partial z_k}{\partial z_i}\right) f_{ki}(z_i, o)\eta_k^\alpha \left(\left(\frac{\partial f_{ik}(z_k, o)}{\partial z_i}\right)\xi_k^\alpha \right. \\ &\qquad \left. + f_{ik}(z_k, o)\left(\frac{\partial z_k}{\partial z_i}\right)\left(\frac{d\xi_k^\alpha}{dz_k}\right)\right) \\ &= \left(\frac{\partial z_k}{\partial z_i}\right)\left(\frac{\partial \log f_{ik}(z_k, o)}{\partial z_i}\right)\left(\sum_\alpha \xi_k^\alpha \eta_k^\alpha\right) + \left(\frac{\partial z_k}{\partial z_i}\right)^2 \left(\sum_\alpha \eta_k^\alpha \left(\frac{d\xi_k^\alpha}{dz_k}\right)\right) \end{aligned}$$

$$= \left(\frac{\partial z_k}{\partial z_i}\right)^2 \left(\sum_\alpha \eta_k^\alpha \left(\frac{d\xi_k^\alpha}{dz_k}\right)\right) .$$

Q.E.D.

Note that Θ-valued (0,1)-form $\bar{\partial}\theta_i$ corresponds to $\theta = \{\theta_{ik}\}$ under Dolbeault's isomorphism (see (4)). Hence (8) shows that, if $\sum_\alpha \xi^\alpha \eta^\alpha = 0$, then

$$\sum_\alpha \hat{\tau}^*_{\xi^\alpha} \eta^\alpha = \{- \sum_\alpha \eta_i^\alpha \frac{d\xi_i^\alpha}{dz_i}\} \in H^0(V_o, \mathcal{O}(K_{V_o}^{\otimes 2})) .$$

Definition 5.2.6. A linear system $L \in G^r(H^0(V_o, \mathcal{O}(F_{s_o})))$ is said to be weakly semi-regular if, for a basis $\{\xi^0, \cdots, \xi^r\}$ of L and $(r+1)$-elements $\eta^0, \cdots, \eta^r$ of $H^0(V_o, \mathcal{O}(K_{V_o} \otimes F_{s_o}^{-1}))$, the equalities:

$$\sum_\alpha \xi^\alpha \eta^\alpha = 0 \quad \text{and} \quad \{\sum_\alpha \eta^\alpha \frac{d\xi_i^\alpha}{dz_i}\} = 0$$

imply $\eta^0 = \cdots = \eta^r = 0$.

If a linear system is semi-regular, then it is weakly semi-regular.

By Theorem 4.1.10,

Theorem 5.2.7. If a linear system $g_n^r \in \mathbb{G}_n^r$ is weakly semi-regular, then $\mathbb{G}_n^r$ is non-singular at g_n^r and

$$\dim_{g_n^r} \mathbb{G}_n^r = (r+1)(n-r) - rg + 3g - 3 .$$

Remark 5.2.8. Severi [78, p.161], says that curves in $\mathbb{P}^r$ of genus g and degree n depend on

$$(r+1)(n-r) - rg + 3g - 3 + \dim \mathrm{Aut}(\mathbb{P}^r)$$

-parameters, provided $(r+1)(n-r) - rg \geqq 0$.

An interesting fact is

Theorem 5.2.9. Every element of $\mathbb{G}_n^1$ is weakly semi-regular. Hence $\mathbb{G}_n^1$ is non-singular and $\dim \mathbb{G}_n^1 = 2n + 2g - 5$.

Proof. Let $\xi^0 = \{\xi_i^0\}$ and $\xi^1 = \{\xi_i^1\}$ be linearly independent elements of $H^0(V_o, \mathcal{O}(F_{s_o}))$. For $\eta^0 = \{\eta_i^0\}$ and $\eta^1 = \{\eta_i^1\}$ of $H^0(V_o, \mathcal{O}(K_{V_o} \otimes F_{s_o}^{-1}))$, assume that

$$\xi^0\eta^0 + \xi^1\eta^1 = 0 ,$$
$$\frac{d\xi_i^0}{dz_i}\eta_i^0 + \frac{d\xi_i^1}{dz_i}\eta_i^1 = 0 .$$

This means that, as holomorphic functions on U_i,

$$\xi_i^0\eta_i^0 + \xi_i^1\eta_i^1 = 0 ,$$
$$\left(\frac{d\xi_i^0}{dz_i}\right)\eta_i^0 + \left(\frac{d\xi_i^1}{dz_i}\right)\eta_i^1 = 0 .$$

Hence

$$\left(\xi_i^0\left(\frac{d\xi_i^1}{dz_i}\right) - \xi_i^1\left(\frac{d\xi_i^0}{dz_i}\right)\right)\eta_i^\alpha = 0, \quad \alpha = 0, 1 .$$

But

$$\xi_i^0\frac{d\xi_i^1}{dz_i} - \xi_i^1\frac{d\xi_i^0}{dz_i} = (\xi_i^0)^2\frac{d}{dz_i}(\xi_i^1/\xi_i^0) \neq 0 ,$$

for ξ_i^1/ξ_i^0 is a non-constant meromorphic function on V_o. Thus $\eta_i^0 = \eta_i^1 = 0$, i.e., $\eta^0 = \eta^1 = 0$. Q.E.D.

This theorem combined with Corollary 5.2.3 again proves Theorem 3.4.17.

Proposition 5.2.10. For $L \in \mathbb{G}_n^2$, assume that (1) Φ_L is a birational map of V_o onto $C = \Phi_L(V_o) \subset \mathbb{P}^2$ and (2) the plane curve C has only ordinary singular points. Then L is weakly semi-regular.

Proof. Let $\{\xi^0,\xi^1,\xi^2\}$ be a basis of L. Put $D^0 = (\xi^0)$ and $F = [D^0]$. Take $\eta^0,\eta^1,\eta^2 \in H^0(V_o, \mathcal{O}(K_{V_o} \otimes F^{-1}))$. Assume that $\eta^2 \not\equiv 0$ and

$$\xi^0\eta^0 + \xi^1\eta^1 = -\xi^2\eta^2 ,$$

$$\frac{d\xi_i^0}{dz_i}\eta_i^0 + \frac{d\xi_i^1}{dz_i}\eta_i^1 = -\frac{d\xi_i^2}{dz_i}\eta_i^2 ,$$

where $\xi^\alpha = \{\xi_i^\alpha\}$, $\eta^\alpha = \{\eta_i^\alpha\}$, $0 \leqq \alpha \leqq 2$. Solving the simultaneous equations, we get

$$\eta^0 = -\frac{df_0}{d(1/f_1)}\eta^2 \quad \text{and} \quad \eta^1 = -\frac{df_2}{df_1}\eta^2 ,$$

where $f_1 = \xi^1/\xi^0$, $f_2 = \xi^2/\xi^0$ and $f_0 = \xi^2/\xi^1 = f_2/f_1$ and df_α, $0 \leqq \alpha \leqq 2$, is the differential of f_α. (df_α is a meromorphic section of K_{V_o}.)

Let $e_P(f_\alpha)$, $0 \leqq \alpha \leqq 2$, be the ramification exponent of the map $f_\alpha : V_o \longrightarrow \mathbb{P}^1$ at $P \in V_o$. Then it is well known that

$$(df_\alpha) = \sum_P (e_P(f_\alpha) - 1)P - 2D_\infty(f_\alpha) .$$

Hence, we get

$$(\eta^1) = (df_2) - (df_1) + (\eta^2)$$

$$= \sum_P (e_P(f_2) - e_P(f_1))P - 2(D_\infty(f_2) - D_\infty(f_1)) + (\eta^2) .$$

Now, we may take ξ^0, ξ^1 and ξ^2 so that

(1) C passes <u>neither</u> of the points $P_0 = (1:0:0)$, $P_1 = (0:1:0)$ and $P_2 = (0:0:1)$,

(2) no singular point of C is on the coordinate axis,

(3) no coordinate axis is tangent to C, and

(4) no tangent line at singular points passes P_0, P_1, P_2.

Then, by (1),

$$D_\infty(f_1) = D_\infty(f_2) = (\xi^0) - \hat{D},$$

where $\hat{D}$ is the fixed part of L. Hence

(5) $(\eta^1) = \sum_P (e_P(f_2) - e_P(f_1))P + (\eta^2)$.

Note that f_1 and f_2 are the projections with the center P_2 and P_1, respectively. We look for points P where $e_P(f_2) - e_P(f_1)$ is negative. If P is such a point, i.e., $e_P(f_1) > e_P(f_2)$, then either $\hat{P} = \Phi_L(P)$ must be a singular point of C or the tangent line to C at $\hat{P}$ passes P_2 (see (4)).

Let $\hat{P}$ be a singular point. Let $\cup C_\nu$ be the irreducible decomposition of C at P. Put

$$m_\nu(\hat{P}) = \text{the multiplicity of } C_\nu \text{ at } \hat{P}.$$

Then, by (4) and by the assumption that Φ_L is birational, there is a unique ν such that

$$e_P(f_1) = e_P(f_2) = m_\nu(\hat{P}).$$

Assume that the tangent line to C at $\hat{P}$ passes P_2. Then, by (2), (3) and (4), $\hat{P}$ is not on the line $\overline{P_1P_2}$ and $e_P(f_2) = 1$.

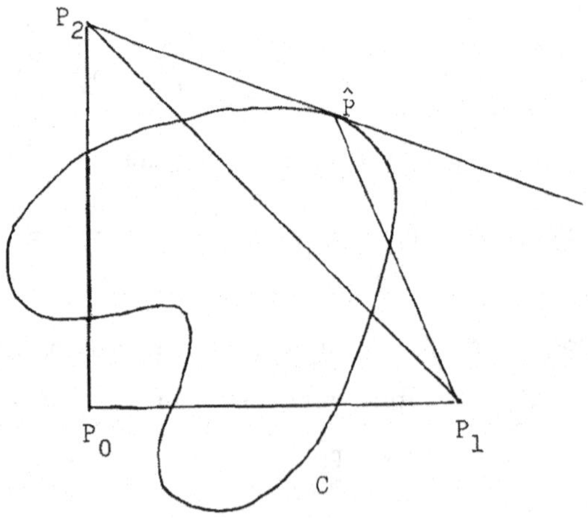

Hence, by (5),

$$D = (\eta^2) - \Sigma'(e_P(f_1) - 1)P$$

is an effective divisor, where Σ' is extended over all $P \in V_o$ such that the tangent line to C at $\Phi_L(P)$ passes P_2. Note that

$$D = (\eta^2) - \sum_P (e_P(f_1) - 1)P + \sum_P (m_\nu(\hat{P}) - 1)P ,$$

where $\hat{P} = \Phi_L(P)$ and ν is the index such that C_ν is the image by Φ_L of an open neighborhood of P in V_o. By Riemann-Hurwitz formula,

$$\deg D = (2g-2) - n - (2g-2+2\,\mathrm{ord}(f_1)) + \sum_P (m_\nu(\hat{P}) - 1)$$

$$= \sum_P (m_\nu(\hat{P}) - 1) - 3n + 2\deg \hat{D} .$$

$(\mathrm{ord}(f_1) = \deg C = n - \deg \hat{D}.)$

Now, if C has only ordinary singular points, then $m_\nu(\hat{P}) = 1$, for all $P \in V_o$. Hence $\deg D < 0$, a contradiction. <u>Q.E.D.</u>

The above proof shows, more generally,

<u>Proposition 5.2.11</u>. For $L \in \mathbb{G}_n^2$, assume that (1) Φ_L is a birational map of V_o onto $C = \Phi_L(V_o) \subset \mathbb{P}^2$ and (2) $3n - 2\deg \hat{D} > \sum_P (m_\nu(\hat{P}) - 1)$, ($\hat{D}$ = the fixed part of L), then L is weakly semi-regular.

Finally, we construct the <u>global moduli space of non-degenerate holomorphic maps of compact Riemann surfaces of genus</u> g $(g \geq 2)$ <u>into</u> $\mathbb{P}^r$ as follows: Let $(X_g, \pi, T_g) = \{V_t\}_{t \in T_g}$ be the Teichmüller family of compact Riemann surfaces of genus g. The <u>Teichmüller modular group</u> Γ_g acts on T_g and X_g properly discontinuously (see, e.g., Mumford [59, p.28]). The quotient space $M_g = T_g/\Gamma_g$ is called the <u>moduli space of compact Riemann surfaces of genus</u> g.

For any $\gamma \in \Gamma_g$ and $t \in T_g$, γ induces a biholomorphic map $\gamma : V_t \longrightarrow V_{\gamma(t)}$. Conversely, it can be shown that Γ_g has the following property:

(*) For two points $t, s \in T_g$ and a biholomorphic map $i : V_t \longrightarrow V_s$, there is a unique $\gamma \in \Gamma_g$ such that $\gamma(t) = s$ and $\gamma : X \longrightarrow X$ induces $i : V_t \longrightarrow V_s$.

Now, for any $\gamma \in \Gamma_g$, any point $t \in T_g$ and any linear system g_n^r of degree n and dimension r on V_t, we define the linear system $\gamma(g_n^r)$ on $V_{\gamma(t)}$ by

$$\gamma(g_n^r) = \{\gamma(P_1)+\cdots+\gamma(P_n) \mid P_1+\cdots P_n \in g_n^r\} .$$

Then, we can easily show that

$$\gamma : \mathbb{G}_n^r \longrightarrow \mathbb{G}_n^r$$

is a biholomorphic map. Thus Γ_g acts on $\mathbb{G}_n^r$. The action is properly discontinuous, as is easily shown. Γ_g also acts on $\mathbb{F}_n^r$ and on $\mathbb{G}_n^r - \mathbb{F}_n^r$ properly discontinuously. By the property (*), the quotient complex space $(\mathbb{G}_n^r - \mathbb{F}_n^r)/\Gamma_g$ can be considered as the global moduli space of non-degenerate holomorphic maps f of compact Riemann surfaces V of genus g $(g \geq 2)$ into $\mathbb{P}^r$ such that

$$n = (\text{ord } f)\cdot(\deg f(V))$$

(c.f., Remark 3.6.14). In particular, $(\mathbb{G}_n^1 - \mathbb{F}_n^1)/\Gamma_g$ can be considered as the global moduli space of algebraic functions of order n and genus g $(g \geq 2)$. Since $\mathbb{G}_n^1 - \mathbb{F}_n^1$ is non-singular by Theorem 5.2.9, $(\mathbb{G}_n^1 - \mathbb{F}_n^1)/\Gamma_g$ is a normal complex space of dimension $2n+2g-5$.

5.3. $\pi(\mathbb{G}_n^r)$.

Consider the projection map

$$\pi : \mathbb{G}_n^r \xrightarrow{\mu} J_n \xrightarrow{\tilde{\pi}} T_g .$$

It is a proper holomorphic map. Hence its image $\pi(\mathbb{G}_n^r)$ is a closed complex subspace of T_g. It is clear that

$$\pi(\mathbb{G}_n^r) = \{t \in T_g \mid V_t \ \underline{\text{has a linear system}} \ g_n^r\} .$$

A famous fact is

Theorem 5.3.1. If $(r+1)(n-r) - rg \geqq 0$, then $\pi(\mathbb{G}_n^r) = T_g$.

Remark 5.3.2. (1) It says in particular $(r = 1)$ that every compact Riemann surface of genus g has a meromorphic function of order $\leqq [\frac{g+3}{2}]$. (2) This theorem is a classically known fact. Its modern (rigorous) proofs were given by Kleiman-Laksov [38] and Kempf [36]. Kleiman-Laksov [39] gave another proof. See also Gunning [28] and Griffiths-Harris [25]. Meis [53] gave an analytic proof for $r = 1$.

We try to give a proof of the theorem along the line of Meis [53]. But it is still incomplete for $r > 1$.

We first recall the proper mapping theorem by Remmert [73]. Let X and Y be complex spaces and let $f : X \longrightarrow Y$ be a holomorphic map. For $x \in X$, put

$$r_x(f) = \operatorname{codim}_x f^{-1}f(x)$$

and call it the rank of f at x. Put

$$r(f) = \max_{x \in X} r_x(f)$$

and call it the (global) rank of f. It can be shown that $r(f) \leqq \dim Y$.

Note that, if X and Y are complex manifolds, then

$$r(f) = \max_{x \in X} \operatorname{rank}(df)_x .$$

Theorem 5.3.3. (Proper mapping theorem by Remmert [73]). Let $f : X \longrightarrow Y$ be a proper holomorphic map of the rank r. Then $f(X)$ is a r-dimensional closed complex subspace of Y. In particular, if Y is irreducible and $r = \dim Y$, then f is surjective.

In order to prove Theorem 5.3.1, it is therefore enough to show that

"If $(r+1)(n-r) - rg \geqq 0$, then $r(\pi) = 3g - 3$."

This holds if the following assertion is satisfied:

"If $(r+1)(n-r) - rg \geqq 0$, then there is $L \in \mathbb{G}_n^r$ such that $\operatorname{codim}_L \pi^{-1}(\pi(L)) = 3g - 3$."

Put $t = \pi(L)$. Then $\pi^{-1}(\pi(L)) = \mathbb{G}_n^r(V_t)$. Assume that L is semi-regular. Then, L is a non-singular point of both $\mathbb{G}_n^r(V_t)$ and $\mathbb{G}_n^r$ and

$$\dim_L \mathbb{G}_n^r(V_t) = (r+1)(n-r) - rg ,$$

$$\dim_L \mathbb{G}_n^r = (r+1)(n-r) - rg + 3g - 3 .$$

Hence

$$\operatorname{codim}_L \pi^{-1}(\pi(L)) = 3g - 3 ,$$

so that the assertion is satisfied.

Note that, if $n < n'$, then $\pi(\mathbb{G}_n^r) \subset \pi(\mathbb{G}_{n'}^r)$. Hence it is enough to assert that $\pi(\mathbb{G}_n^r) = T_g$ for the smallest n, i.e.,

$$n = g + r - [\tfrac{g}{r+1}] \quad ([\ ] \text{ is the Gauss' notation}) .$$

Hence, we conclude that

<u>Assertion</u>. Theorem 5.3.1 is proved, if one finds a compact Riemann surface V of genus g and a semi-regular linear system g_n^r on it, where $n = g + r - [\frac{g}{r+1}]$.

This is actually what Meis [53] did for $r = 1$. In fact, he found such V and g_n^1 as follows:

<u>Case 1. g is even.</u>

In this case, $n = \frac{g+2}{2}$. We may assume that $g \geqq 4$. Let V be a non-singular model (of the closure in $\mathbb{P}^2$) of the curve:

$$y^n = (x-1)(x-2)(x-3)(x-4)^{n-1}(x-5)^{n-1}(x-6)^{n-1} .$$

Then V has the genus g. The meromorphic function x satisfies $h^1(2D_\infty(x)) = 0$. Hence, by (2) of Corollary 5.1.10, the pencil g_n^1 determined by x is semi-regular.

<u>Case 2. g is odd.</u>

In this case, $n = \frac{g+3}{2}$. Let V be a non-singular model (of the closure in $\mathbb{P}^2$) of the curve:

$$y^3 = \frac{(x-1)\cdots(x-n)}{(x-(n+1))\cdots(x-(2n-2))} .$$

Then V has the genus g. The meromorphic function y satisfies $h^1(2D_\infty(y)) = 0$, so that the pencil g_n^1 determined by y is semi-regular.

<u>Problem</u>. Find such V and g_n^r for $r > 1$.

<u>Proposition 5.3.4</u>. If $2n < g+2$, then $\pi : \mathbb{G}_n^1 \longrightarrow T_g$ is not

surjective. $\pi(\mathbb{G}_n^1)$ is a nowhere dense closed complex subspace of T_g of dimension at most $2n+2g-5$.

Proof. By Theorem 5.2.9, $\mathbb{G}_n^1$ is non-singular and of dimension $2n+2g-5$. Since $2n < g+2$, any $L \in \mathbb{G}_n^1$ can not be semi-regular and

$$\operatorname{codim}_L \pi^{-1}(\pi(L)) \leqq 2n+2g-5 < 3g-3 . \qquad \text{Q.E.D.}$$

Now, using the previous notations, $\mathbb{G}_n^r$ is locally (around $L \in \mathbb{G}_n^r$) given by

$$\{(L',x,t) \in U \times D \times W \mid \tilde{u}(x,t)\xi_\alpha(L') = 0, \quad 0 \leqq \alpha \leqq r\},$$

where W, D and U are open neighborhoods of o in T_g, of x_o in $J(V_o)$ and of L in $G^r(H^0(V_o\, \mathcal{O}(F_{s_o})))$, respectively. ($s_o = (x_o, o)$.) (See §4.1.)

Assume that L is weakly semi-regular. Then $\mathbb{G}_n^r$ is non-singular at L and the tangent space $T_L\mathbb{G}_n^r$ is given by

$$T_L\mathbb{G}_n^r \cong T_L G^r(H^0(V_o, \mathcal{O}(F_{s_o}))) \times K ,$$

where

$$K = \{(h,\theta) \in H^1(V_o, \Theta) \times H^1(V_o, \mathbb{H}) \mid \tau(\xi_\alpha, h) + \tau_{\xi_\alpha}\theta = 0, \quad 0 \leqq \alpha \leqq r\}.$$

The differential $(d\pi)_L : T_L\mathbb{G}_n^r \longrightarrow T_o(T_g)$ is given by

$$(d\pi)_L : (a,h,\theta) \in T_L G^r(H^0(V_o, \mathcal{O}(F_{s_o}))) \times K \longmapsto \theta \in H^1(V_o, \mathbb{H}) .$$

We look for the kernel of $(d\pi)_L$.

"$(a,h,o) \in \ker(d\pi)_L$."

$\Longleftrightarrow$

"$\tau(\xi_\alpha, h) = 0, \quad 0 \leqq \alpha \leqq r.$"

$\Longleftrightarrow$

"$\langle h, \xi_\alpha \eta \rangle = 0$ for all $\eta \in H^0(V_0, \mathcal{O}(K_{V_0} \otimes F_{s_0}^{-1})), \quad 0 \leqq \alpha \leqq r.$"

$\Longleftrightarrow$

"$\langle h, \sum_\alpha H^0(V_0, \mathcal{O}(K_{V_0} - D_\alpha)) \rangle = 0, \quad (D_\alpha = (\xi_\alpha)).$"

Hence

<u>Lemma 5.3.5</u>. Let $L \in \mathbb{G}_n^r$ be weakly semi-regular. Then

$$\operatorname{Ker}(d\pi)_L \cong T_L G^r(H^0(V_0, \mathcal{O}(F_{s_0}))) \times (\sum_\alpha H^0(V_0, \mathcal{O}(K_{V_0} - D_\alpha)))^{\perp},$$

where $(\sum_\alpha H^0(V_0, \mathcal{O}(K_{V_0} - D_\alpha)))^{\perp}$ is the annihilator of $\sum_\alpha H^0(V_0, \mathcal{O}(K_{V_0} - D_\alpha))$ in $H^1(V_0, \mathcal{O})$.

<u>Theorem 5.3.6</u>. Assume that $2n \geqq g+2$. Then $\pi : \mathbb{G}_n^1 \longrightarrow T_g$ is of maximal rank at $L \in \mathbb{G}_n^1$ if and only if L is semi-regular.

<u>Proof</u>. Note that π is surjective. By Lemma 5.3.5,

$$\begin{aligned} \dim \operatorname{Ker}(d\pi)_L &= 2(h^0(D) - 2) + g - \dim(H^0(V_0, \mathcal{O}(K - D)) + H^0(V_0, \mathcal{O}(K - D'))) \\ &= 2(h^0(D) - 2) + g - (2h^1(D) - h^1(2D - \hat{D})) \\ &= 2n - 2 - g + h^1(2D - \hat{D}), \end{aligned}$$

where $D, D' \in L$, $(D \neq D')$, and $\hat{D}$ is the fixed part of L.

On the other hand, π is of maximal rank at L if and only if

$$\dim \operatorname{Ker}(d\pi)_L = \dim \mathbb{G}_n^1 - \dim T_g = 2n - 2 - g.$$

Hence, π is of maximal rank if and only if $h^1(2D - \hat{D}) = 0$, i.e., L is semi-regular (see Corollary 5.1.10). <u>Q.E.D.</u>

Corollary 5.3.7. (c.f., Farkas [17]). For a fixed $n \geqq 2$, put

$$S_n = \{t \in T_g \mid V_t \text{ has a pencil } g_n^1 \text{ which is \underline{not} semi-regular}\}.$$

Then S_n is a nowhere dense closed complex subspace of T_g.

Proof. If $2n < g+2$, this follows from Proposition 5.3.4. If $2n \geqq g+2$, then this follows from Theorem 5.3.6 and Sard's theorem.

Q.E.D.

Example 5.3.8. $g = 4$ and $n = 3$. In this case, $\mathbb{G}_3^1$ and T_4 are both 9-dimensional. $\pi : \mathbb{G}_3^1 \longrightarrow T_4$ is a ramified covering map. (1) If V_t is hyperelliptic, then $\pi^{-1}(t) = \{g_2^1+P \mid P \in V_t\}$ $(\cong V_t)$ and no element of $\pi^{-1}(t)$ is semi-regular. (2) If V_t is not hyperelliptic, then the canonical curve C_t of V_t is the complete intersection of a quadric Q_t and a cubic surface in $\mathbb{P}^3$. (2-i) If Q_t is non-singular, then $\pi^{-1}(t)$ consists of 2 points, which are semi-regular. (2-ii) If Q_t is singular, then $\pi^{-1}(t)$ consists of 1 point, which is not semi-regular.

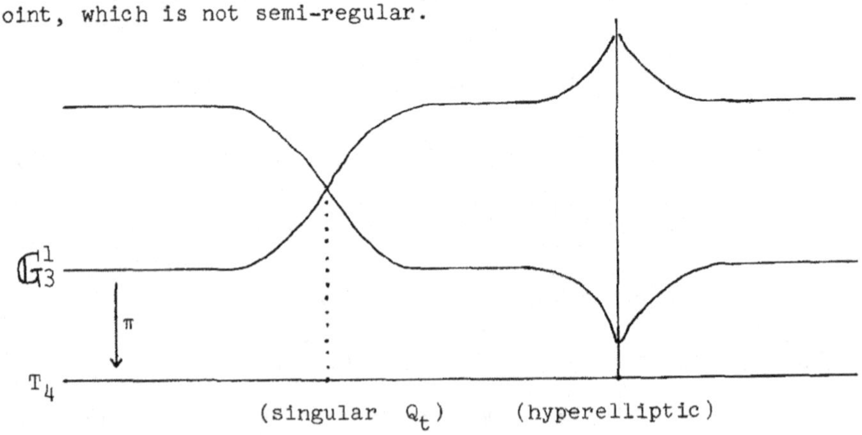

Remark 5.3.9. If g is even and $n = \frac{g+2}{2}$, then $\pi : \mathbb{G}_n^1 \longrightarrow T_g$ is a ramified covering map. The covering order k is given by the following table:

g	2	4	6	8	...	g
k	1	2	5	14	...	g!/(n!(n-1)!)

(see Griffiths-Harris [25, p.299 and p.359]).

Another direct consequence of Lemma 5.3.5 is

Proposition 5.3.10. For a weakly semi-regular $L \in \mathbb{G}^r_n$, $(d\pi)_L$ is injective if and only if (1) L is complete and (2) $\sum_\alpha H^0(V_o, \mathcal{O}(K-D_\alpha)) = H^0(V_o, \mathcal{O}(K))$.

We apply this proposition to $\mathbb{G}^1_n$. Note that every $L \in \mathbb{G}^1_n$ is weakly semi-regular (Theorem 5.2.9). Assume that L is complete. Take distinct $D_0, D_1 \in L$. The equality

$$H^0(V_o, \mathcal{O}(K-D_0)) + H^0(V_o, \mathcal{O}(K-D_1)) = H^0(V_o, \mathcal{O}(K))$$

holds if and only if

$$2h^1(D_0) = g + h^1(2D_0 - \hat{D}) .$$

($\hat{D}$ = the fixed part of L.) By Riemann-Roch theorem, this holds if and only if

(*) $$h^0(2D_0 - \hat{D}) = 3 - \deg \hat{D} .$$

Put $D_0 = D_0(f)+\hat{D}$ and $D_1 = D_\infty(f)+\hat{D}$, where f is a meromorphic function on V_o. Then

$$2D_0 - \hat{D} \sim 2D_\infty(f) + \hat{D} \quad \text{(linearly equivalent)} .$$

Note that $L(2D_\infty(f))$ has at least 3 linearly independent elements: $1, f, f^2$. Hence

$$h^0(2D_0 - \hat{D}) \geqq h^0(2D_\infty(f)) \geqq 3 .$$

Thus (*) holds if and only if L has no fixed point and $h^0(2D_0) = 3$. Thus we get

<u>Theorem 5.3.11</u>. $\pi : \mathbb{G}_n^1 \longrightarrow T_g$ is a holomorphic local imbedding around $L \in \mathbb{G}_n^1$ if and only if (1) L is complete, (2) L has no fixed point and (3) $h^0(2D) = 3$ for $D \in L$.

By Corollary 5.2.3, there is a biholomorphic map

$$\iota : R_n/\mathrm{Aut}(\mathbb{P}^1) \longrightarrow \mathbb{G}_n^1 - \mathbb{F}_n^1 ,$$

which makes the following diagram commutative:

$$\begin{array}{ccc} R_n/\mathrm{Aut}(\mathbb{P}^1) & \xrightarrow{\ \iota\ } & \mathbb{G}_n^1 - \mathbb{F}_n^1 \\ & {\scriptstyle\pi}\searrow \quad \swarrow{\scriptstyle\pi} & \\ & T_g & \end{array}$$

ι is, in fact, given by

$$\iota : f \ (\mathrm{mod}\ \mathrm{Aut}(\mathbb{P}^1)) \longmapsto \text{the pencil given by } f.$$

Hence

$$\pi(\mathbb{G}_n^1 - \mathbb{F}_n^1) = \{t \in T_g \mid V_t \text{ has a meromorphic function of order } n\} .$$

We denote this set by $T_g(n)$, i.e.,

$$T_g(n) = \{t \in T_g \mid V_t \ \underline{\text{has a meromorphic function of order}}\ n\} .$$

<u>Lemma 5.3.12</u>. $\pi(\mathbb{F}_n^1) = \pi(\mathbb{G}_{n-1}^1)$.

<u>Proof</u>. If $t \in \pi(\mathbb{F}_n^1)$, then there is $g_n^1 \in \mathbb{G}_n^1(V_t)$ with a fixed point P. Then, $g_n^1 - P \in \mathbb{G}_{n-1}^1(V_t)$, so that $t \in \pi(\mathbb{G}_{n-1}^1)$.

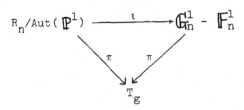

Conversely, if $g^1_{n-1} \in \mathbb{G}^1_{n-1}(V_t)$, then $g^1_{n-1}+P \in \mathbb{F}^1_n(V_t)$ for any (fixed) point $P \in V_t$. Q.E.D.

Now, let p be a prime number such that

$$(p-1)^2 \leqq g-1 .$$

Then

$$T_g(p) \cap \pi(\mathbb{F}^1_p) = \emptyset \text{ (empty)} .$$

In fact, if $t \in T_g(p)$, then there is $f \in R_p(V_t)$. By Corollary 2.4.5, $R_n(V_t) = \emptyset$ for all $n \leqq p-1$. This means that $\mathbb{F}^1_p(V_t) = \emptyset$, i.e., $t \notin \pi(\mathbb{F}^1_p)$. Hence

$$T_g(p) = \pi(\mathbb{G}^1_p - \mathbb{F}^1_p) = \pi(\mathbb{G}^1_p) - \pi(\mathbb{F}^1_p) = \pi(\mathbb{G}^1_p) - \pi(\mathbb{G}^1_{p-1}) .$$

Note that $h^0(D_\infty(f)) = 2$ for any $f \in R_p(V_t)$ (see Theorem 2.1.9). Hence every $L \in \mathbb{G}^1_p - \mathbb{F}^1_p$ is complete. In conclusion, we get following theorem, which was announced in Namba [66].

Theorem 5.3.13. Let p be a prime number such that $(p-1)^2 \leqq g-1$. Then

(1) $T_g(p)$ is an open subspace of a closed complex subspace of T_g and has dimension $2p+2g-5$.

(2) $T_g(p)$ is singular at $t \in T_g(p)$ if and only if $h^0(2D_\infty(f)) > 3$ for $f \in R_p(V_t)$.

Proof. We have gotten: $T_g(p) = \pi(\mathbb{G}^1_p) - \pi(\mathbb{G}^1_{p-1})$. Note also that $\pi : \mathbb{G}^1_p - \mathbb{F}^1_p \longrightarrow T_g(p)$ is a bijective holomorphic map (see Corollary 2.4.5). Hence (1) holds. (2) follows from Theorem 5.3.11.

Q.E.D.

Corollary 5.3.14.

(1) (Rauch [71]) If $g \geqq 2$, then $T_g(2)$, the hyperelliptic locus, is a non-singular closed complex subspace of T_g of dimension $2g-1$

(2) If $g \geqq 4$, then $T_g(3)$, the locus of trigonal compact Riemann surfaces, is non-singular and of dimension $2g+1$.

(3) If p ($\geqq 5$) is a prime number such that $(p-1)(2p-3) \leqq g-1$, then $T_g(p)$ is non-singular.

Proof. In order to apply (2) of the theorem, we first give a general remark.

For a compact Riemann surface V of genus g, put

$$m = \min \{n \mid R_n(V) \neq \emptyset\} .$$

For $f \in R_m(V)$, put $D = D_\infty(f)$. Then, by Theorem 2.1.9, $h^0(D) = 2$. Hence $\dim S_D = m-2$, where

$$S_D = \bigcap_{(\omega) \geqq D} \langle \omega \rangle . \quad \text{(see §5.1).}$$

Now,

"$h^0(2D) = 3.$"

$$\Longleftrightarrow$$

"$h^1(2D) = g+2-2m.$"

$$\Longleftrightarrow$$

"$\dim S_{2D} = 2m-3.$"

$$\Longleftrightarrow$$

"If $D' \sim D$ $(D' \neq D)$, then $S_D \cap S_{D'} = \emptyset.$"

Now, let p be a prime number such that $(p-1)^2 \leqq g-1$. Let $t \in T_g(p)$ and $f \in R_p(V_t)$. Then

"$h^0(2D_\infty(f)) = 3.$"

$\Longleftrightarrow$

"$S_{D_\infty(f)} \cap S_{D_0(f)} = \emptyset.$"

If $p = 2$ and $g \geqq 2$, then $\dim S_{D_\infty(f)} = p - 2 = 0$, i.e., $S_{D_\infty(f)}$ is one point in $\mathbb{P}^{g-1}$. (It is the canonical image of $D_\infty(f)$.) Since $S_{D_\infty(f)} \cap S_{D_0(f)} = \emptyset$, (1) is proved. ($\mathbb{F}_2^1 = \emptyset$.)

If $p = 3$ and $g = 4$, then $T_4(3) = T_4 - T_4(2)$ is non-singular and of dimension 9.

If $p = 3$ and $g \geqq 5$, then the canonical curve C of V_t is contained in a rational ruled surface M (see §2.5). If $\sigma : M \longrightarrow \mathbb{P}^1$ is the ruling, then f is given by

$$D_\lambda(f) = \sigma^{-1}(\lambda) \cap C \quad \text{for} \quad \lambda \in \mathbb{P}^1 .$$

Note that $\sigma^{-1}(\lambda)$ is a line in $\mathbb{P}^{g-1}$. Since $\dim S_{D_\infty(f)} = 3 - 2 = 1$,

$$S_{D_\lambda(f)} = \sigma^{-1}(\lambda) \quad \text{for} \quad \lambda \in \mathbb{P}^1.$$

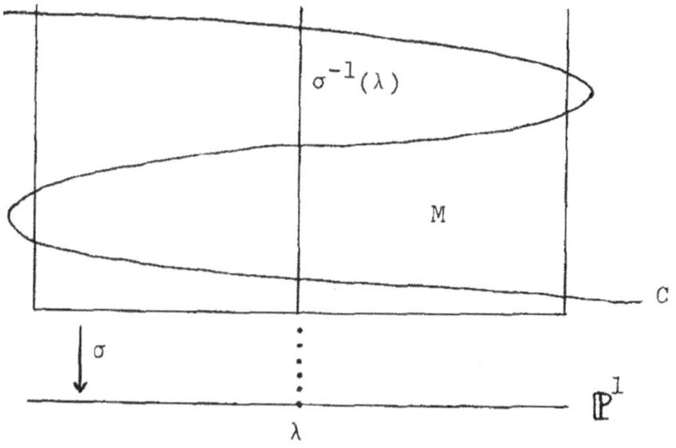

Note that $\sigma^{-1}(0)$ and $\sigma^{-1}(\infty)$ do not intersect. Hence (2) is proved.

Finally, we show (3). Let $p \geqq 5$ and $(p-1)(2p-3) \leqq g-1$. Take $t \in T_g(p)$ and $f \in R_p(V_t)$. Then, by Corollary 2.4.4,

$$R_n(V_t) = \emptyset \quad \text{for} \quad 1 \leqq n \leqq 2p-2 \quad \text{with} \quad n \neq p.$$

Assume that $h^0(2D_\infty(f)) > 3$. Consider the meromorphic map

$$\Phi_{|2D_\infty(f)|} : V_t \longrightarrow \hat{C} = \Phi_{|2D_\infty(f)|}(V_t) \subset \mathbb{P}^N ,$$

where $N = h^0(2D_\infty(f)) - 1 \geqq 3$. Then $\hat{C}$ is not a plane curve. Note that $|2D_\infty(f)|$ has no fixed point. In fact,

$$2D_\infty(f) = D_\infty(f^2) \sim D_0(f^2) .$$

Hence $\Phi_{|2D_\infty(f)|}$ is a holomorphic map.

Since $\dim |2D_\infty(f)-D_\infty(f)| = 1$, there is a unique quadric Q of rank $\leqq 4$ in $\mathbb{P}^N$ containing $\hat{C}$ and determines f (see Proposition 2.5.2). It is in fact of rank 3 (see Proposition 2.5.4). Its vertex $V(Q)$ does not intersect with $\hat{C}$. This is because

$$2D_\infty(f) \sim D_\infty(f) + D_0(f) .$$

Put

$$d = \deg \hat{C} ,$$

$$k = \text{the mapping order of } \Phi_{|2D_\infty(f)|} : V_t \longrightarrow \hat{C} .$$

Then $dk = 2p$. Note that $d \geqq N \geqq 3$. (In fact, there is a hyperplane in $\mathbb{P}^N$ containing at least N points of $\hat{C}$.) Hence d must be either p or $2p$.

Let

$$u : \hat{C} \longrightarrow \tilde{C} \subset \mathbb{P}^2$$

be the projection with the center the vertex $V(Q)$. Then $\tilde{C}$ is a conic in $\mathbb{P}^2$.

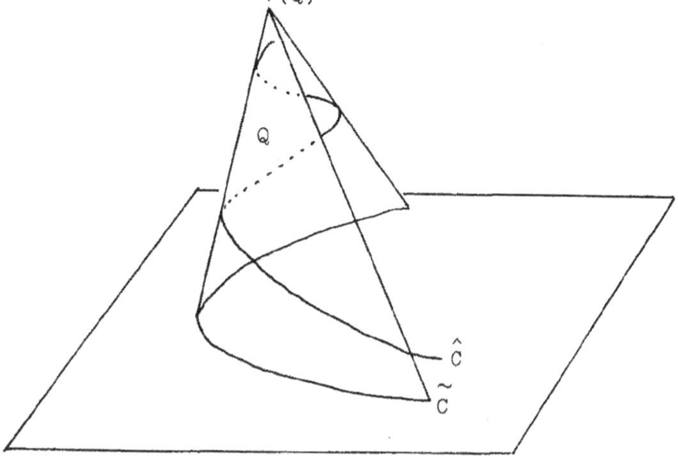

Put

$$k' = \text{the mapping order of } u .$$

Then $d = 2k'$, so that $d = 2p$ and $k = 1$. Thus $\Phi_{|2D_\infty(f)|} : V_t \longrightarrow \hat{C}$ is a birational morphism.

We show that Q is a unique quadric of rand $\leqq 4$ in $\mathbb{P}^N$ containing $\hat{C}$. In fact, if Q' is another such quadric, then one of the projections

$$\pi_S : V_t \longrightarrow \mathbb{P}^1 \quad (\pi_S = \pi_S \cdot \Phi_{|2D_\infty(f)|} \quad \text{(identified).})$$

with the center , S, a linear subspace of dimension $N-2$ in Q', has the order $\leqq \frac{1}{2} \deg(2D_\infty(f)) = p$. By the assumption, it must be f (mod $\mathrm{Aut}(\mathbb{P}^1)$), so that $Q' = Q$.

Let S be a $(N-2)$-dimensional linear subspace of $\mathbb{P}^N$ such that $S \not\subset Q$ and such that

$$\deg(\hat{C} \cap S) \geqq N-1, \quad (\hat{C} \cap S \text{ is regarded as a divisor on } V_t).$$

(For example, take independent points $x_1, x_2, \cdots, x_{N-1}$ on $\hat{C}$ so that x_1 and x_2 belong to different linear subspaces of dimension $N-2$

in Q. Then the linear subspace S of dimension N-2 generated by these points satisfies the conditions.)

Consider the projection $\pi_S : C \longrightarrow \mathbb{P}^1$. Put

$$h = \pi_S \cdot \Phi_{|2D_\infty(f)|} .$$

Then

$$\mathrm{ord}(h) \leqq 2p - (N-1) \leqq 2p-2 .$$

Hence

$$h \equiv f \pmod{\mathrm{Aut}(\mathbb{P}^1)} .$$

Then

$$\dim |2D_\infty(f) - D_\infty(h)| = \dim |2D_\infty(f) - D_\infty(f)| = 1 .$$

Hence there must be a quadric of rank $\leqq 4$ in $\mathbb{P}^N$ containing $\hat{C}$ and S, a contradiction. Q.E.D.

Problem. Sharpen the condition of (3) of Corollary 5.3.14.

BIBLIOGRAPHY

[1] Ahlfors, L.V.: The complex analytic structure of the space of closed Riemann surfaces. Analytic Functions, Princeton Univ. Press, Princeton, 1960.

[2] Akahori, T. and Namba, M.: Examples of obstructed holomorphic maps. Proc. Japan Acad., 54 (1978), 189-191.

[3] Andreotti, A. and Mayer, A.: On period relations for abelian integrals on algebraic curves. Ann. Scuola Norm. Sup. Pisa Series 3, 21 (1967), 189-238.

[4] Bers, L.: Uniformization and moduli. Contrib. to Function Theory, Tata Institute, Bombay, 1960, 41-49.

[5] Bliss, G.A.: Algebraic Functions. Dover, New York, 1966.

[6] Bloch, S.: Semi-regularity and de Rham cohomology. Inv. Math., 17 (1972), 51-66.

[7] Bochner, S. and Montgomery, D.: Groups on analytic manifolds. Ann. Math., 48 (1947), 659-669.

[8] Bott, R.: Homogeneous vector bundles. Ann. Math., 66 (1957), 203-248.

[9] Burns, D.M.Jr. and Wahl, J.M.: Local contributions to global deformations of surfaces. Inv. Math., 26 (1974), 67-88.

[10] Commichau, M.: Deformationen kompakter komplexer Mannigfaltigkeiten. Math. Ann., 213 (1975), 43-96.

[11] Donin, I.F.: Complete families of deformations of germs of complex spaces. Math. USSR Sbornik, 18 (1972), 397-406.

[12] Douady, A.: Le problème des modules pour les variétés analytiques complexes. Séminaire Bourbaki, No. 277, 1964/5.

[13] __________: Le problème des modules pour les sous-espaces analytiques compacts d'un espace analytiques donné. Ann. Inst. Fourier, Grenoble, 16 (1966), 1-98.

[14] __________: Flatness and Privilege. Topics in Several Complex Variables, Finland, 1967, 47-74.

[15] __________: Le problème des modules locaux pour les espaces $\mathbb{C}$-analytiques compacts. Ann. Sci. École Norm. Sup., Sér. IV 4 (1974), 567-602.

[16] Earle, C.J.: Families of Riemann surfaces and Jacobi varieties. Ann. Math., 107 (1978), 255-286.

[17] Farkas, H.M.: Special divisors and analytic subloc of Teichmüller space. Amer. J. Math., 88 (1966), 881-901.

[18] Fischer, G.: Complex Analytic Geometry. Lecture Notes in Math., 538, Springer-Verlag, 1976.

[19] Frölicher, A. and Nijenhuis, A.: A theorem on stability of complex structures. Proc. Nat. Acad. Sci. U.S.A., 43 (1957), 239-241.

[20] Fujita, T.: Defining equations for certain types of polarized varieties. Complex Analysis and Algebraic Geometry, edited by W.L. Baily, Jr. and T. Shioda, Iwanami Shoten-Cambridge Univ. Press, 1977.

[21] Fulton, W.: Algebraic Curves. Benjamin, 1969.

[22] Grauert, H.: Ein Theorem der analytischen Garbentheorie und die Modulräume komplexer Strukturen. Publ. Math. IHES., No. 5, 1960.

[23] ________: Über Modifikationen und exzeptionelle analytische Mengen. Math. Ann., 146 (1962), 331-368.

[24] ________: Der Satz von Kuranishi für kompakte komplexe Räume. Inv. Math., 25 (1974), 107-142.

[25] Griffiths, P. and Harris, J.: Principles of Algebraic Geometry. A Wiley-Interscience Publ., 1978.

[26] Grothendieck, A.: Techniques de construction en géométrie analytique. Séminaire H. Cartan, 13ème année, 1960/61, Exp. 16 et 17.

[27] ____________ and Dieudonné, J.: Eléménts de géométrie algébrique III: Etude cohomologique des faisceaux cohérents (seconde partie). Publ. Math. IHES., No. 17, 1963.

[28] Gunning, R.C.: Lectures on Riemann Surfaces: Jacobi Varieties. Princeton Univ. Press, Princeton, N.J., 1972.

[29] Gunning, R.C. and Rossi, H.: Analytic Functions of Several Complex Variabels, Prentice-Hall, Englewood Cliffs, N.J., 1965.

[30] Holmann, H.: Quotienten komplexer Räume. Math. Ann., 142 (1961), 407-440.

[31] ________: Komplexe Räume mit komplexen transformationsgruppen. Math. Ann., 150 (1963), 327-360.

[32] Horikawa, E.: On deformations of holomorphic maps, I, II, III. J. Math. Soc. Japan, 25 (1973), 647-666; ibid, 26 (1974), 372-396; Math. Ann., 222 (1976), 275-282.

[33] ________: On deformations of quintic surfaces, Inv. Math., 31 (1975), 43-85.

[34] Kas, A.: On obstructions to deformations of complex analytic surfaces. Proc. Nat. Acad. Sci. U.S.A., 58 (1967), 402-404.

[35] Kaup, W.: Infinitesimale Transformationengruppen komplexer Räume. Math. Ann., 160 (1965), 72-92.

[36] Kempf, G.: Schubert methods with an application to algebraic curves. Publ. Math. Centrum, Amsterdam, 1971.

[37] Kempf, G.: On the geometry of a theorem of Riemann. Ann. Math., 98 (1973), 178-185.

[38] Kleiman, S.L. and Laksov, D.: On the existence of special divisors. Amer. J. Math., 94 (1972), 431-436.

[39] ______________________: Another proof of the existence of special divisors. Acta Math., 132 (1974), 163-176.

[40] Kodaira, K.: Characteristic linear systems of complete continuous systems. Amer. J. Math., 78 (1956), 716-744.

[41] ________: A theorem of completeness of characteristic systems for analytic families of compact submanifolds of complex manifolds. Ann. Math., 75 (1962), 146-162.

[42] Kodaira, K. and Spencer, D.C.: Groups of complex line bundles over compact Kähler varieties. Proc. Nat. Acad. Sci. U.S.A., 39 (1953), 868-872.

[43] ______________________: On deformations of complex analytic structures, I-II. Ann. Math., 67 (1958), 328-466.

[44] ______________________: A theorem of completeness for complex analytic fibre spaces. Acta Math., 100 (1958), 281-294.

[45] ______________________: A theorem of completeness of characteristic systems of complete continuous systems. Amer. J. Math., 81 (1959), 477-500.

[46] Kodaira, K. and Nirenberg, L. and Spencer, D.C.: On the existence of deformations of complex analytic structures. Ann. Math., 68 (1958), 450-459.

[47] Kouchiyama, N.: On deformations of holomorphic maps, (in Japanese). Master's thesis, Kyoto Univ., 1978.

[48] Kuranishi, M.: On the locally complete families of complex analytic structures. Ann. Math., 75 (1962), 536-577.

[49] __________: New proof for the existence of locally complete families of complex structures. Proc. Conf. on Complex Analysis, Minneapolis, 1964; Springer, 1965.

[50] __________: Deformations of compact complex manifolds. Proc. International Seminar, Univ. Montreal, Montreal, 1969.

[51] Lefschetz, S.: Algebraic Geometry. Princeton Univ. Press, Princeton, N.J., 1953.

[52] Martens, H.: On the varieties of special divisors on a curve, I, II. Jour. reine Angew. Math., 227 (1967), 111-120; ibid, 233 (1968), 89-100.

[53] Meis, T.: Die minimale Blätterzahl der Konkretisierung einer kompakten Riemannischen Fläche. Schr. Math. Inst. Univ. Münster, 1960.

[54] Miyajima, K.: On the existence of Kuranishi family for deformations of holomorphic maps. Science Rep. Kagoshima Univ.,

27 (1978), 43-76.

[55] Morrow, J. and Kodaira, K.: Complex manifolds. Holt Reinhalt and Winston, New York, 1971.

[56] Mumford, D.: Further pathologies in algebraic geometry. Amer. J. Math., 84 (1962), 642-647.

[57] __________: Lectures on curves on an algebraic surface. Ann. Math. Studies, 59, Princeton Unov. Press, 1966.

[58] __________: Varieties defined by quadratic equations. C.I.M.E. (1969)-III, 29-100.

[59] __________: Curves and their Jacobians. Univ. Michigan Press, 1975.

[60] Nakamura, I.: Complex parallelizable manifolds and their small deformations. J. Diff. Geom., 10 (1975), 85-112.

[61] Namba, M.: On maximal families of compact complex submanifolds of complex manifolds. Tohoku Math. J., 24 (1972), 581-609.

[62] __________: On deformations of automorphism groups of compact complex manifolds. Tohoku Math. J., 26 (1974), 237-283.

[63] __________: Notes on complex Lie semigroups. Proc. Japan Acad., 51 (1975), 362-364.

[64] __________: Moduli of open holomorphic maps of compact complex manifolds. Math. Ann., 220 (1976), 65-76.

[65] __________: On families of effective divisors on algebraic manifolds. Proc. Japan Acad., 53 (1977), 206-209.

[66] __________: Meromorphic functions on compact Riemann surfaces. Proc. Japan Acad., 54 (1978), 192-193.

[67] __________: Deformations of compact complex manifolds and some related topics, (survey article), to appear.

[68] Narasimhan, R.: Introduction to the Theory of Analytic Spaces. Lecture Notes in Math., 25, Springer-Verlag, 1966.

[69] Petri, K.: Über die invariante Darstellung algebraischer Funktionen einer Veränderlichen. Math. Ann., 88 (1922), 242-289.

[70] Pourcin, G.: Théorème de Douady an-dessus de S. Ann. Scuola Norm. Sup. Pisa, 23 (1969), 451-459.

[71] Rauch, H.E.: Weierstrass points, branch points and the moduli of Riemann surfaces. Comm. Pure Appl. Math., 12 (1959), 543-560.

[72] __________: Variational methods in the problem of the moduli of Riemann surfaces. Contrib. to Function Theory, Tata Institute Bombay, 1960, 17-40.

[73] Remmert, R.: Holomorphe und meromorphe Abbildungen komplexer Räume. Math. Ann., 133 (1957), 328-370.

[74] Riemenschneider, O.: Über die Anwedung algebraischer Methoden in der Deformationstheorie komplexer Räume. Math. Ann., 187 (1970), 40-55.

[75] Saint-Donat, B.: Sur les équations définissant une courbe algébrique. C. R. Acad. Sci. Paris, 274 (1972), 324-327 et 487-489.

[76] ______________: On Petri's analysis of the linear system of quadrics through a canonical curve. Math. Ann., 206 (1973), 157-175.

[77] Schuster, H.: Zur Theorie der Deformationen kompakter komplexer Räume. Inv. Math., 9 (1970), 284-294.

[78] Severi, F.: Vorlesungen über Algebraische Geometrie, (tr. by E. Löffler). Leipzig, Teubner, 1921.

[79] Šokurov, V.V.: The Noether-Enriques theorem on canonical curves. Math. USSR Sbornik, 15 (1971), 361-403.

[80] Suwa, T.: Stratification of local moduli spaces of Hirzebruch manifolds. Rice Univ. Studies, Complex Analysis II, 1972, 129-146.

[81] Teichmüller, O.: Veränderliche Riemannsche Flächen. Deutsche Math., 7 (1944), 344-359.

[82] Ueno, K.: Classification Theory of Algebraic Varieties and Compact Complex Spaces. Lecture Notes in Math., 439, Springer-Verlag, 1975.

[83] Walker, R.: Algebraic Curves. Dover, New York, 1962.

[84] Wavrik, J.: Obstructions to the existence of a space of moduli. Global Analysis, Papers in honor of K. Kodaira, Princeton, 1969, 403-414.

[85] Weil, A.: On Picard varieties. Amer. J. Math., 74 (1952), 865-894.

[86] Zappa, G.: Sull'estistenza, sopra le superficie algebriche, di sistemi continui completi infiniti, la cui curva generica è a serie caratteristica incompleta. Pont. Acad. Acta., 9 (1945), 91-93.

INDEX

Vol. 609: General Topology and Its Relations to Modern Analysis and Algebra IV. Proceedings 1976. Edited by J. Novák. XVIII, 225 pages. 1977.

Vol. 610: G. Jensen, Higher Order Contact of Submanifolds of Homogeneous Spaces. XII, 154 pages. 1977.

Vol. 611: M. Makkai and G. E. Reyes, First Order Categorical Logic. VIII, 301 pages. 1977.

Vol. 612: E. M. Kleinberg, Infinitary Combinatorics and the Axiom of Determinateness. VIII, 150 pages. 1977.

Vol. 613: E. Behrends et al., L^P-Structure in Real Banach Spaces. X, 108 pages. 1977.

Vol. 614: H. Yanagihara, Theory of Hopf Algebras Attached to Group Schemes. VIII, 308 pages. 1977.

Vol. 615: Turbulence Seminar, Proceedings 1976/77. Edited by P. Bernard and T. Ratiu. VI, 155 pages. 1977.

Vol. 616: Abelian Group Theory, 2nd New Mexico State University Conference, 1976. Proceedings. Edited by D. Arnold, R. Hunter and E. Walker. X, 423 pages. 1977.

Vol. 617: K. J. Devlin, The Axiom of Constructibility: A Guide for the Mathematician. VIII, 96 pages. 1977.

Vol. 618: I. I. Hirschman, Jr. and D. E. Hughes, Extreme Eigen Values of Toeplitz Operators. VI, 145 pages. 1977.

Vol. 619: Set Theory and Hierarchy Theory V, Bierutowice 1976. Edited by A. Lachlan, M. Srebrny, and A. Zarach. VIII, 358 pages. 1977.

Vol. 620: H. Popp, Moduli Theory and Classification Theory of Algebraic Varieties. VIII, 189 pages. 1977.

Vol. 621: Kauffman et al., The Deficiency Index Problem. VI, 112 pages. 1977.

Vol. 622: Combinatorial Mathematics V, Melbourne 1976. Proceedings. Edited by C. Little. VIII, 213 pages. 1977.

Vol. 623: I. Erdelyi and R. Lange, Spectral Decompositions on Banach Spaces. VIII, 122 pages. 1977.

Vol. 624: Y. Guivarc'h et al., Marches Aléatoires sur les Groupes de Lie. VIII, 292 pages. 1977.

Vol. 625: J. P. Alexander et al., Odd Order Group Actions and Witt Classification of Innerproducts. IV, 202 pages. 1977.

Vol. 626: Number Theory Day, New York 1976. Proceedings. Edited by M. B. Nathanson. VI, 241 pages. 1977.

Vol. 627: Modular Functions of One Variable VI, Bonn 1976. Proceedings. Edited by J.-P. Serre and D. B. Zagier. VI, 339 pages. 1977.

Vol. 628: H. J. Baues, Obstruction Theory on the Homotopy Classification of Maps. XII, 387 pages. 1977.

Vol. 629: W. A. Coppel, Dichotomies in Stability Theory. VI, 98 pages. 1978.

Vol. 630: Numerical Analysis, Proceedings, Biennial Conference, Dundee 1977. Edited by G. A. Watson. XII, 199 pages. 1978.

Vol. 631: Numerical Treatment of Differential Equations. Proceedings 1976. Edited by R. Bulirsch, R. D. Grigorieff, and J. Schröder. X, 219 pages. 1978.

Vol. 632: J.-F. Boutot, Schéma de Picard Local. X, 165 pages. 1978.

Vol. 633: N. R. Coleff and M. E. Herrera, Les Courants Résiduels Associés à une Forme Méromorphe. X, 211 pages. 1978.

Vol. 634: H. Kurke et al., Die Approximationseigenschaft lokaler Ringe. IV, 204 Seiten. 1978.

Vol. 635: T. Y. Lam, Serre's Conjecture. XVI, 227 pages. 1978.

Vol. 636: Journées de Statistique des Processus Stochastiques, Grenoble 1977, Proceedings. Edité par Didier Dacunha-Castelle et Bernard Van Cutsem. VII, 202 pages. 1978.

Vol. 637: W. B. Jurkat, Meromorphe Differentialgleichungen. VII, 194 Seiten. 1978.

Vol. 638: P. Shanahan, The Atiyah-Singer Index Theorem, An Introduction. V, 224 pages. 1978.

Vol. 639: N. Adasch et al., Topological Vector Spaces. V, 125 pages. 1978.

Vol. 640: J. L. Dupont, Curvature and Characteristic Classes. X, 175 pages. 1978.

Vol. 641: Séminaire d'Algèbre Paul Dubreil, Proceedings Paris 1976–1977. Edité par M. P. Malliavin. IV, 367 pages. 1978.

Vol. 642: Theory and Applications of Graphs, Proceedings, Michigan 1976. Edited by Y. Alavi and D. R. Lick. XIV, 635 pages. 1978.

Vol. 643: M. Davis, Multiaxial Actions on Manifolds. VI, 141 pages. 1978.

Vol. 644: Vector Space Measures and Applications I, Proceedings 1977. Edited by R. M. Aron and S. Dineen. VIII, 451 pages. 1978.

Vol. 645: Vector Space Measures and Applications II, Proceedings 1977. Edited by R. M. Aron and S. Dineen. VIII, 218 pages. 1978.

Vol. 646: O. Tammi, Extremum Problems for Bounded Univalent Functions. VIII, 313 pages. 1978.

Vol. 647: L. J. Ratliff, Jr., Chain Conjectures in Ring Theory. VIII, 133 pages. 1978.

Vol. 648: Nonlinear Partial Differential Equations and Applications, Proceedings, Indiana 1976–1977. Edited by J. M. Chadam. VI, 206 pages. 1978.

Vol. 649: Séminaire de Probabilités XII, Proceedings, Strasbourg, 1976–1977. Edité par C. Dellacherie, P. A. Meyer et M. Weil. VIII, 805 pages. 1978.

Vol. 650: C*-Algebras and Applications to Physics. Proceedings 1977. Edited by H. Araki and R. V. Kadison. V, 192 pages. 1978.

Vol. 651: P. W. Michor, Functors and Categories of Banach Spaces. VI, 99 pages. 1978.

Vol. 652: Differential Topology, Foliations and Gelfand-Fuks-Cohomology, Proceedings 1976. Edited by P. A. Schweitzer. XIV, 252 pages. 1978.

Vol. 653: Locally Interacting Systems and Their Application in Biology. Proceedings, 1976. Edited by R. L. Dobrushin, V. I. Kryukov and A. L. Toom. XI, 202 pages. 1978.

Vol. 654: J. P. Buhler, Icosahedral Golois Representations. III, 143 pages. 1978.

Vol. 655: R. Baeza, Quadratic Forms Over Semilocal Rings. VI, 199 pages. 1978.

Vol. 656: Probability Theory on Vector Spaces. Proceedings, 1977. Edited by A. Weron. VIII, 274 pages. 1978.

Vol. 657: Geometric Applications of Homotopy Theory I, Proceedings 1977. Edited by M. G. Barratt and M. E. Mahowald. VIII, 459 pages. 1978.

Vol. 658: Geometric Applications of Homotopy Theory II, Proceedings 1977. Edited by M. G. Barratt and M. E. Mahowald. VIII, 487 pages. 1978.

Vol. 659: Bruckner, Differentiation of Real Functions. X, 247 pages. 1978.

Vol. 660: Equations aux Dérivée Partielles. Proceedings, 1977. Edité par Pham The Lai. VI, 216 pages. 1978.

Vol. 661: P. T. Johnstone, R. Paré, R. D. Rosebrugh, D. Schumacher, R. J. Wood, and G. C. Wraith, Indexed Categories and Their Applications. VII, 260 pages. 1978.

Vol. 662: Akin, The Metric Theory of Banach Manifolds. XIX, 306 pages. 1978.

Vol. 663: J. F. Berglund, H. D. Junghenn, P. Milnes, Compact Right Topological Semigroups and Generalizations of Almost Periodicity. X, 243 pages. 1978.

Vol. 664: Algebraic and Geometric Topology, Proceedings, 1977. Edited by K. C. Millett. XI, 240 pages. 1978.

Vol. 665: Journées d'Analyse Non Linéaire. Proceedings, 1977. Edité par P. Bénilan et J. Robert. VIII, 256 pages. 1978.

Vol. 666: B. Beauzamy, Espaces d'Interpolation Réels: Topologie et Géometrie. X, 104 pages. 1978.

Vol. 667: J. Gilewicz, Approximants de Padé. XIV, 511 pages. 1978.

Vol. 668: The Structure of Attractors in Dynamical Systems. Proceedings, 1977. Edited by J. C. Martin, N. G. Markley and W. Perrizo. VI, 264 pages. 1978.

Vol. 669: Higher Set Theory. Proceedings, 1977. Edited by G. H. Müller and D. S. Scott. XII, 476 pages. 1978.